BARRON'S

Regents Exams and Answers

Physics—Physical Setting

Revised Edition

ALBERT S. TARENDASH
Assistant Principal, Supervision (Retired),
Department of Chemistry and Physics
Stuyvesant High School, New York, New York

The section on how to answer Part C questions and the glossary were adapted from *Let's Review: Physics* by Miriam A. Lazar and Albert S. Tarendash, published by Kaplan North America, LLC d/b/a Barron's Educational Series, 1996.

Acknowledgments

The 2014 Answers Explained were written by Gregory Guido.

The 2015–2019 Answers Explained were written by Miriam A. Lazar.

Published by Kaplan North America, LLC d/b/a Barron's Educational Series
1515 West Cypress Creek Road
Fort Lauderdale, Florida 33309
www.barronseduc.com

ISBN: 978-1-5062-6637-4

10 9 8 7 6 5

Contents

Preface v

How to Use This Book 1

Test-Taking Tips 3

General Helpful Tips 3

How to Answer Part C Questions 8

What to Expect on the Regents Examination in Physics 13

Format of the Physics Examination 13

Topics Covered on the Regents Examination in Physics 16

New York State Physical Setting/Physics Core 17

Topic Outline 17

Question Index 31

Glossary of Important Terms 34

Reference Tables for Physics 45

A. List of Physical Constants 45
B. Prefixes for Powers of 10 46
C. Approximate Coefficients of Friction 46
D. The Electromagnetic Spectrum 47
E. Absolute Indices of Refraction 47
F. Energy Level Diagrams for Hydrogen and Mercury 48
G. Classification of Matter 49
H. Particles of the Standard Model 50
I. Circuit Symbols 51
J. Resistivities at 20°C 51
K. Equations for Physics 52

Using the Equations to Solve Physics Problems 57

Mechanics 57
Electricity 64
Waves 69
Modern Physics 71

Regents Examinations, Answers, and Self-Analysis Charts 73

June 2013 75
June 2014 127
June 2015 180
June 2016 235
June 2017 288
June 2018 345
June 2019 404

Preface

A HELPFUL WORD TO THE STUDENT

As you are aware, the purpose of this book is to help you review for the New York State Regents examination in physics. You can also use it effectively to prepare for classroom, midterm, and final examinations. The book contains a number of special sections and other features designed to aid you in achieving good grades on your examinations. Included are the following:

- ***How to Use This Book.*** Your journey begins here. This section explains how to use the entire book as an effective test-preparation tool. It provides a method of identifying those areas you have mastered and those that will require additional work.
- ***Test-Taking Techniques.*** In this section, you will learn how to take the Regents examination effectively. Included are a list of materials you will need to bring to the examination and directions on how to fill out your answer sheet, how to read the questions, and how to deal with difficult items. In addition, it provides some suggestions on how to approach the extended-response questions found in Part C of the examination.
- ***What to Expect on the Regents Examination in Physics.*** This section explains the format, content, and grading of the examination.
- ***New York State Physical Setting/Physics Core:*** Topic Outline and Question Index. The Topic Outline provides a detailed description of the Regents Physics syllabus. The Question Index keys the Regents examination questions that appear in this book to the topic outline. The index will help you find questions on similar topics and will provide information about the areas that are stressed most heavily on the exam.
- ***Glossary of Important Terms.*** The glossary will help you understand the technical terms that may appear on the examination.

- ***Reference Tables for Physics.*** A significant part of the Regents examination requires you to be able to find and use the information contained in these reference tables. This section describes each table and the types of information that may be obtained from it.
- ***Using the Equations to Solve Physics Problems.*** This section provides examples that show you how to use each of the equations given in Reference Table *K*.
- ***Recent Regents Examinations.*** These questions will provide the bulk of your study and review. When you have answered all of the questions that appear in this book, you will be able to approach the Regents examination with confidence!
- ***Explanation of Answers and Self-Analysis Chart for Each Examination.*** The detailed answer explanations will help you to understand why certain choices are correct while others are incorrect. The self-analysis chart will pinpoint your strengths and weaknesses on each practice examination you take.

This book was written to provide you with the basic tools you need to perform well on the Regents Physics examination. If you follow its advice and prepare diligently, you will achieve your goal.

Best wishes for success!

Albert S. Tarendash

How to Use This Book

1. Read the section entitled *Test-Taking Techniques* to learn how to prepare properly for an examination and how to take it with maximum efficiency.
2. Read the section entitled *What to Expect on the Regents Examination in Physics* to familiarize yourself with the structure and contents of this examination.
3. Read the section entitled *Reference Tables for Physics* to familiarize yourself with the contents and use of these tables.
   Note: On occasion, the New York State Education Department may change the content or format of the Regents examination and/or the reference tables. Your classroom teacher is your best source of information about such changes.
4. Read the section entitled *Using the Equations to Solve Physics Problems.*
5. Take the first Regents examination in this book, answering *all* of the questions.
6. Refer to the *Glossary of Important Terms* to learn the meanings of words and terms you do not understand.
7. Check your answers, and then complete the self-analysis chart at the end of the examination to pinpoint your strengths and weaknesses.
8. Read the detailed explanation of *all of the questions,* paying closest attention to the questions you answered incorrectly. Occasionally, the *wrong choices* are explained, and these explanations may help you understand why you chose an incorrect answer.
9. When you have determined your areas of weakness, refer to the *New York State Regents Physics Core: Topic Outline and Question Index* to locate similar questions on other recent examinations. (You can also use this outline and index to determine which areas have been stressed in recent years.)
10. Repeat steps 5–9 for the other examinations, *with the exception of the most recent test.*

11. When you have completed your studying, but no more than 1 or 2 days before the actual examination, take the most recent examination in this book *under strict examination conditions.*

After you have checked your answers to this last examination, you will have a rough idea of how you will perform on the Regents examination you will take.

IMPORTANT NOTE ABOUT THE REFERENCE TABLES AND EQUATIONS

In this book, the reference tables and equations beginning on page 45 have been *slightly modified* from the official New York State Reference Tables for Physics (2006 edition). The tables in this book are indexed by letter, and the equations are indexed by letter and number. Some of the tables have been moved in order to separate them from the equations, and each table is accompanied by a brief explanation. Otherwise, the material on these tables and the official tables is *identical*.

All of the explanations and answers in this book refer to the tables and equations *contained within this book*.

Test-Taking Techniques

The following pages contain several tips to help you achieve a good grade on the Physics Regents exam.

GENERAL HELPFUL TIPS

TIP 1

Be confident and prepared.

SUGGESTIONS

- Review previous tests.
- Use a clock or watch, and take previous exams at home under examination conditions (i.e., don't have the radio or television on).
- Get a review book. (One review book is Barron's *Let's Review: Physics*.)
- Talk over the answers to questions on these tests with someone else, such as another student in your class or someone at home.
- Finish all your homework assignments.
- Look over classroom exams that your teacher gave during the term.
- Take class notes carefully.
- Practice good study habits.
- Know that there is an answer for every question.
- Be aware that the people who made up the Regents exam want you to pass.
- Remember that thousands of students over the last few years have taken and passed a Physics Regents. You can pass too!
- Complete your study and review at least one day before the examination. Last-minute cramming does not help and may hurt your performance.

- On the night prior to the exam day: lay out all the things you will need, such as clothing, pens, and admission cards.
- Go to bed early; eat wisely.
- Bring the required materials to the examination. This generally means a pen, two sharpened pencils, and a good quality eraser. In addition, Parts A, B, and C may require the use of a ruler and a protractor. If your school does not supply a calculator, be certain to bring one to the examination. Some schools also require a signed Regents admission card for identification.
- Good advice: Assume your school will *not* supply you with any materials!
- Once you are in the exam room, arrange things, get comfortable, be relaxed, attend to personal needs (the bathroom).
- Keep your eyes on your own paper; do not let them wander over to anyone else's paper.
- Be polite in making any reasonable requests of the exam room proctor, such as changing your seat or having window shades raised or lowered.

TIP 2

Read test instructions carefully.

SUGGESTIONS

- Be familiar with the format of the examination.
- Know how the test will be graded.
- If your school supplies an electronic scoring sheet, be certain you are familiar with the additional directions for recording and changing answers.
- If you decide to change an answer, be certain that you erase your original response completely.
- Any stray marks on your answer sheet should be erased completely.
- Be familiar with the directions for Parts B and C. Answer each question completely. Explanations should be written as *whole sentences* and substitutions into equations *must include units*. Be certain that your answers are clearly labeled and well organized. Place a box around numerical answers. Be neat!
- Ask for assistance from the exam room proctor if you do not understand the directions.

TIP 3

Read each question carefully before you record your answer.

SUGGESTIONS

- Be sure you understand *what* the question is asking; circle key words and numbers.
- Try to recognize information that is *given* in the question.
- On extended response problems, check the blank answer sheet; it will provide clues to answering the questions.
- Will a physics formula help you find the answer to the question?
- Some choices in Parts A and B may look appealing yet be incorrect. (These traps are known as *distractors*.)
- Try to eliminate those choices that are *obviously* incorrect.

TIP 4

Budget your test time (3 hours).

SUGGESTIONS

- Bring a watch or clock to the test.
- The Regents examination is designed to be completed in 1½ to 2 hours.
- If you are absolutely uncertain of the answer to a question, mark your question booklet and move on to the next question.
- If you persist in trying to answer every difficult question *immediately*, you may find yourself rushing or unable to finish the remainder of the examination.
- When you have finished the examination, return to those unanswered questions.
- Good advice: If at all possible, reread the *entire* examination—and your responses—at least one more time. (This will help you eliminate those errors that result from misreading questions.)

TIP 5

Use your reasoning skills.

SUGGESTIONS

- Answer *all* questions.
- Relate (connect) the question to anything that you studied, wrote in your notebook, or heard your teacher say in class.
- Relate (connect) the question to any film, demonstration, or experiment you saw in class, any project you did, or to anything you may have learned from newspapers, magazines, or television.
- Look over the entire test to see whether one part of it can help you answer another part.

TIP 6

Use your reference tables and refer frequently to the "Equations for Physics."

SUGGESTIONS

- You should be familiar with the *content* of each table.
- Frequently, the answers to questions can be found from information contained within the table.
- The equations are grouped according to where they appear in the Core.
- The definition of each symbol in the equation is also provided.
- These equations will aid you in answering many of the questions on the examination.

TIP 7

Do not be afraid to guess on multiple-choice questions.

SUGGESTIONS

- In general, go with your first answer choice.
- Eliminate obvious incorrect choices.
- If still unsure of an answer, make an educated guess.
- There is no penalty for guessing; therefore, answer ALL questions. An omitted answer gets no credit.

TIP 8

Sign the declaration found on your answer sheet.

SUGGESTION

- Unless this declaration is signed, your paper cannot be scored.

SUMMARY OF TIPS

1. Be confident and prepared.
2. Read test instructions carefully.
3. Read each question carefully and read each choice before you record your answer.
4. Budget your test time (3 hours).
5. Use your reasoning skills.
6. Use your reference tables and refer frequently to the "Summary of Equations."
7. Don't be afraid to guess.
8. Sign the declaration found on your answer sheet.

HOW TO ANSWER PART C QUESTIONS

An *extended-response question* is an examination question that requires the test taker to do more than to choose among several responses or to fill in a blank. You may need to perform numerical calculations, draw and interpret graphs, and provide extended written responses to a question or problem.

Part C of the New York State Regents Examination in Physics contains free-response questions. This appendix is designed to provide you with a number of general guidelines for answering them.

SOLVING PROBLEMS INVOLVING NUMERICAL CALCULATIONS

To receive full credit you must:

- Provide the appropriate equation(s).
- Substitute values and units into the equation(s).
- Display the answer, with appropriate units and to the correct number of significant figures.
- If the answer is a vector quantity, include its direction.

Although SI units are used on the Regents examination, you are expected to have some familiarity with other metric units such as the gram and the kilometer.

You should write as legibly as possible. Teachers are human, and nothing irks them more than trying to decipher a careless, messy scrawl. It is also a good idea to identify your answer clearly, either by placing it in a box or by writing the word "answer" next to it.

A final word: If you provide the correct answer but do not show any work, you will not receive any credit for the problem!

The following is a sample problem and its model solution.

Problem

A 5.0-kilogram object has a velocity of 10 meters per second [east]. Calculate the momentum of this object.

Solution

$$\mathbf{p} = m\mathbf{v}$$

$$\mathbf{p} = (5.0\text{ kg})(10\text{ m/s [east]})$$

$$\boxed{\mathbf{p} = 50\text{ kg.m/s [east]}}$$

GRAPHING EXPERIMENTAL DATA

To receive full credit you must:

- Label both axes with the appropriate variables and units.
- Divide the axes so that the data ranges fill the graph as nearly as possible.
- Plot all data points accurately.
- Draw a best-fit line carefully with a straightedge. The line should pass through the origin *only if the data warrant it.*
- If a part of the question requires that the slope be calculated, calculate the slope *from the line,* not from individual data points.

A graph should have a title, and the *independent variable* is usually drawn along the x-axis.

The following is a sample problem and its model solution.

Problem

The weights of various masses, measured on Planet X, are given in the table below.

Mass (kg)	Weight (N)
15	21
20	32
25	35
30	48
35	56

1. Draw a graph that illustrates these data.
2. Use the *graph* to calculate the acceleration due to gravity on Planet X.

Solution

1. The graph shown below incorporates the essential items that were listed in the table.

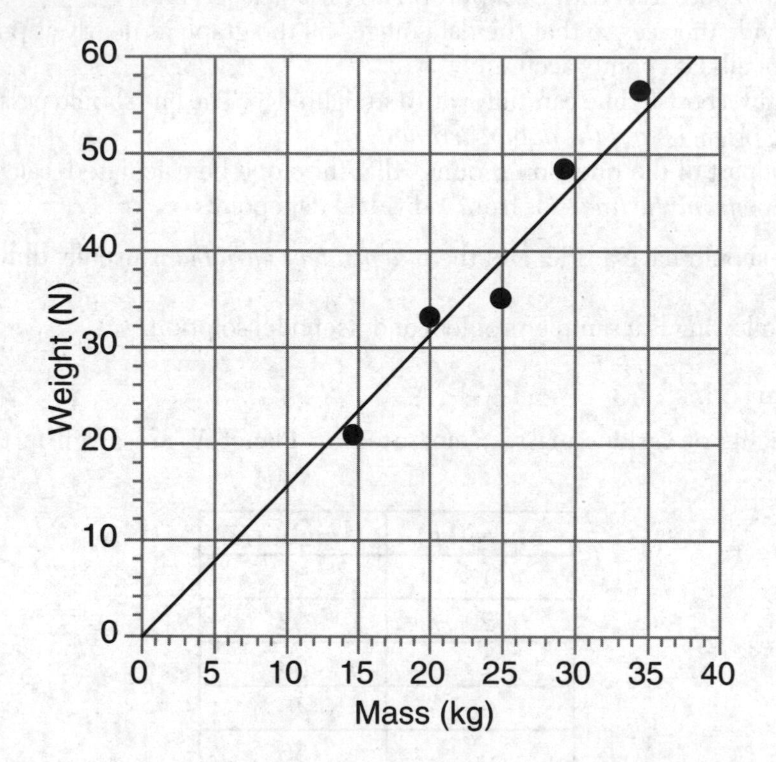

2. Since the magnitude of the gravitational acceleration can be determined by calculating the ratio of weight to mass ($g = W/m$), we can calculate the value of g from the slope of the graph.

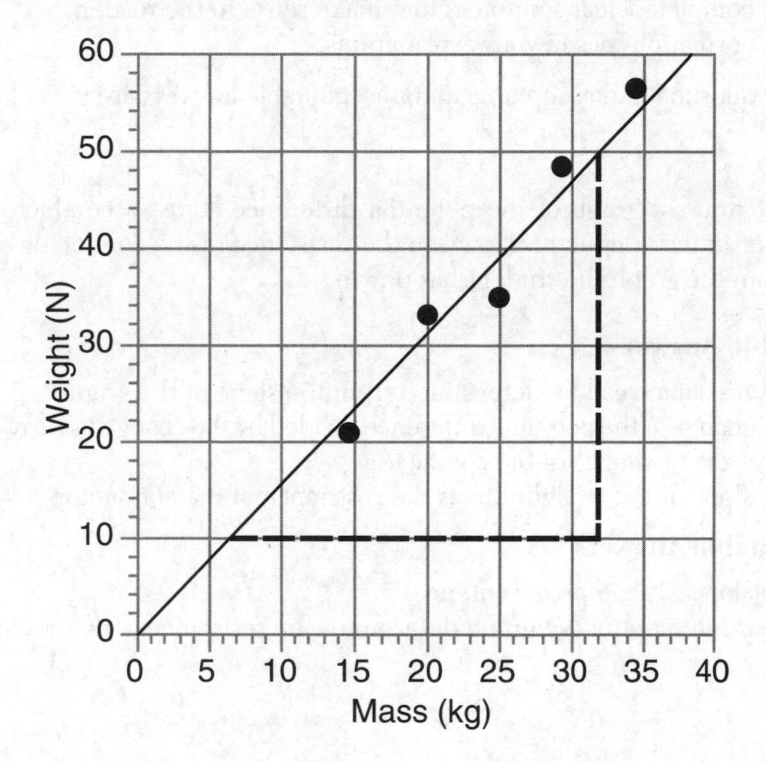

We choose two points on the line; we do not use the data points themselves:

$$g = \frac{\Delta W}{\Delta m} = \frac{50\ \text{N} - 10\ \text{N}}{32\ \text{kg} - 6\ \text{kg}} = 1.5\frac{\text{N}}{\text{kg}} = 1.5\,\text{m/s}^2$$

DRAWING DIAGRAMS

To receive full credit you must:

- Draw your diagrams neatly, and label them clearly.
- Draw vectors *to scale* and *in the correct direction*. If you are given a scale, you must draw your vectors to that scale.
- Bring a straightedge and a protractor with you so that you can draw neat, accurate diagrams.

WRITING AN EXTENDED-RESPONSE ANSWER

To receive full credit you must:

- Use complete, clear sentences that make sense to the reader.
- Use correct physics in your explanations.

A sample question and acceptable and unacceptable answers are given below.

Question

A student draws a graph of the potential difference across a conductor versus the current in the conductor. How can the resistance of the conductor be determined from the graph the student has drawn?

Acceptable Answers

- The resistance can be determined from the slope of the graph.
- The change in the potential difference divided by the change in current determines the resistance of the conductor.
- The slope of the straight line is the resistance of the conductor.

Unacceptable Answers

- The slope. (Incomplete sentence)
- The y-intercept of the graph determines the resistance. (Incorrect physics)

What to Expect on the Regents Examination in Physics

FORMAT OF THE PHYSICS EXAMINATION

The physics examination will be three hours long and will include four parts: A, B–1, B–2, and C. You should be prepared to answer multiple-choice questions as well as questions that require an extended written response.

In general, questions will fall into three categories:

Content questions will test your knowledge and understanding of the material contained within the New York State Physics Core. You may be asked to provide definitions of physical phenomena, interpret diagrams, and solve simple problems.

Skills questions will test your ability to apply, analyze, and evaluate the material contained within the Core. You may be asked to draw and/or interpret graphs and diagrams and solve problems of a more complex nature.

Applications questions will test your ability to apply your scientific knowledge and skills to real-world situations.

Note: The topic outline found on pages 18–30 contains the content, skills, and real-world applications that make up the New York State Physics Core.

Some of the questions on the examination will require use of the 2006 edition of the *Reference Tables for Physical Setting/Physics* (see pages 45–56).

You will be required to answer ALL of the questions on the Physical Setting/Physics Regents examination.

Analysis of a Recent Physical Setting/ Physics Regents Examination

Part	Question Format	Types of Questions	*Percent of Examination
A	Multiple-choice questions	Content questions	41
B–1	Multiple-choice questions	Content and skills questions	14
B–2	Multiple-choice and extended response questions	Content and skills questions	21
C	Extended response questions	Content, skills, and applications questions	24

*These percentages may vary slightly in future examinations.

The maximum *raw* score on the examination is 85 points. A teacher's chart will be provided for converting your *raw* score to a *scaled* score that has a maximum of 100 points. A sample conversion table taken from a recent Regents Physics Examination is shown on the next page. (Remember: Conversion tables can vary slightly from year to year.)

Sample Conversion Table

Raw Score	Scale Score	Raw Score	Scale Score	Raw Score	Scale Score	Raw Score	Scale Score
85	**100**	63	**78**	41	**57**	19	**31**
84	**99**	62	**77**	40	**56**	18	**29**
83	**98**	61	**76**	39	**55**	17	**28**
82	**97**	60	**76**	38	**54**	16	**26**
81	**96**	59	**75**	37	**53**	15	**25**
80	**95**	58	**74**	36	**52**	14	**23**
79	**94**	57	**73**	35	**50**	13	**22**
78	**93**	56	**72**	34	**49**	12	**20**
77	**92**	55	**71**	33	**48**	11	**19**
76	**91**	54	**70**	32	**47**	10	**17**
75	**90**	53	**69**	31	**46**	9	**16**
74	**89**	52	**68**	30	**45**	8	**14**
73	**88**	51	**67**	29	**43**	7	**12**
72	**87**	50	**66**	28	**42**	6	**11**
71	**86**	49	**65**	27	**41**	5	**9**
70	**85**	48	**64**	26	**40**	4	**7**
69	**84**	47	**63**	25	**38**	3	**5**
68	**83**	46	**62**	24	**37**	2	**4**
67	**82**	45	**61**	23	**36**	1	**2**
66	**81**	44	**60**	22	**35**	0	**0**
65	**80**	43	**59**	21	**33**		
64	**79**	42	**58**	20	**32**		

The table is used to convert the number of points you actually received on the examination (your "raw" score) to your final score on the examination (your "scaled" score). **Note that this table will change from one examination to another.**

TOPICS COVERED ON THE REGENTS EXAMINATION IN PHYSICS

All of the questions on the Physics examination will test major understandings, skills, and real-world applications drawn from following five subject areas:

M. Math Skills
I. Mechanics
II. Energy
III. Electricity and Magnetism
IV. Waves
V. Modern Physics

It is suggested that you read the *Topic Outline* found on pages 18–30 in order to learn the exact nature of the material that is subject to testing.

New York State Physical Setting/ Physics Core

TOPIC OUTLINE

The Topic Outline on pages 18–30 is taken from Appendices C and D of the New York State Physical Setting/Physics Core. All Regents physics examinations are based on this Core. The Topic Outline is divided into six sections:

M. Math Skills
I. Mechanics
II. Energy
III. Electricity and Magnetism
IV. Waves
V. Modern Physics

Each course area contains one or more of the following items:

- *Performance indicators*, that is, the major understandings that you must have mastered for the examination.
- *Process skills* that you need to be able to demonstrate during the examination.
- *Real-world applications* that relate physics concepts to the world around you.

Note: When an asterisk (*) is associated with a performance indicator, it means that you need to be able to solve problems using one or more of the equations given in the reference tables on pages 45–56.

M. Math Skills

Sequence	Process Skills (The student will be able to . . .)	Real-World Application
M.1	use algebraic and geometric representations to describe and compare data	use scaled diagrams to represent and manipulate vector quantities, represent physical quantities in graphical form, construct graphs of real-world data (scatter plots, line or curve of best fit), manipulate equations to solve for unknowns, use dimensional analysis to confirm algebraic solutions
M.2	use deductive reasoning to construct and evaluate arguments	interpret graphs to determine the mathematical relationship between the variables
M.3	apply algebraic and geometric concepts and skills in the solution of problems	explain the physical relevance of the properties of a graph of real-world data using slope, intercepts, and area under a curve

I. Mechanics

Sequence	Performance Indicators
I.1	Measured quantities can be classified as either vector or scalar.
I.2	An object in linear motion may travel with a constant velocity* or with acceleration*. *(Note: Testing of acceleration will be limited to cases in which acceleration is constant.)*
I.3	An object in free fall accelerates due to the force of gravity*. Friction and other forces cause the actual motion of a falling object to deviate from its theoretical motion. *(Note: Initial velocities of objects in free fall may be in any direction.)*
I.4	The resultant of two or more vectors, acting at any angle, is determined by vector addition.
I.5	A vector may be resolved into perpendicular components.*
I.6	The path of a projectile is the result of the simultaneous effect of the horizontal and vertical components of its motion; these components act independently.
I.7	A projectile's time of flight is dependent upon the vertical components of its motion.
I.8	The horizontal displacement of a projectile is dependent upon the horizontal component of its motion and its time of flight.
I.9	According to Newton's First Law, the inertia of an object is directly proportional to its mass. An object remains at rest or moves with constant velocity, unless acted upon by an unbalanced force.
I.10	When the net force on a system is zero, the system is in equilibrium.
I.11	According to Newton's Second Law, an unbalanced force causes a mass to accelerate*.
I.12	Weight is the gravitational force with which a planet attracts a mass.* The mass of an object is independent of the gravitational field in which it is located.

I. Mechanics (*Continued*)

I.13	Kinetic friction* is a force that opposes motion.
I.14	Centripetal force* is the net force which produces centripetal acceleration*. In uniform circular motion, the centripetal force is perpendicular to the tangential velocity.
I.15	The impulse* imparted to an object causes a change in its momentum*.
I.16	The elongation or compression of a spring depends upon the nature of the spring (its spring constant) and the magnitude of the applied force*.
I.17	According to Newton's Third Law, forces occur in action/reaction pairs. When one object exerts a force on a second, the second exerts a force on the first that is equal in magnitude and opposite in direction.
I.18	Momentum is conserved in a closed system.* *(Note: Testing will be limited to momentum in one dimension.)*
I.19	Gravitational forces are only attractive, whereas electrical and magnetic forces can be attractive or repulsive.
I.20	The inverse square law applies to electrical* and gravitational* fields produced by point sources.
I.21	Field strength* and direction are determined using a suitable test particle. *(Notes: 1) Calculations are limited to electrostatic and gravitational fields. 2) The gravitational field near the surface of Earth and the electrical field between two oppositely charged parallel plates are treated as uniform.)*

Is. Mechanics Skills

Sequence	Process Skills (The student will be able to . . .)	Real-World Application
Is.1	construct and interpret graphs of position, velocity, or acceleration versus time	Global Positioning Systems (GPS), track and field
Is.2	determine and interpret slopes and areas of motion graphs	mathematical slopes, calculus
Is.3	determine the acceleration due to gravity near the surface of the Earth	weights, bungee jumping, skydiving
Is.4	determine the resultant of two or more vectors graphically or algebraically	navigation (e.g., boats, planes)
Is.5	draw scaled force diagrams, using a ruler and a protractor	building design (stress analysis), cranes, picture hangers
Is.6	resolve a vector into perpendicular components: graphically and algebraically	push lawn mower, amusement park wave swing
Is.7	sketch the theoretical path of a projectile	tennis, soccer, golf, archery
Is.8	use vector diagrams to analyze mechanical systems (equilibrium and nonequilibrium)	cars, elevators, tightrope walker, apparent weightlessness (micro-gravity)
Is.9	verify Newton's Second Law for linear motion	space shuttle, cruise control
Is.10	determine the coefficient of friction for two surfaces	skidding on driving surfaces, ice skating, Teflon surfaces, sledding
Is.11	verify Newton's Second Law for uniform circular motion	amusement park rides (e.g., merry-go-rounds)
Is.12	verify conservation of momentum	car crashes, balls, bats
Is.13	determine a spring constant	car suspension systems, rubber bands, spring scales

II. Energy

Sequence	Performance Indicators
II.1	When work* is done on or by a system, there is a change in the total energy* of the system.
II.2	Work done against friction results in an increase in the internal energy of the system.
II.3	Power* is the time-rate at which work is done or energy is expended.
II.4	All energy transfers are governed by the law of conservation of energy.*
II.5	Energy may be converted among mechanical, electromagnetic, nuclear, and thermal forms.
II.6	Potential energy is the energy an object possesses by virtue of its position or condition. Types of potential energy are gravitational* and elastic*.
II.7	Kinetic energy* is the energy an object possesses by virtue of its motion.
II.8	In an ideal mechanical system, the sum of the macroscopic kinetic and potential energies (mechanical energy) is constant.*
II.9	In a nonideal mechanical system, as mechanical energy decreases there is a corresponding increase in other energies such as internal energy.*

IIs. Energy Skills

Sequence	Process Skills (The student will be able to . . .)	Real-World Application
IIs.1	describe and explain the exchange between potential energy, kinetic energy, and internal energy for simple mechanical systems, such as a pendulum, a roller coaster, a spring, a freely falling object	skiing, skateboarding
IIs.2	predict velocities, heights, and spring compressions based on energy conservation	diving board, trampoline
IIs.3	determine the energy stored in a spring	ballpoint pen, pop-up toys
IIs.4	observe and explain energy conversions in real-world situations	hydroelectric power, solar power, Sun, engines
IIs.5	recognize and describe conversions among different forms of energy in real or hypothetical devices such as a motor, a generator, a photocell, a battery	solar-powered calculator, electric fan, heat pumps, air conditioners, Peltier devices
IIs.6	compare the power developed when the same work is done at different rates	elevators, running versus walking up stairs, motorcycles versus tractor-trailers
IIs.7	determine the factors that affect the period of a pendulum	Pirate Ship and Sky Coaster (amusement park rides), grandfather clock, swing

III. Electricity and Magnetism

Sequence	Performance Indicators
III.1	Gravitational forces are only attractive, whereas electrical and magnetic forces can be attractive or repulsive.
III.2	The inverse square law applies to electrical* and gravitational* fields produced by point sources.
III.3	Energy may be stored in electric* or magnetic fields. This energy may be transferred through conductors or space and may be converted to other forms of energy.
III.4	The factors affecting resistance in a conductor are length, cross-sectional area, temperature, and resistivity.*
III.5	All materials display a range of conductivity. At constant temperature, common metallic conductors obey Ohm's Law*.
III.6	A circuit is a closed path in which a current* can exist. *(Note: Use conventional current.)*
III.7	Electrical power* and energy* can be determined for electric circuits.
III.8	Circuit components may be connected in series* or in parallel.* Schematic diagrams are used to represent circuits and circuit elements.
III.9	Moving electric charges produce magnetic fields. The relative motion between a conductor and a magnetic field may produce a potential difference in the conductor.

IIIs. Electricity and Magnetism Skills

Sequence	Process Skills (The student will be able to . . .)	Real-World Application
IIIs.1	measure current and voltage in a circuit	transformers, power supplies, battery testers, power meters, multi-meters
IIIs.2	use measurements to determine the resistance of a circuit element	dimmer switches, volume controls, temperature controls (potentiometers)
IIIs.3	interpret graphs of voltage versus current	power meters, sound-board meters
IIIs.4	measure and compare the resistance of conductors of various lengths and cross-sectional areas	toasters, hair dryers, power transmission lines
IIIs.5	construct simple series and parallel circuits	household wiring, jumper cables, fuses, and circuit breakers
IIIs.6	draw and interpret circuit diagrams which include voltmeters and ammeters	schematic plans
IIIs.7	predict the behavior of lightbulbs in series and parallel circuits	holiday lights, flashlights
IIIs.8	map the magnetic field of a permanent magnet, indicating the direction of the field between the N (north-seeking) and S (south-seeking) poles	compass, magnets, magnetic storage media (e.g., floppy disks, hard drives, tapes)

IV. Waves

Sequence	Performance Indicators
IV.1	An oscillating system produces waves. The nature of the system determines the type of wave produced.
IV.2	Waves carry energy and information without transferring mass. This energy may be carried by pulses or periodic waves.
IV.3	Waves are categorized by the direction in which particles in a medium vibrate about an equilibrium position relative to the direction of propagation of the wave such as transverse and longitudinal waves.
IV.4	Mechanical waves require a material medium through which to travel.
IV.5	The model of a wave incorporates the characteristics of amplitude, wavelength*, frequency*, period*, wave speed*, and phase.
IV.6	Electromagnetic radiation exhibits wave characteristics. Electromagnetic waves can propagate through a vacuum.
IV.7	All frequencies of electromagnetic radiation travel at the same speed in a vacuum.*
IV.8	When a wave strikes a boundary between two media, reflection*, transmission, and absorption occur. A transmitted wave may be refracted.
IV.9	When a wave moves from one medium into another, the wave may refract due to a change in speed. The angle of refraction (measured with respect to the normal) depends on the angle of incidence and the properties of the media (indices of refraction).*
IV.10	The absolute index of refraction is inversely proportional to the speed of a wave.*

IV. Waves (*Continued*)

IV.11	When waves of a similar nature meet, the resulting interference may be explained using the Principle of Superposition. Standing waves are a special case of interference.
IV.12	Resonance occurs when energy is transferred to a system at its natural frequency.
IV.13	Diffraction occurs when waves pass by obstacles or through openings. The wavelength of the incident wave and the size of the obstacle or opening affect how the wave spreads out.
IV.14	When a wave source and an observer are in relative motion, the observed frequency of the waves traveling between them is shifted (Doppler effect).

IVs. Waves Skills

Sequence	Process Skills (The student will be able to . . .)	Real-World Application
IVs.1	compare the characteristics of two transverse waves such as amplitude, frequency, wavelength, speed, period, and phase	stadium waves, electromagnetic waves, S-waves (secondary earthquake waves)
IVs.2	draw wave forms with various characteristics	oscilloscopes
IVs.3	identify nodes and antinodes in standing waves	guitar string (vibrating stretched wire), pipe organ (vibrating air column)
IVs.4	differentiate between transverse and longitudinal waves	polarized sunglasses, liquid crystal displays (e.g., computer screens, watches, calculator), speakers, 3-D movies
IVs.5	determine the speed of sound in air	echoes
IVs.6	predict the superposition of two waves interfering constructively and destructively (indicating nodes, antinodes, and standing waves)	stereo speakers, surround sound, iridescence (e.g., butterfly wings, soap bubbles), Tacoma Narrows Bridge, beats, electronic tuners
IVs.7	observe, sketch, and interpret the behavior of wave fronts as they reflect, refract, and diffract	ocean waves, amusement park wave pools, harbor waves, pond ripples, ultrasonic cleaners (standing waves)
IVs.8	draw ray diagrams to represent the reflection and refraction of waves	barcode scanners, mirrors
IVs.9	determine empirically the index of refraction of a transparent medium	diamonds, spearfishing, lenses

V. Modern Physics

Sequence	Performance Indicators
V.1	States of matter and energy are restricted to discrete values (quantized).
V.2	Charge is quantized on two levels. On the atomic level, charge is restricted to the elementary charge (charge on an electron or proton). On the subnuclear level charge appears as fractional values of the elementary charge (quarks).
V.3	On the atomic level, energy is emitted or absorbed in discrete packets called photons.*
V.4	The energy of a photon is proportional to its frequency.*
V.5	On the atomic level, energy and matter exhibit the characteristics of both waves and particles.
V.6	Among other things, mass-energy and charge are conserved at all levels (from subnuclear to cosmic).
V.7	The Standard Model of Particle Physics has evolved from previous attempts to explain the nature of the atom and states that: 1. Atomic particles are composed of subnuclear particles. 2. The nucleus is a conglomeration of quarks which manifest themselves as protons and neutrons. 3. Each elementary particle has a corresponding antiparticle.
V.8	Behaviors and characteristics of matter, from the microscopic to the cosmic levels, are manifestations of its atomic structure. The macroscopic characteristics of matter, such as electrical and optical properties, are the result of microscopic interactions.
V.9	The total of the fundamental interactions is responsible for the appearance and behavior of the objects in the universe.
V.10	The fundamental source of all energy in the universe is the conversion of mass into energy.*

Vs. Modern Physics Skills

Sequence	Process Skills (The student will be able to . . .)	Real-World Application
Vs.1	interpret energy-level diagrams	black light posters, lasers
Vs.2	correlate spectral lines with energy-level diagram	neon lights, street lights

QUESTION INDEX

What follows is an index to the examination questions that are explained in this book. The questions are indexed according to the sequence numbers given in the core and in the skills section.

Some questions embrace more than one topic and/or skill. These questions are marked with a dagger (†).

SEQUENCE	JUNE 2013	JUNE 2014	JUNE 2015	JUNE 2016	JUNE 2017	JUNE 2018	JUNE 2019
M. Math Skills							
M.1	66, 70	†51, †52, †53, †54	16, 67, 73, 74, 75	†2, †5, 7–†8, †12–15, 17–18, 23, 30–33, 38–39, 41–43, †59–60, 68, †70–†73, 76–84	†4–†7, 9–†13, 15–17, 19, 21–22, 28–29, 32, 35, 39–43, 48, 49, †55–57, 60–61, †63–64, †66–68, 71–75, 77–82, 84, 85	†3, †4, †6, 9, †13, †14, 15, 19, 20, 26, 29, 30, 32, 36–39, 41–44, 50, 51, †58–†70, 76–78, 82, 84	†2, 7, †8, 10, †12–16, 23, 26, 32, 34, †36, 39, 41, 45, 46, 48, 49, 51–53, †57–75, 78–80
M.2	57	†41, †46	24, 43, 48	37, 40, †44–45	†44		11, 33, 37
M.3	58, 59	48, †64, †65	61, 62	46, 62, †65	3, 36, 38	53, †59, 71, 72	56
I. Mechanics							
I.1	1, 36	1, 36	1	1	1	1	1
I.2	3, 4, 72, 73	2, 7	2, 39, 40	31, 37	†4, †6	†3, †4, †6	23, 34
I.3		4	22		†7	7	3, 59, 60
I.4	2		3, 41			53	†2, 19
I.5		3	37, 41		2	5, 71, 92	7, †40
I.6	5		†38		37		†40
I.7	6		†38, 61, 62	56–58	†55	8	†40
I.8		47	†38		†55	5	†40
I.9	7	16	†4, 33	4	8, 32, 33	2	
I.10	9	44	†4	3	8		
I.11	12, 38, 40	8, 73, 74	†4	51, 52	43	40	41
I.12	10, 63, 64, 65	75, 77, 78	5, 11, 66	†2, †65	†5		5, 66, 67
I.13	8	76	†65, 68, 70	9, 53		11, 73, 74, 75	
I.14		38, 45	57, 58, 59		†12, 79, 80	12, 74, 76, 77, 79, 80	6, 42
I.15		9, 10	7	†59, 60, 61		9	61–65
I.16			6	†70, 71		32	†8
I.17	11	5	9	†5		10	18, 33
I.18	37, 54, 55		10	†8	11, 40, 41	39	
I.19							
I.20	47	71, 72	12			†14, 20	†36
I.21	13	6	11	25			
Is. Mechanics Skills							
Is.1		63, †64	60	63, 64			37, 55
Is.2		†65		62		45	
Is.3							
Is.4	71, 74, 75	†52, †53	3, 41, 69	67, 68, 69		52	2

SEQUENCE	JUNE 2013	JUNE 2014	JUNE 2015	JUNE 2016	JUNE 2017	JUNE 2018	JUNE 2019
Is.5	†51, 54						
Is.6			37	39		71, 72	7
Is.7		57					
Is.8		55, 56					
Is.9							
Is.10		79, 80					
Is.11							
Is.12							
Is.13					60	32	
II. Energy							
II.1	81		42, 48	†12, †13	19, 66, 70	54, 55	10
II.2	84					56	
II.3	39	58, 59	44, 51, 52	17, 18		15	†12
II.4	21	84	†15, †45				39
II.5	15, 27, 85	15	14	74			
II.6	14, 82, 83		†45	14	67, 68	†38, 60, 61	9, 39, 52, 53
II.7	45	81, 82	†45	15	77, 78	†38	39
II.8							11, 52, 53
II.9		42, 62	†15		39, 69		39
IIs. Energy Skills							
IIs.1		31					
IIs.2							39
IIs.3		19		72, †73			
IIs.4							51
IIs.5							
IIs.6							†12
IIs.7		50					
III. Electricity and Magnetism							
III.1	23	32	30, 71, 72		14, 30, 34	16, 34	4, 35
III.2	43	12, 33		49	15		
III.3	33, 41, 46, 56	13, 37, 39	16, 31	32	10, †13, 42, 48	†13, 18, 28, 37	†13, †14, 47, 68–70
III.4	†51, 52	14	17	45, 46	18, 47	17, 19, 66, 67, 68	†15
III.5			81, 82			43	73, 74
III.6	42	22	18	36, †44, 77, 78	63, 64, 73	33, †58, †59	
III.7	30, 32	40, 69, 70	83	38	21, 22, 56, 57, 74, 75	69, †70	17, 48, 75
III.8			19, 81, 82	33, 76, 77, 78, 79, 80	20, 71, 72, 76	33, 50	†16, 43, 71–72
III.9	28	11		10, 22			
IIIs. Electricity and Magnetism Skills							
IIIs.1	31						
IIIs.2	60, 61, 62	67, 68	13				
IIIs.3							84, 85

SEQUENCE	JUNE 2013	JUNE 2014	JUNE 2015	JUNE 2016	JUNE 2017	JUNE 2018	JUNE 2019
IIIs.4				42			
IIIs.5		66		33, 50, 75			54
IIIs.6					35, 71–76	33, 43	
IIIs.7			83, 84				
IIIs.8				11, 35	51		19
IV. Waves							
IV.1			26		26		
IV.2	16		20	20	23		
IV.3	19	61	26	6, 34	29	47	20
IV.4					25		20
IV.5	18, 22, 24, 29, 80	20, 21, 23, 35	20, 21, 24, 53, 54	21, 24	17	26, 35, 82, 83, 84, 85	21, 22, 78–80
IV.6	17		23				
IV.7		24	23	16			
IV.8	76, 77	18	34		9, 27		24
IV.9	49, 78,79	49	8		27, 80, 82	23	25
IV.10	20	†41	46	23	28	24	26, †57, 58
IV.11	48		35	26, 49	45	27, 48	
IV.12	25	27	27	48	24	22	
IV.13			28	47		25	29
IV.14	26	26	32	19	50	49	28
IVs. Waves Skills							
IVs.1		34	25, 49		62	41	
IVs.2	53				58, 59		76, 77
IVs.3					46	81	27
IVs.4		17					
IVs.5							
IVs.6		25	35		65		50
IVs.7		60					29, 56
IVs.8			63		83	57	
IVs.9			64, †65	54, 55	84, 85	62, 63	
V. Modern Physics							
V.1	69		29, 76				
V.2	35	28	50		31, 52		†44, †38
V.3			77, 80	81, 82			
V.4	67, 68	†46, 83	47, 78, 79	27, 83, 84	44	29, 64, 65	
V.5							
V.6		85					
V.7	44			28, 29	52	42	31, †38, †44
V.8							30
V.9		29			54		
V.10	50	30	55, 56	30	61	31	45
Vs. Modern Physics Skills							
Vs.1	34		29, 76	81	16		32, †40
Vs.2		43		85			46

Glossary of Important Terms

absolute index of refraction The ratio of the speed of light in a vacuum to the speed of light in a medium.

absorption spectrum A series of dark spectral lines or bands formed by the absorption of specific wavelengths of light by atoms or molecules.

acceleration The time rate of change in velocity. The SI unit is meters per second^2.

accuracy The agreement of a measured value with an accepted standard.

alpha particle A helium nucleus; a particle consisting of two protons and two neutrons.

alternating current An electric current that varies in magnitude and alternates in direction.

ammeter A device used to measure electric current. It is constructed by placing a low-resistance shunt across the coil of a galvanometer.

ampere (A) The SI unit of electric current, equivalent to the unit coulomb per second.

amplitude The maximum displacement in periodic phenomena such as wave motion, pendulum motion, and spring oscillation.

angle of incidence The angle made by the incident wave with the surface of a medium; the angle made by the incident ray with the normal to the surface of the medium.

angle of reflection The angle made by the reflected wave with the surface of a medium; the angle made by the reflected ray with the normal to the surface of the medium.

angle of refraction The angle made by the refracted wave with the surface of a medium; the angle made by the refracted ray with the normal to the surface of the medium.

anode The positive terminal of a DC source of potential difference.

antimatter One or more atoms composed entirely of antiparticles.

antinode The point or locus of points on an interference pattern (such as a standing wave or double slit pattern) that results in maximum constructive interference.

antiparticle The counterpart of a subatomic particle. An antiparticle has the same mass as its companion particle, but its electric charge is opposite in sign.

atomic number The number of protons in the nucleus of an atom. The atomic number defines the element.

Balmer series The visible-ultraviolet line spectrum of atomic hydrogen. It is the result of electrons falling from higher levels to the $n = 2$ state.

baryon A "heavy" particle, such as a proton or a neutron. Baryons are composed of three quarks. For example, a neutron is composed of an up quark and two down quarks (*udd*).

battery A combination of two or more electric cells.

beta (−) particle An electron formed in the nucleus by the disintegration of a neutron.

beta (+) particle A positron, the antiparticle of the electron, formed in the nucleus by the disintegration of a proton.

binding energy The energy equivalent of the mass defect of a nucleus.

cathode The negative terminal of a DC source of potential difference.

cathode ray tube A device for visualizing an electron beam. It consists of an evacuated tube with a source of electrons at one end and a fluorescent screen at the other end. The electron beam is controlled by electric and magnetic fields.

Celsius scale (°C) The temperature scale that fixes the (atmospheric) freezing point of water at 0° and the boiling point of water at 100°.

centripetal acceleration The acceleration that is directed along the radius and toward the center of a curved path in which an object is moving.

centripetal force The force that causes centripetal acceleration. It is responsible for changing an object's direction, not its speed.

chromatic aberration A lens defect in which different colors of light are focused at different points.

circuit A closed loop formed by a source of potential difference connected to one or more resistances.

coefficient of friction The ratio of the force of friction on an object to the normal force on it.

coherent light A series of light waves that have a fixed phase relationship; the type of light produced by a laser. Lasers produce beams of monochromatic coherent light.

component One of the two or more vectors into which a given vector may be resolved.

concurrent forces Two or more forces acting at the same point.

conductor A material that allows electrons to flow through it freely. Metals such as copper and silver are conductors.

constructive interference The combination of two in-phase wave disturbances to produce a single wave disturbance whose amplitude is the sum of the amplitudes of the individual disturbances.

converging lens A lens that focuses its transmitted light to a point. Generally, convex lenses are converging lenses.

convex lens A lens that is thicker in the middle than at the edges.

coulomb (C) The SI unit of electric charge, approximately equal to 6.25×10^{18} elementary charges.

Coulomb's law The electrostatic force between two point charges is directly proportional to the product of the charges and inversely proportional to the square of the distance between the charges.

critical angle The angle of incidence for which the corresponding angle of refraction is 90°.

cycle One complete repetition of the pattern in any periodic phenomenon.

de Broglie wavelength The wavelength of a matter wave.

destructive interference The combination of two out-of-phase wave disturbances to produce a single wave disturbance whose amplitude is the difference of the amplitudes of the individual disturbances.

diffraction The bending of a wave around a barrier.

diffuse reflection The reflection of parallel light rays by irregular surfaces.

direct current An electric current that flows in one direction only.

dispersion The separation of polychromatic light into its individual colors.

displacement A change of position in a specific direction.

Doppler effect An apparent change in frequency that results when a wave source and an observer are in relative motion with respect to each other.

elastic potential energy The energy stored in a spring when it is compressed or stretched.

electric current The time rate of flow of charged particles. The SI unit of electric current is the ampere (A).

electric field The region of space around a charged object that affects other charges.

electric field intensity The ratio of the force that an electric field exerts on a charge to the magnitude of the charge.

electric motor A device that converts electrical energy into mechanical energy.

electric potential The total work done by an electric field in bringing 1 coulomb of positive charge from infinity to a specific point. The potential is a positive number if the charge is repelled by the field and a negative number if the charge is attracted by the field. At infinity, the potential is taken to be zero. Electric potential is measured in volts.

electromagnetic force The fundamental force that governs the attraction or repulsion among charged particles, whether at rest or in motion.

electromagnetic induction The process by which the magnetic field and the mechanical energy are used to generate a potential difference.

electromagnetic radiation The propagation of electromagnetic waves in space.

electromagnetic spectrum The entire range of electromagnetic waves from the lowest to the highest frequencies.

electromagnetic wave A periodic wave, consisting of mutually perpendicular electric and magnetic fields, that is radiated away from the vicinity of an accelerating charge.

electromotive force The potential difference produced as a result of the conversion of other forms of energy into electrical energy.

electron A fundamental, negatively charged, subatomic particle; a lepton.

electron-volt (eV) A unit of energy equal to the work needed to move an elementary charge across a potential difference of 1 volt.

electroscope A device used to detect the presence of electric charges.

elementary charge The magnitude of charge present on a proton or an electron. An elementary charge is approximately equal to 1.6×10^{-19} coulomb.

emission spectrum A series of bright spectral lines or bands formed by the emission of certain wavelengths of light by excited atoms falling to lower energy states.

energy A quantity related to work.

equilibrant A single balancing force that maintains the static equilibrium of an object.

equivalent resistance A single resistance that can be substituted for a group of resistances in series or in parallel.

excited state A condition in which the energy of an atom is greater than its lowest energy state.

ferromagnetic Referring to a material, such as iron, that has the ability to strengthen greatly the magnetic field of a current-carrying coil.

field lines A series of lines used to represent the magnitude and direction of a field.

fission The process of splitting a heavy nucleus, such as uranium-235, into lighter fragments. Fission is accompanied by the release of large quantities of energy.

force A push or a pull on an object. If the force is unbalanced, an acceleration will result.

frame of reference A coordinate grid and a set of synchronized clocks that can be used to determine the position and time of an event.

free fall A motion in the Earth's gravitational field without regard to air resistance.

frequency The number of repetitions produced per unit time by periodic phenomena.

friction The force present as the result of contact between two surfaces. The direction of a frictional force is opposite to the direction of motion.

fundamental forces The four forces in nature responsible for all interactions among matter. (See also *electromagnetic force, gravitational force, strong force, weak force*.)

fusion (1) The process of uniting lighter nuclei, such as deuterium, into a heavier nucleus. Fusion is accompanied by the release of large quantities of energy. (2) In the study of heat and thermodynamics, a synonym for *melting*.

galvanometer A device, consisting of a coil-shaped wire placed between the opposite poles of a permanent magnet, that is used to detect small amounts of electric current.

gamma radiation Very high energy photons of electromagnetic radiation. Gamma photons have the highest frequencies in the electromagnetic spectrum.

geiger counter A device that detects charged nuclear particles.

generator A device that uses a magnetic field and mechanical energy to induce a source of electromotive force.

gravitational field The region of space around a mass that affects other masses.

gravitational field intensity The ratio of the force that a gravitational field exerts on a mass to the magnitude of the mass, numerically equal to the acceleration due to gravity.

gravitational force The fundamental universal attraction between masses.

gravitational potential energy The energy that an object acquires as a result of the work done in moving the object against a gravitational field.

ground An extremely large source or reservoir of electrons, which can supply or accept electrons as the need arises.

ground state The lowest energy state of an atom.

hadron Any particle that interacts through the strong force. Hadrons are classified as baryons or as mesons.

heat energy The energy that is transferred from warmer objects to cooler ones because of a temperature difference between them.

hertz (Hz) The SI unit of frequency, equivalent to the unit second^{-1}.

image An optical reproduction of an object.

impulse The product of the net force acting on an object and the time during which the force acts. The impulse delivered to an object is equal to its change in momentum. The direction of the impulse is the direction of the force. The SI unit of impulse is the newton • second, which is equivalent to the kilogram • meter per second.

incident ray A ray of a wave impinging on a surface.

incident wave A wave impinging on a surface.

induced current An electric current that is the result of an induced electromotive force.

induced emf A potential difference created when a magnetic field is interrupted over a time period.

induction (1) A method of charging a neutral object by using a charged object and a ground. The induced charge is always opposite to the charge on the charged object. (2) See *induced current* and *induced emf*.

inertia The property of matter that resists changes in motion. Mass is the quantitative measure of inertia.

instantaneous velocity The ratio of displacement to time at any given instant; the slope of a line tangent to a displacement-time graph at any given point.

insulator A material that is a very poor conductor because it has few conduction electrons. Wood and glass are examples of insulators.

interference pattern Regions of constructive and destructive interference that are present in a medium as a result of the combination of two or more waves.

internal energy The total kinetic and potential energy associated with the atoms and molecules of an object.

joule (J) The SI unit of work and energy, equivalent to the unit newton • meter.

kilogram (kg) The SI unit of mass; a fundamental unit.

kinetic energy The energy that an object possesses because of its motion.

laser An acronym for light amplification by the stimulated emission of radiation. A laser is a device that emits extremely intense, monochromatic, coherent light.

lepton Any particle that participates in the weak force: electron, muon, tau, and their neutrinos.

longitudinal wave A wave in which the disturbance is parallel to the direction of the wave's motion. Sound waves are longitudinal.

magnet Any material that aligns itself, when free to do so, in an approximate north-south direction. Magnets exert forces on one another and on charged particles in motion.

magnetic field The region of space around a magnet or charge in motion that exerts a force on magnets or other moving charges.

magnification The ratio of image size to object size.

mass (1) The measure of an object's ability to obey Newton's second law of motion. (2) The measure of an object's ability to obey Newton's law of universal gravitation. The SI unit of mass is the kilogram.

mass defect The mass lost by a nucleus when it is assembled from its nucleons. (See also *binding energy*.)

mass number The sum of the number of protons and neutrons in a nucleus; the number of nucleons the nucleus contains.

matter waves According to quantum theory, the waves associated with moving particles.

medium A material through which a disturbance, such as a wave, travels.

meson Any particle that is composed of a quark and an antiquark. For example, a π^+ meson is composed of an up quark and an antidown quark $\left(u\bar{d}\right)$.

momentum The product of mass and velocity. The direction of an object's momentum is the direction of its velocity. The SI unit of momentum is the kilogram • meter per second.

natural frequency A specific frequency with which an elastic body may vibrate if disturbed.

net force The unbalanced force present on an object; the accelerating force.

neutrino A lepton with no charge and questionable mass. It and its antiparticle are products of beta-decay reactions.

Newton (N) The SI unit of force, equivalent to the unit kilogram • meter per second^2.

Newton's first law of motion Objects remain in a state of uniform motion unless acted upon by an unbalanced force.

Newton's law of universal gravitation Any two bodies in the universe are attracted to each other with a force that is directly proportional to their masses and inversely proportional to the square of the distance between them.

Newton's second law of motion The unbalanced force on an object is equal to the product of its mass and acceleration.

Newton's third law of motion If object A exerts a force on object B, then object B exerts an equal and opposite force on object A.

node The point or locus of points on an interference pattern, such as a standing wave or double-slit pattern, that results in total destructive interference.

normal A line perpendicular to a surface.

normal force The force that keeps two surfaces in contact. If an object is on a *horizontal* surface, the normal force on the object is equal to its weight.

nucleon A proton or a neutron.

nucleus The dense, positively charged core of an atom.

ohm (Ω) The SI unit of electrical resistance, equivalent to the unit volt per ampere.

Ohm's law A relationship in which the ratio of the potential difference across certain conductors to the current in them is constant at constant temperature.

parallel circuit An electric circuit with more than one current path.

particle accelerator A device used to accelerate charged nuclear particles.

period The time for one complete repetition of a periodic phenomenon. The SI unit of period is the second.

periodic wave A regularly repeating series of waves.

phase (1) A form in which matter can exist, including liquid, solid, gas, and plasma. (2) In wave motion, the points on the wave that have specific time and space relationships.

photoelectric effect A phenomenon in which light causes electrons to be ejected from certain materials.

photon The fundamental particle of electromagnetic radiation.

Planck's constant A universal constant (h) relating the energy of a photon to its frequency; its approximate value is 6.62×10^{-34} joule • second.

point charge A charge with negligible physical dimensions.

polarization A process that produces transverse waves that vibrate in only one plane. Polarization is limited to transverse waves: light can be polarized; sound cannot.

polychromatic Referring to light waves of different colors (frequencies).

potential difference The ratio of the work required to move a test charge between two points in an electric field to the magnitude of the test charge. The unit of potential difference is the volt.

potential energy The energy that a system has because of its relative position or condition.

power The time rate at which work is done or energy is expended. The SI unit of power is the watt, which is equivalent to the unit joule per second.

precision The limit of the ability of a measuring device to reproduce a measurement.

pressure The force on a surface per unit area. The SI unit of pressure is the pascal (Pa).

proton A positively charged subatomic particle with a charge equal in magnitude to that of the electron; a baryon.

pulse A nonperiodic disturbance in a medium.

quantum A discrete quantity of energy.

quarks The particles of which protons, neutrons, baryons, and mesons are composed. Quarks carry a charge of either one-third or two-thirds of an elementary charge and come in six "flavors": top, bottom, up, down, charm, and strange. Each quark has a companion antiquark.

radioactive decay A spontaneous change in the nucleus of an atom.

radioactivity Changes in the nucleus of an atom that produce the emission of subatomic particles or photons.

ray A straight line indicating the direction of travel of a wave.

real image An image created by the actual convergence of light waves. Real images from single mirrors and single lenses are inverted and can be projected on a screen.

refraction The change in the direction of a wave when it passes obliquely from one medium to another in which it moves at different speed.

regular reflection The reflection of parallel light rays incident on a smooth plane surface.

resistance The opposition of a material to the flow of electrons through it; the ratio of potential difference to current.

resistivity A quantity that allows the resistance of substances to be compared. Numerically, it is the resistance of a 1-meter conductor with a cross-sectional area of 1 square meter. The SI unit of resistivity is the ohm • meter.

resistor A device that supplies resistance to a circuit.
resolution The process of determining the magnitude and direction of the components of a vector.
resonance The spontaneous vibration of an object at a frequency equal to that of the wave that initiates the resonant vibration.
resultant A vector sum.

satellite A body that revolves around a larger body as a result of a gravitational force.
scalar quantity A quantity, such as mass or work, that has magnitude but not direction.
series circuit A circuit with only one current path.
significant digits The digits that are part of any measurement.
solenoid A coil of wire wound as a helix. When a current is passed through the solenoid, it becomes an electromagnet.
speed The time rate of change of distance; the magnitude of velocity. The SI unit of speed is the meter per second.
spring constant The ratio of the force required to stretch or compress a spring to the magnitude of the stretch or compression.
standard model A model of matter and the fundamental interactions that govern it.
standing wave A wave pattern created by the continual interference of an incident wave with its reflected counterpart. The standing wave does not travel, but oscillates about an equilibrium position.
static equilibrium The condition of a body when a net force of zero is acting on it.
superconductor A material with no electrical resistance.
strong force The strongest of the four fundamental forces. It mediates interactions among certain nuclear particles.

temperature The "hotness" of an object, measured with respect to a chosen standard.
torque A force, applied perpendicularly to a designated line, that tends to produce rotational motion.
total internal reflection The reflection of a wave inside a relatively dense medium produced when the angle of the wave with the boundary exceeds the critical angle.
total mechanical energy The sum of the potential and kinetic energies of a mechanical system.
transverse wave A wave in which the disturbance is perpendicular to the direction of the wave's motion. Light waves are transverse.

uniform In the study of motion, a term that is equivalent to *constant*.

vector A representation of a vector quantity; an arrow in which the length represents the magnitude of the quantity and the arrowhead points in the direction of its orientation.

vector quantity A quantity, such as force or velocity, that has both magnitude and direction.

velocity The time rate of change of displacement.

virtual image An image formed by projecting diverging light behind a mirror or a lens.

volt (V) The SI unit of potential difference, equivalent to the unit joule per coulomb.

voltage Another term for *potential difference*.

voltmeter A device used to measure potential difference and constructed by placing a large resistor in series with the coil of a galvanometer.

watt (W) The SI unit of power, equivalent to the unit joule per second.

wave A series of periodic oscillations of a particle or a field both in time and in space.

wave front All points on a wave that are in phase with each other.

wavelength The length of one complete wave cycle.

weak force The fundamental force that arises in certain types of radioactive decay.

weight The gravitational force present on an object.

work The product of the force on an object and its displacement. The SI unit of work is the joule.

Reference Tables for Physics

When you take the New York State Regents examination in physics, you will be provided with a set of reference tables to aid you in answering the questions on the examination. For those students not taking the examination, the following tables will provide a convenient reference when answering the questions and problems presented in this book. Each of these tables is given below along with a brief description of it. Please read the Important Note on page 2 concerning these tables.

A. LIST OF PHYSICAL CONSTANTS

List of Physical Constants

Name	Symbol	Value
Universal gravitational constant	G	6.67×10^{-11} N•m^2/kg^2
Acceleration due to gravity	g	9.81 m/s^2
Speed of light in a vacuum	c	3.00×10^{8} m/s
Speed of sound in air at STP		3.31×10^{2} m/s
Mass of Earth		5.98×10^{24} kg
Mass of the Moon		7.35×10^{22} kg
Mean radius of Earth		6.37×10^{6} m
Mean radius of the Moon		1.74×10^{6} m
Mean distance—Earth to the Moon		3.84×10^{8} m
Mean distance—Earth to the Sun		1.50×10^{11} m
Electrostatic constant	k	8.99×10^{9} N•m^2/C^2
1 elementary charge	e	1.60×10^{-19} C
1 coulomb (C)		6.25×10^{18} elementary charges
1 electronvolt (eV)		1.60×10^{-19} J
Planck's constant	h	6.63×10^{-34} J•s
1 universal mass unit (u)		9.31×10^{2} MeV
Rest mass of the electron	m_e	9.11×10^{-31} kg
Rest mass of the proton	m_p	1.67×10^{-27} kg
Rest mass of the neutron	m_n	1.67×10^{-27} kg

The most important physical constants and their symbols, where appropriate, are given in this table. The value of each constant is given to three significant digits along with its units.

For example, if you were asked to calculate the gravitational force between the Moon and Earth, you would use this table to find the masses of the Moon and Earth and the mean Earth–Moon distance.

B. PREFIXES FOR POWERS OF 10

Prefixes for Powers of 10

Prefix	Symbol	Notation
tera	T	10^{12}
giga	G	10^{9}
mega	M	10^{6}
kilo	k	10^{3}
deci	d	10^{-1}
centi	c	10^{-2}
milli	m	10^{-3}
micro	μ	10^{-6}
nano	n	10^{-9}
pico	p	10^{-12}

The metric prefixes, along with their symbols and values, are given for numbers between 10^{-12} and 10^{12}.

For example, 10^{-1} meter is known as a *decimeter* (dm).

C. APPROXIMATE COEFFICIENTS OF FRICTION

Approximate Coefficients of Friction

	Kinetic	Static
Rubber on concrete (dry)	0.68	0.90
Rubber on concrete (wet)	0.58	
Rubber on asphalt (dry)	0.67	0.85
Rubber on asphalt (wet)	0.53	
Rubber on ice	0.15	
Waxed ski on snow	0.05	0.14
Wood on wood	0.30	0.42
Steel on steel	0.57	0.74
Copper on steel	0.36	0.53
Teflon on Teflon	0.04	

The coefficients of friction (μ) for a number of pairs of surfaces are provided. The *kinetic* coefficients of friction are used when the surfaces are in relative motion. The *static* coefficients of friction are used when the surfaces are at rest with respect to each other.

D. THE ELECTROMAGNETIC SPECTRUM

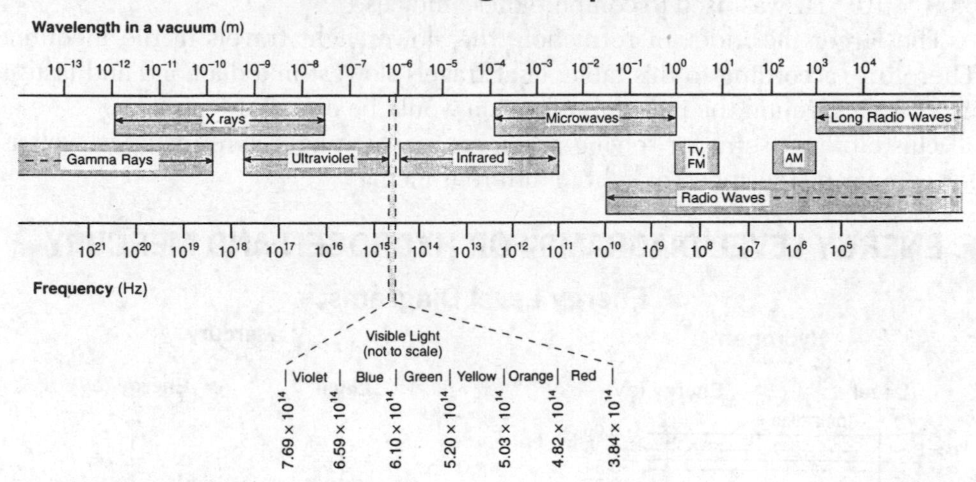

The wavelength and frequency ranges for the principal types of electromagnetic radiation are given in this chart. Note that the type of radiation is principally determined by its *source* and not necessarily by its wavelength (or frequency). As a result, there is some overlap between certain types of radiation. In addition, the frequency ranges for visible light are provided at the bottom of the chart.

E. ABSOLUTE INDICES OF REFRACTION

Absolute Indices of Refraction
($f = 5.09 \times 10^{14}$ Hz)

Material	Index
Air	1.00
Corn oil	1.47
Diamond	2.42
Ethyl alcohol	1.36
Glass, crown	1.52
Glass, flint	1.66
Glycerol	1.47
Lucite	1.50
Quartz, fused	1.46
Sodium chloride	1.54
Water	1.33
Zircon	1.92

The absolute index of refraction is defined as the ratio of the speed of light in a vacuum to the speed of light in a medium. (Monochromatic yellow light of 5.09×10^{14} Hz was used to compute these indices.)

The larger the index of refraction, the slower light travels in the medium. Therefore, according to this table, light travels slowest in a diamond and fastest in air. In a vacuum, the index of refraction would be exactly 1.

This table is useful for solving problems in which light is refracted as well as for comparing the speed of light in different media.

F. ENERGY LEVEL DIAGRAMS FOR HYDROGEN AND MERCURY

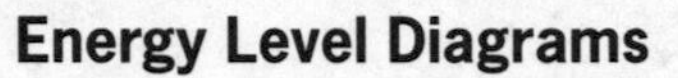

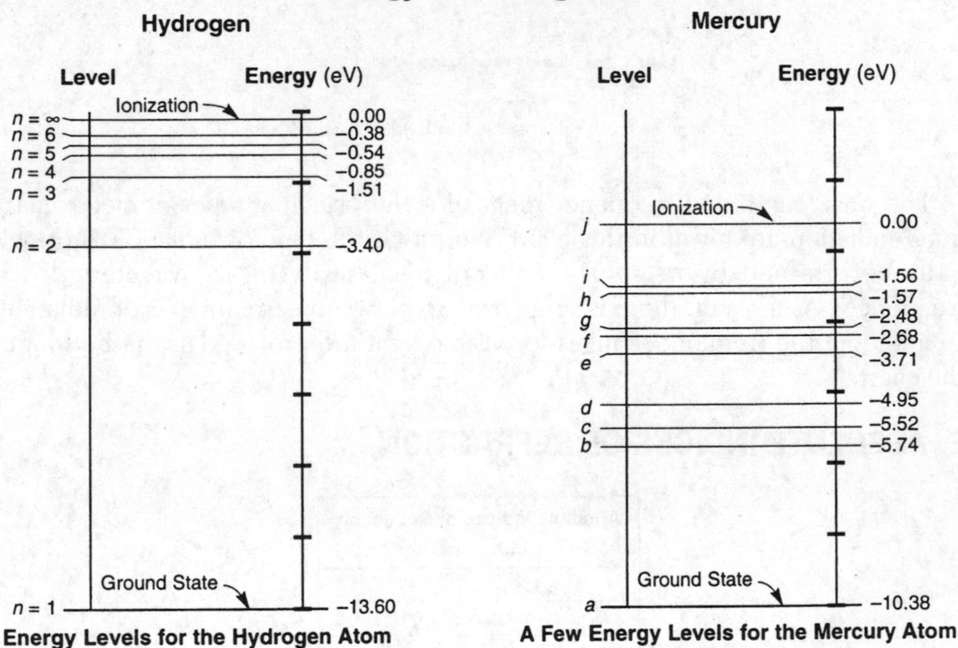

Energy Levels for the Hydrogen Atom

A Few Energy Levels for the Mercury Atom

Each energy state is represented by a horizontal line. Its energy value, in eV, is given at the right. Note that these values are negative numbers that increase to a maximum value of 0.00. The lowest energy state is known as the *ground state*. Every other state is known as an *excited state*. At a value of 0.00 eV, the electron is no longer associated with the atom, a condition known as *ionization*.

At the left is the label corresponding to the energy state. For hydrogen, these states are integers known as *principal quantum numbers*. For mercury, the atom is much more complex and the various states are represented as letters.

In order to calculate the energy involved in a particular transition, one *subtracts* the final energy value from the initial value. If the difference is negative, then energy is released by the atom as a photon. If it is positive, then the energy is absorbed by the atom.

G. CLASSIFICATION OF MATTER

Classification of Matter

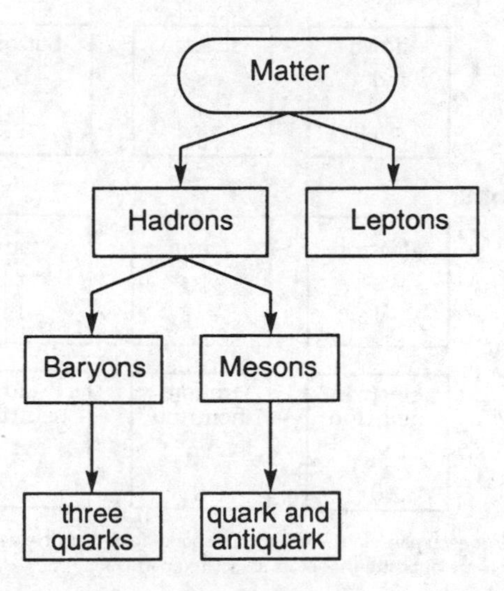

This chart classifies matter according to the *standard model*. For example, all nuclear particles are classified as either hadrons or leptons. Leptons (and their antiparticles) are not composed of smaller units and have a charge of ± 1 or 0. Hadrons are composed of smaller units and are ultimately composed of quarks and/or antiquarks.

H. PARTICLES OF THE STANDARD MODEL

Particles of the Standard Model

Quarks

Name	up	charm	top
Symbol	u	c	t
Charge	$+\frac{2}{3}$ e	$+\frac{2}{3}$ e	$+\frac{2}{3}$ e
	down	strange	bottom
	d	s	b
	$-\frac{1}{3}$ e	$-\frac{1}{3}$ e	$-\frac{1}{3}$ e

Leptons

electron	muon	tau
e	μ	τ
−1e	−1e	−1e
electron neutrino	muon neutrino	tau neutrino
ν_e	ν_μ	ν_τ
0	0	0

Note: For each particle there is a corresponding antiparticle with a charge opposite that of its associated particle.

The names, symbols, and charges of the six quarks and six leptons are given in this chart. Electric charges are given in terms of the elementary charge (e). Each particle has a corresponding *antiparticle* that has an electric charge opposite that of the particle. For example, the antimuon ($\bar{\mu}$) has an electric charge of $+1e$.

I. CIRCUIT SYMBOLS

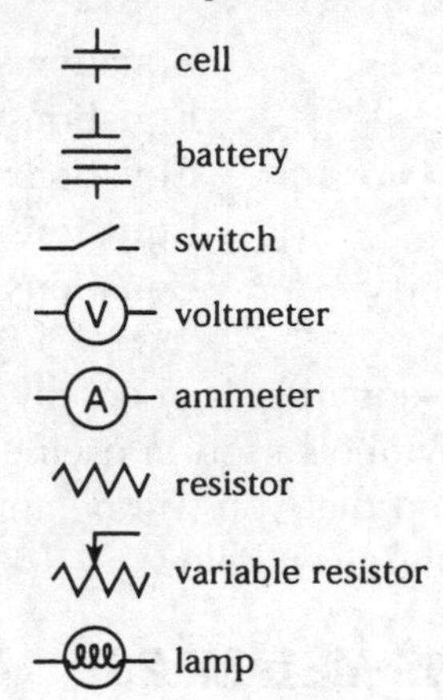

The symbols used in simple electric circuits are provided in this chart. It is useful in interpreting the elements that make up a particular type of circuit.

J. RESISTIVITIES AT 20°C

Resistivities at 20°C	
Material	Resistivity (Ω•m)
Aluminum	2.82×10^{-8}
Copper	1.72×10^{-8}
Gold	2.44×10^{-8}
Nichrome	$150. \times 10^{-8}$
Silver	1.59×10^{-8}
Tungsten	5.60×10^{-8}

The resistivities, in ohm • meters, are given for a number of conductors at 20°C. Resistivities are useful for comparing the relative abilities of various materials to conduct an electric current. They are also used to calculate the resistance of a conductor whose length and cross-sectional area are known.

K. EQUATIONS FOR PHYSICS

Reference Table *K* is divided into the following five areas:

- Mechanics: Equations ME1–ME23
- Electricity: Equations EL1–EL10
- Waves: Equations W1–W6
- Modern Physics: Equations MP1–MP3
- Geometry and Trigonometry: Equations GT1–GT4

In each area, the relevant equation is generally given on the left side of the page. The definition of the symbols used in each equation is given on the right side of the page. It is expected that you will be able to solve quantitative problems using any of the equations given below.

MECHANICS (EQUATIONS ME1–ME23)

ME1. $\overline{v} = \frac{d}{t}$

ME2. $a = \frac{\Delta v}{t}$

ME3. $v_f = v_i + at$

ME4. $d = v_i t + \frac{1}{2}at^2$

ME5. $v_f^2 = v_i^2 + 2ad$

ME6. $A_y = A \sin \theta$

ME7. $A_x = A \cos \theta$

ME8. $a = \frac{F_{net}}{m}$

ME9. $F_f = \mu F_N$

ME10. $F_g = \frac{Gm_1m_2}{r^2}$

ME11. $g = \frac{F_g}{m}$

a = acceleration

a_c = centripetal acceleration

A = any vector quantity

d = displacement or distance

E_T = total energy

F = force

F_c = centripetal force

F_f = force of friction

F_g = weight or force due to gravity

F_N = normal force

F_{net} = net force

F_s = force on a spring

g = acceleration due to gravity or gravitational field strength

G = universal gravitational constant

h = height

J = impulse

k = spring constant

ME12. $p = mv$	KE = kinetic energy
	m = mass
ME13. $p_{before} = p_{after}$	p = momentum
ME14. $J = F_{net}t = \Delta p$	P = power
	PE = potential energy
ME15. $F_s = kx$	PE_s = potential energy stored in a spring
ME16. $PE_s = \frac{1}{2}kx^2$	Q = internal energy
	r = radius or distance between centers
ME17. $F_c = ma_c$	t = time interval
ME18. $a_c = \frac{v^2}{r}$	v = velocity or speed
	$\overline{v}$ = average velocity or average speed
ME19. $\Delta PE = mg\Delta h$	W = work
ME20. $KE = \frac{1}{2}mv^2$	x = change in spring length from the equilibrium position
ME21. $W = Fd = \Delta E_T$	Δ = change
ME22. $E_T = PE + KE + Q$	θ = angle
ME23. $P = \frac{W}{t} = \frac{Fd}{t} = F\overline{v}$	μ = coefficient of friction

Equations are provided for motion, vectors, Newton's laws, momentum and impulse, work, energy, and power.

ELECTRICITY (EQUATIONS EL1–EL10)

EL1. $F_e = \frac{kq_1q_2}{r^2}$

EL2. $E = \frac{F_e}{q}$

EL3. $V = \frac{W}{q}$

EL4. $I = \frac{\Delta q}{t}$

EL5. $R = \frac{V}{I}$

EL6. $R = \frac{\rho L}{A}$

EL7. $P = VI = I^2R = \frac{V^2}{R}$

EL8. $W = Pt = VIt = I^2Rt = \frac{V^2t}{R}$

A = cross-sectional area
E = electric field strength
F_e = electrostatic force
I = current
k = electrostatic constant
L = length of conductor
P = electrical power
q = charge
R = resistance
R_{eq} = equivalent resistance
r = distance between centers
t = time
V = potential difference
W = work (electrical energy)
Δ = change
ρ = resistivity

EL9. **Series Circuits**

EL9A. $I = I_1 = I_2 = I_3 = \ldots$

EL9B. $V = V_1 + V_2 + V_3 + \ldots$

EL9C. $R_{eq} = R_1 + R_2 + R_3 + \ldots$

EL10. **Parallel Circuits**

EL10A. $I = I_1 + I_2 + I_3 + \ldots$

EL10B. $V = V_1 = V_2 = V_3 = \ldots$

EL10C. $\frac{1}{R_{eq}} = \frac{1}{R_1} + \frac{1}{R_2} + \frac{1}{R_3} + \ldots$

Equations are provided for electrostatic forces, electric fields, potential difference, current, resistance, and power and energy in electric circuits. In addition, the current, potential difference, and resistance relationships are provided for series and parallel circuits.

WAVES (EQUATIONS W1–W6)

W1. $v = f\lambda$

W2. $T = \frac{1}{f}$

W3. $\theta_i = \theta_r$

W4. $n = \frac{c}{v}$

W5. $n_1 \sin \theta_1 = n_2 \sin \theta_2$

W6. $\frac{n_2}{n_1} = \frac{v_1}{v_2} = \frac{\lambda_1}{\lambda_2}$

c = speed of light in a vacuum
f = frequency
n = absolute index of refraction
T = period
v = velocity
λ = wavelength
θ = angle
θ_i = incident angle
θ_r = reflected angle

Equations are provided for all types of waves (relationships among speed, wavelength, frequency, and period) as well as for reflection and refraction.

MODERN PHYSICS (EQUATIONS MP1–MP3)

MP1. $E_{photon} = hf = \frac{hc}{\lambda}$

MP2. $E_{photon} = E_i - E_f$

MP3. $E = mc^2$

c = speed of light in a vacuum
E = energy
f = frequency
h = Planck's constant
m = mass
λ = wavelength

Three energy equations are included in this table. MP1 is the Planck equation for relating the energy of a photon to its frequency (or wavelength). MP2 is the energy of a photon that is emitted or absorbed as the result of a transition between two energy levels in an atom. MP3 is the Einstein equation that relates energy and mass.

GEOMETRY AND TRIGONOMETRY (EQUATIONS GT1–GT4)

GT1. **Rectangle**

$A = bh$

GT2. **Triangle**

$A = \frac{1}{2}bh$

GT3. **Circle**

GT3A. $A = \pi r^2$

GT3B. $C = 2\pi r$

GT4. **Right Triangle**

GT4A. $c^2 = a^2 + b^2$

GT4D. $\sin \theta = \frac{a}{c}$

GT4C. $\cos \theta = \frac{b}{c}$

GT4D. $\tan \theta = \frac{a}{b}$

A = area

b = base

C = circumference

h = height

r = radius

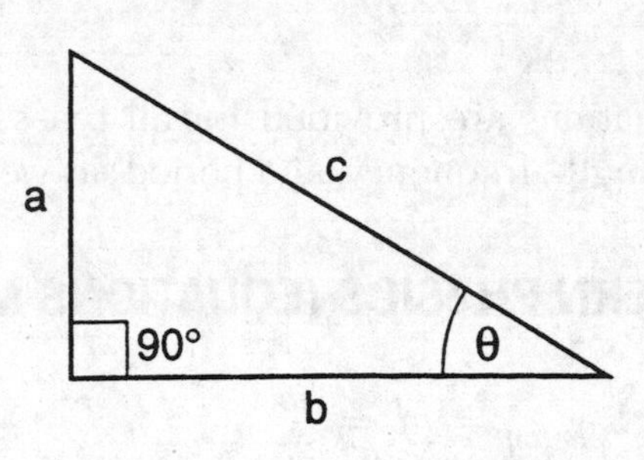

The geometric relationships for the areas of rectangles, triangles, and circles are given. In addition, the relationship for the circumference of a circle is provided. This table also gives the Pythagorean theorem and several trigonometric relationships for a right triangle.

Using the Equations to Solve Physics Problems

Note to the student: This section contains simple examples that use each of the equations given in *Reference Table K.* On the Regents examination in physics, you may be required to rearrange these equations or apply more than one equation in order to solve a problem.

MECHANICS

ME1. $\overline{v} = \frac{d}{t}$

EXAMPLE: It takes a car 0.20 hour to travel a distance of 12 kilometers.

$$\begin{aligned}\overline{v} &= \frac{d}{t} \\ &= \frac{12 \text{ km}}{0.20 \text{ h}} \\ &= 60 \text{ km/h}\end{aligned}$$

ME2. $a = \frac{\Delta v}{t}$

EXAMPLE: An object is uniformly accelerated from rest to a speed of 25 meters per second in 10 seconds.

$$\begin{aligned}\text{acceleration } a &= \frac{(25 - 0) \text{ m/s}}{10 \text{ s}} \\ &= 2.5 \text{ m/s}^2\end{aligned}$$

ME3. $v_f = v_i + at$

EXAMPLE: An object moving with an initial velocity of 3.0 meters per second accelerates for 15 seconds at 2.0 meters per second. What is the final velocity of the object?

$$\begin{aligned} v_f &= 3.0 \text{ m/s} + \left(2.0 \text{ m/s}^2\right)(15 \text{ s}) \\ &= 33 \text{ m/s} \end{aligned}$$

ME4. $d = v_i t + \frac{1}{2}at^2$

EXAMPLE: A cart with an initial velocity of 6.00 meters per second rolls down an inclined plane with an acceleration of 4.00 meters per second squared. What distance will it travel in 3.00 seconds?

$$\begin{aligned} d &= (6.00 \text{ m/s})(3.00 \text{ s}) + \frac{1}{2}\left(4.00 \text{ m/s}^2\right)(3.00 \text{ s})^2 \\ &= 36.0 \text{ m} \end{aligned}$$

ME5. $v_f^{\,2} = v_i^{\,2} + 2ad$

EXAMPLE: A car moving at a constant speed of 4.10 meters per second then accelerates uniformly at 3.20 meters per second squared. What will be its speed when it has traveled a distance of 40.0 meters?

$$\begin{aligned} v_f^{\,2} &= \left(4.10 \text{ m/s}^2\right) + (2)\left(3.20 \text{ m/s}^2\right)(40.0 \text{ m}) \\ &= 16.8 \text{ m}^2/\text{s}^2 + 256 \text{ m}^2/\text{s}^2 \\ v_f &= 16.5 \text{ m/s} \end{aligned}$$

ME6. $A_y = A \sin \theta$

ME7. $A_x = A\cos\theta$

EXAMPLE: A projectile is fired at an angle of 30° to the horizontal with an initial velocity of 100 meters per second. Calculate the vertical and horizontal components of the projectile's initial velocity.

$$\begin{aligned} A_y &= A\sin\theta \\ &= \left(100\ \frac{\text{m}}{\text{s}}\right)(0.500) \\ &= 50.0\ \frac{\text{m}}{\text{s}} \\ A_x &= A\cos\theta \\ &= \left(100\ \frac{\text{m}}{\text{s}}\right)(0.866) \\ &= 86.6\ \frac{\text{m}}{\text{s}} \end{aligned}$$

ME8. $a = \dfrac{F_{net}}{m}$

EXAMPLE: A 3.0-kilogram mass is being moved along a horizontal surface by a force of 6.0 newtons. If the surface is frictionless, what is the acceleration produced by the 6.0-newton force?

$$\begin{aligned} F_{net} &= ma \\ 6.0\ \text{N} &= (3.0\ \text{kg})(a) \\ a &= 2.0\ \text{m/s}^2 \end{aligned}$$

ME9. $F_f = \mu F_N$

EXAMPLE: A 10-newton rubber object is dragged across a dry, horizontal concrete floor at constant speed. Calculate the force of friction on the object.

Since the object is on a horizontal surface, the normal force is equal to its weight (10 N). Use *Reference Table C* to obtain the *kinetic* coefficient of friction between rubber and dry concrete (0.68).

$$\begin{aligned} F_f &= (0.68)(10\ \text{N}) \\ &= 6.8\ \text{N} \end{aligned}$$

ME10. $F_g = \frac{Gm_1m_2}{r^2}$

EXAMPLE: Two masses of 10.0 kg and 1.0 kg, respectively, are located 1.0 meter apart. How large a gravitational force does each mass exert on the other?

$$F_g = \left(6.67 \times 10^{-11}\text{N} \bullet \text{m}^2/\text{kg}^2\right)\left(10\ \text{kg}\right)\frac{1.0\ \text{kg}}{\left(1.0\ \text{m}\right)^2}$$

$$= 6.7 \times 10^{-10}\ \text{N}$$

ME11. $g = \frac{F_g}{m}$

EXAMPLE: The acceleration due to gravity on planet A is 20 meters per second squared. On this planet what is the gravitational force on an object whose mass is 2.0 kilograms?

$$F_g = mg$$

$$= \left(2.0\ \text{kg}\right)\left(20\ \text{m/s}^2\right)$$

$$= 40\ \text{N}$$

ME12. $p = mv$

EXAMPLE: A cart whose mass is 20 kg moves with a velocity of 5.0 meters per second eastward. What is its momentum?

$$p = mv$$

$$= \left(20\ \text{kg}\right)\left(5.0\ \text{m/s}\right)$$

$$= 1.0 \times 10^2\ \text{kg} \bullet \text{m/s in the direction of the velocity}$$

ME13. $p_{before} = p_{after}$

EXAMPLE: A 5.0-kilogram object traveling at 5.0 meters per second east collides with a 2.0-kilogram object traveling at 2.0 meters per second east. After the collision, the 2.0-kilogram object is moving at 4.0 meters per second east. What is the velocity of the 5.0-kilogram object? [Neglect friction.]

Designate east by using a positive sign.

$$(5.0\text{ kg})(+5.0\text{ m/s}) + (2.0\text{ kg})(+2.0\text{ m/s}) = (5.0\text{ kg})(v) + (2.0\text{ kg})(+4.0\text{ m/s})$$

$$v = +4.2\text{ m/s (east)}$$

ME14. $J = Ft = \Delta p$

EXAMPLE: A spring exerts a force of 50.0 newtons on a cart located on a frictionless plane. The cart has a mass of 2.0 kilograms and the force acts for 0.20 second.

$$\begin{aligned} \text{J} = \text{Impulse } Ft &= (50.0\text{ N})(0.20\text{ s}) \\ &= 10\text{ N} \bullet \text{s} \end{aligned}$$

The change of momentum of the cart is also 10 N • s, and that is the same thing as 10 kilogram-meters per second.

$$\Delta p = 10\text{ kg} \bullet \text{m/s}$$

ME15. $F_s = kx$

EXAMPLE: A block is suspended from a spring which has a spring constant, k, of 200 newtons per meter. What is the force on the spring if the spring is stretched 0.20 meter?

$$\begin{aligned} F_s &= kx \\ &= (200\text{ N/m})(0.20\text{ m}) \\ &= 40.0 \end{aligned}$$

ME16. $PE_s = \frac{1}{2}kx^2$

EXAMPLE: A block is suspended from a spring and as a result the spring is stretched 0.20 meter. The spring constant is 200 newtons per meter. How much potential energy is stored in the spring?

$$\begin{aligned} PE_s &= \frac{1}{2}kx^2 \\ &= \frac{1}{2}(200\text{ N/m})(0.20\text{ m})^2 \\ &= 8.0\text{ J} \end{aligned}$$

ME17. $Fc = ma_c$

ME18. $a_c = \frac{v^2}{r}$

EXAMPLE: An object moves around a circle whose radius is 2.0 meters. The constant speed of the object is 6.0 meters per second and the mass of the object is 0.20 kilogram.

$$\begin{aligned}\text{centripetal acceleration } a_c &= \frac{v^2}{r} \\ &= \frac{(6.0 \text{ m/s})^2}{2.0 \text{ m}} \\ &= 18. \text{ m/s}^2\end{aligned}$$

$$\begin{aligned}\text{centripetal force } F_c &= ma_c \left(= \frac{mv^2}{r} \right) \\ &= (0.20 \text{ kg})(18 \text{ m/s}^2) \\ &= 3.6 \text{ N}\end{aligned}$$

ME19. $\Delta PE = mg\Delta h$

EXAMPLE: A 3.0-kilogram mass is raised 4.0 meters from a surface. Calculate its gain in gravitational potential energy.

$$\begin{aligned}\Delta PE &= (3.0 \text{ kg})(9.8 \text{ m/s}^2)(4.0 \text{ m}) \\ &= 1.2 \times 10^2 \text{ J}\end{aligned}$$

ME20. $KE = \frac{1}{2}mv^2$

EXAMPLE: A car with a mass of 1000 kilograms travels with a speed of 20 meters per second.

$$\begin{aligned}E_k &= \frac{1}{2}(1000 \text{ kg})(20 \text{ m/s})^2 \\ &= 200{,}000 \text{ J}\end{aligned}$$

ME21. $W = Fd = \Delta E_T$

EXAMPLE: If a horizontal force of 30 newtons is used to push an object 40 meters along a horizontal surface, the work done on the object

$$= Fd$$
$$= (30\ \text{N})(40\ \text{m})$$
$$= 1200\ \text{J}$$

Note: 1 newton-meter = 1 joule

ME22. $E_T = PE + KE + Q$

EXAMPLE: As an object, initially at rest, slides down a plane, it loses 25 joules of potential energy. If the object gains 18 joules of kinetic energy, what is the change in the internal energy of the object? How is the internal energy change used?

At the top of the incline, the total energy of the object is present as potential energy and internal energy. At the bottom of the incline, the lost potential energy is converted to kinetic energy and additional internal energy. The total energy of the object remains constant throughout the slide. Modify equation ME22 to read:

$$\Delta E_T = \Delta PE + \Delta KE + \Delta Q$$
$$\Delta E_T = 0\ \text{J}$$
$$= (-25\ \text{J}) + (+18\ \text{J}) + \Delta Q$$
$$\Delta Q = +7\ \text{J}$$

The change in the internal energy is used to do work against the friction between the object and the incline.

ME23. $P = \frac{W}{t} = \frac{Fd}{t} = F\overline{v}$

Power is the rate of doing work.

EXAMPLE: An object whose mass is 3.0 kilograms is moved at constant speed by a force of 5.0 newtons. If the object is moved 40 meters in 20 seconds, then therequired power $= \frac{\text{work}}{\text{time}}$.

But the work which is done is equal to the product of the applied force and the distance that the object is moved:

$$\begin{aligned} W &= Fd \\ &= (5.0\text{ N})(40\text{ m}) \\ &= 2.0 \times 10^2 \text{ joules} \end{aligned}$$

$$\begin{aligned} \text{power} &= \frac{\text{work done}}{\text{time required}} \\ &= \frac{200\text{ joules}}{20\text{ s}} \\ &= 10\text{ watts.} \end{aligned}$$

$$\begin{aligned} \text{Also, power} &= \frac{\text{force} \times \text{distance moved}}{\text{time required}} \\ &= \frac{5.0\text{ N} \times 40\text{ m}}{20\text{ s}} \\ &= 10\text{ watts} \end{aligned}$$

ELECTRICITY

EL1. $\frac{F = kq_1q_2}{r^2}$

EXAMPLE: Two small charged spheres are 3.0 meters apart, and each sphere has a charge of 2.0×10^{-6} coulomb. Calculate the magnitude of the force exerted by one sphere on the other.

$$\begin{aligned} F &= \frac{\left(9.00 \times 10^9 \text{ N m}^2/\text{C}^2\right)\left(2.0 \times 10^{-6}\text{ C}\right)^2}{(3.0\text{ m})^2} \\ &= 4.0 \times 10^{-3}\text{ N} \end{aligned}$$

EL2. $E = \dfrac{F_e}{q}$

EXAMPLE: When a charge of 0.040 coulomb is placed at a point in the electric field, the force on the charge is 100 newtons. What is the magnitude of the electric field at that point?

$$E = \frac{100\ \text{N}}{0.040\ \text{C}}$$
$$= 2500\ \text{N/C}$$

EL3. $V = \dfrac{W}{q}$

EXAMPLE: The work required to move a charge of 0.04 coulomb from one point to another in an electric field is 200 joules.

$$\text{potential difference} = \frac{200\ \text{J}}{0.04\ \text{C}}$$
$$= 5000\ \text{V}$$

EL4. $I = \dfrac{\Delta q}{t}$

EXAMPLE: Calculate the current in a conductor if 130 coulombs of charge is transferred in 20.0 seconds.

$$I = \frac{130\ \text{C}}{20.0\ \text{s}}$$
$$= 6.50\ \text{A}$$

EL5. $R = \dfrac{V}{I}$

EXAMPLE: A resistor has a resistance of 10 ohms and a current in it of 0.50 ampere. What is the potential difference across the resistor?

$$\text{V} = (0.50\ \text{A})(10\ \Omega)$$
$$= 5.0\ \text{V}$$

EL6. $R = \dfrac{\rho L}{A}$

EXAMPLE: Calculate the resistance of a bar of silver if its length is 2.00 meters and its cross-sectional area is 0.0100 meter^2.

Use *Reference Table J* to obtain the resistivity of silver ($1.59 \times 10^{-8}\ \Omega \cdot \text{m}$).

$$
\begin{aligned}
R &= \left(1.59 \times 10^{-8}\ \Omega \bullet \text{m}\right)\left(\frac{2.00\ \text{m}}{0.0100\ \text{m}^2}\right) \\
&= 3.18 \times 10^{-6}\ \Omega
\end{aligned}
$$

EL7. $P = VI = I^2R = \dfrac{V^2}{R}$

a. $P = VI$

EXAMPLE: The potential difference across a resistor is 15 volts and the current in it is 0.50 ampere.

$$
\begin{aligned}
P &= (15\ \text{V})(0.50\ \text{A}) \\
&= 7.5\ \text{W}
\end{aligned}
$$

b. $P = I^2R$

EXAMPLE: The current in a resistor is 3.0 ampere, and its resistance is 10 ohms.

$$
\begin{aligned}
P &= (30\ \text{A})^2(10\ \Omega) \\
&= 90\ \text{W}
\end{aligned}
$$

c. $P = \dfrac{V^2}{R}$

EXAMPLE: The potential difference across a resistor is 100 volts, and its resistance is 50.0 ohms.

$$
\begin{aligned}
P &= \frac{(100\ \text{V})^2}{50.0\ \Omega} \\
&= 200\ \text{W}
\end{aligned}
$$

EL8. $W = Pt = VIt = I^2Rt = \frac{V^2t}{R}$

a. $W = Pt$

EXAMPLE: An electric toaster rated at 1500 watts is used for 40 seconds. How much energy does it use?

$$W = (1500\ \text{J})(40\ \text{s}) = 6.0 \times 10^4\ \text{J}$$

b. $W = VIt$

EXAMPLE: The potential difference across a resistor is 15 volts and the current in it is 0.50 ampere. How much energy is used by the resistor in 5.0 minutes when operating at the rated power?

$$W = (15\ \text{V})(0.50\ \text{A})(300\ \text{s}) = 2.3 \times 10^3\ \text{J}$$

c. $W = I^2Rt$

EXAMPLE: The current in a resistor is 3.0 ampere, and its resistance is 10. ohms. How much energy is used by the resistor in 5.0 minutes?

$$W = (3.0\ \text{A})^2\,(10\ \Omega)(300\ \text{s}) = 2.7 \times 10^4\ \text{J}$$

EL9A. Series Circuit: $I = I_1 = I_2 = I_3 = \ldots$

EXAMPLE: In a series circuit containing three resistors having resistances of 5, 10, and 15 ohms, respectively, we know that the current in the 15-ohm resistor is 0.5 ampere. This tells us that the current is also 0.5 ampere in the 5-ohm and 10-ohm resistors as well as in the generator connected to the circuit.

EL9B. Series Circuit: $V = V_1 + V_2 + V_3 + \ldots$

EXAMPLE: In the circuit described in the previous question, if we know that the voltages across the three resistors are 2.5, 5.0, and 7.5 volts, respectively, then the voltage supplied by the source is the sum of these three values, namely 15 volts.

EL9C. Series Circuit: $R_{eq} = R_1 + R_2 + R_3 + \ldots$

EXAMPLE: In the circuit described in the previous two questions, we can add the individual resistances to get the total resistance.

$$\begin{aligned} \text{total resistance} &= (5 + 10 + 15)\ \Omega \\ &= 30\ \Omega \end{aligned}$$

EL10A. Parallel Circuit: $I = I_1 + I_2 + I_3 + \ldots$

EXAMPLE: A parallel circuit has two branches, a 10-ohm resistor connected parallel to a 20-ohm resistor. If the current in the 10-ohm resistor is 6.0 amperes, and the current in the 20-ohm resistor is 3.0 amperes, how much current is supplied by the source?

$$\begin{aligned} I &= 6.0\ \text{A} + 3.0\ \text{A} \\ &= 9.0\ \text{A} \end{aligned}$$

EL10B. Parallel Circuit: $V = V_1 = V_2 = V_3 = \ldots$

In a parallel circuit the potential difference of the source is equal to the potential difference across each of the branches.

EXAMPLE: If the potential difference across the 20-ohm resistor is 60 volts, what is the potential difference supplied by the source?

$$\begin{aligned} V &= V_1 \\ &= 60\ \text{V} \end{aligned}$$

EL10C. Parallel Circuit: $\frac{1}{R_{eq}} = \frac{1}{R_1} + \frac{1}{R_2} + \frac{1}{R_3} + \ldots$

In a parallel circuit the reciprocal of the total resistance (or equivalent resistance) is equal to the sum of the reciprocals of the individual resistances.

EXAMPLE: In the circuit described in the previous two questions, what is the total resistance of the circuit?

$$\frac{1}{R_{eq}} = \frac{1}{10\ \Omega} + \frac{1}{20\ \Omega}$$

$$\frac{1}{R_{eq}} = \frac{3}{20\ \Omega}$$

$$R_{eq} = 6.7\ \Omega$$

WAVES

W1. $v = f\lambda$

EXAMPLE: The frequency of a wave is 2.0 hertz, and its wavelength is 0.030 meter.

$$v = (2.0\ \text{Hz})(0.030\ \text{m})$$
$$= 0.060\ \text{m/s}$$

W2. $T = \frac{1}{f}$

EXAMPLE: If the frequency of a wave is 0.25 cycle per second, what is the period of the wave?

$$T = \frac{1}{f}$$
$$= \frac{1}{0.25\ \text{Hz}}$$
$$= 4.0\ \text{s}$$

W3. $\theta_i = \theta_r$

EXAMPLE: When a ray of monochromatic light strikes a plane mirror, the incident ray makes an angle of 30° with the surface of the mirror. Calculate the angle of reflection for the reflected ray.

Angles of incidence and reflection are measured with respect to a *normal* drawn to the surface of the mirror. Therefore, the angle of incidence is the *complement* of 30°, that is, 60°.

$$\theta_r = \theta_i = 60°$$

W4. $n = \frac{c}{v}$

EXAMPLE: A ray of monochromatic light travels from a vacuum into a material whose index of refraction is 1.4. Calculate the speed of light in this material.

$$v = \frac{3.0 \times 10^8 \text{ m/s}}{1.4}$$
$$= 2.1 \times 10^8 \text{ m/s}$$

W5. $n_1 \sin \theta_1 = n_2 \sin \theta_2$

EXAMPLE: If the angle of incidence in air is 30°, what is the sine of the angle of refraction in crown glass?

$$1.00 \sin 30° = 1.52 \sin \theta_2$$
$$\sin \theta_2 = \frac{0.50}{1.52}$$
$$\sin \theta_2 = 0.33$$

W6. $\frac{n_2}{n_1} = \frac{v_1}{v_2} = \frac{\lambda_1}{\lambda_2}$

EXAMPLE: If the speed of light in a medium is 1.5×10^8 meters per second, what is the index of refraction of the medium?

MODERN PHYSICS

MP1. $E_{photon} = hf = \dfrac{hc}{\lambda}$

a. **EXAMPLE:** If the frequency of a monochromatic light beam is 1.2×10^{15} cycles per second, what is the energy of each of its photons?

$$E = \left(6.6 \times 10^{-34}\ \text{J} \bullet \text{s}\right)\left(1.2 \times 10^{15}\ \text{Hz}\right)$$
$$= 7.9 \times 10^{-19}\ \text{J}$$

b. **EXAMPLE:** Calculate the energy of a photon of monochromatic orange light whose wavelength is 6.0×10^{-7} m.

$$E_{photon} = \left(6.63 \times 10^{-34}\ \text{J} \bullet \text{s}\right)\left(3.00 \times 10^{8}\ \text{m/s}\right)/6.0 \times 10^{-7}\ \text{m}$$
$$= 3.3 \times 10^{-19}\ \text{J}$$

MP2. $E_{photon} = E_i - E_f$

EXAMPLE: A hydrogen atom jumps from an energy level of -0.85 eV to an energy level of -13.60 eV. What is the energy of the photon emitted as a result of this transition?

$$E_{photon} = (-0.85\ \text{eV}) - (-13.60\ \text{eV})$$
$$= 12.75\ \text{eV}$$

MP3. $E = mc^2$

EXAMPLE: What is the energy equivalence, in joules, of 1.0×10^{-3} kilogram of matter?

$$E = \left(1.0 \times 10^{-3}\ \text{kg}\right)\left(3.0 \times 10^{8}\ \text{m/s}\right)$$
$$= 9.0 \times 10^{13}\ \text{J}$$

Regents Examinations, Answers, and Self-Analysis Charts

IMPORTANT NOTE ABOUT THE REFERENCE TABLES AND EQUATIONS

In this book, the reference tables and equations beginning on page 45 have been *slightly modified* from the official *New York State Reference Tables for Physics (2006 edition).* The tables in this book are indexed by letter, and the equations are indexed by letter and number. Some of the tables have been moved in order to separate them from the equations, and each table is accompanied by a brief explanation. Otherwise, the material on these tables and the official tables is *identical*.

All of the explanations and answers in this book refer to the tables and equations *contained within this book*.

Examination June 2013

Physics: The Physical Setting

PART A

Answer all questions in this part.

Directions (1–35): For *each* statement or question, select the *number* of the word or expression that, of those given, best completes the statement or answers the question. Some questions may require the use of the *2006 Edition Reference Tables for Physical Setting/Physics*. Record your answers in the spaces provided.

1 Which term identifies a scalar quantity?

(1) displacement
(2) momentum
(3) velocity
(4) time

1 _____

2 Two 20-newton forces act concurrently on an object. What angle between these forces will produce a resultant force with the greatest magnitude?

(1) 0°
(2) 45°
(3) 90°
(4) 180°

2 _____

3 A car traveling west in a straight line on a highway decreases its speed from 30.0 meters per second to 23.0 meters per second in 2.00 seconds. The car's average acceleration during this time interval is

(1) 3.5 m/s^2 east
(2) 3.5 m/s^2 west
(3) 13 m/s^2 east
(4) 13 m/s^2 west

3 _____

4 In a race, a runner traveled 12 meters in 4.0 seconds as she accelerated uniformly from rest. The magnitude of the acceleration of the runner was

(1) 0.25 m/s^2
(2) 1.5 m/s^2
(3) 3.0 m/s^2
(4) 48 m/s^2

4 ____

5 A projectile is launched at an angle above the ground. The horizontal component of the projectile's velocity, v_x, is initially 40 meters per second. The vertical component of the projectile's velocity, v_y, is initially 30 meters per second. What are the components of the projectile's velocity after 2.0 seconds of flight? [Neglect friction.]

(1) $v_x = 40$ m/s and $v_y = 10$ m/s
(2) $v_x = 40$ m/s and $v_y = 30$ m/s
(3) $v_x = 20$ m/s and $v_y = 10$ m/s
(4) $v_x = 20$ m/s and $v_y = 30$ m/s

5 ____

6 A ball is thrown with an initial speed of 10 meters per second. At what angle above the horizontal should the ball be thrown to reach the greatest height?

(1) 0°
(2) 30°
(3) 45°
(4) 90°

6 ____

7 Which object has the greatest inertia?

(1) a 0.010-kg bullet traveling at 90 m/s
(2) a 30-kg child traveling at 10 m/s on her bike
(3) a 490-kg elephant walking with a speed of 1.0 m/s
(4) a 1500-kg car at rest in a parking lot

7 ____

8 An 8.0-newton wooden block slides across a horizontal wooden floor at constant velocity. What is the magnitude of the force of kinetic friction between the block and the floor?

(1) 2.4 N
(2) 3.4 N
(3) 8.0 N
(4) 27 N

8 ____

9 Which situation represents a person in equilibrium?

(1) a child gaining speed while sliding down a slide
(2) a woman accelerating upward in an elevator
(3) a man standing still on a bathroom scale
(4) a teenager driving around a corner in his car 9 ____

10 A rock is thrown straight up into the air. At the highest point of the rock's path, the magnitude of the net force acting on the rock is

(1) less than the magnitude of the rock's weight, but greater than zero
(2) greater than the magnitude of the rock's weight
(3) the same as the magnitude of the rock's weight
(4) zero 10 ____

11 The diagram below shows a compressed spring between two carts initially at rest on a horizontal, frictionless surface. Cart *A* has a mass of 2 kilograms and cart *B* has a mass of 1 kilogram. A string holds the carts together.

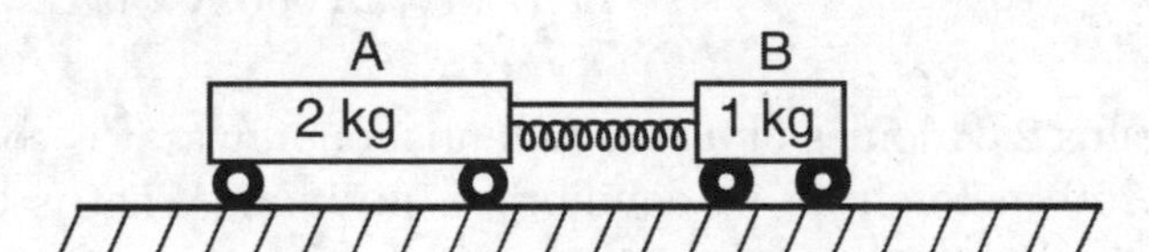

The string is cut and the carts move apart. Compared to the magnitude of the force the spring exerts on cart *A*, the magnitude of the force the spring exerts on cart *B* is

(1) the same
(2) half as great
(3) twice as great
(4) four times as great 11 ____

12 An 8.0-newton block is accelerating down a frictionless ramp inclined at 15° to the horizontal, as shown in the diagram below.

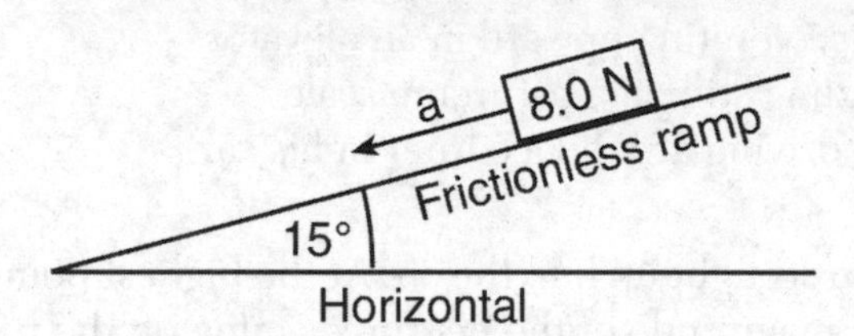

What is the magnitude of the net force causing the block's acceleration?

(1) 0 N
(2) 2.1 N
(3) 7.7 N
(4) 8.0 N

12 ____

13 At a certain location, a gravitational force with a magnitude of 350 newtons acts on a 70-kilogram astronaut. What is the magnitude of the gravitational field strength at this location?

(1) 0.20 kg/N
(2) 5.0 N/kg
(3) 9.8 m/s^2
(4) 25 000 N • kg

13 ____

14 A spring gains 2.34 joules of elastic potential energy as it is compressed 0.250 meter from its equilibrium position. What is the spring constant of this spring?

(1) 9.36 N/m
(2) 18.7 N/m
(3) 37.4 N/m
(4) 74.9 N/m

14 ____

15 When a teacher shines light on a photocell attached to a fan, the blades of the fan turn. The brighter the light shining on the photocell, the faster the blades turn. Which energy conversion is illustrated by this demonstration?

(1) light → thermal → mechanical
(2) light → nuclear → thermal
(3) light → electrical → mechanical
(4) light → mechanical → chemical

15 ____

16 Which statement describes a characteristic common to all electromagnetic waves and mechanical waves?

(1) Both types of waves travel at the same speed.
(2) Both types of waves require a material medium for propagation.
(3) Both types of waves propagate in a vacuum.
(4) Both types of waves transfer energy.

16 ____

17 An electromagnetic wave is produced by charged particles vibrating at a rate of 3.9×10^8 vibrations per second. The electromagnetic wave is classified as

(1) a radio wave
(2) an infrared wave
(3) an x ray
(4) visible light

17 ____

18 The energy of a sound wave is most closely related to the wave's

(1) frequency
(2) amplitude
(3) wavelength
(4) speed

18 ____

19 A sound wave traveling eastward through air causes the air molecules to

(1) vibrate east and west
(2) vibrate north and south
(3) move eastward, only
(4) move northward, only

19 ____

20 What is the speed of light ($f = 5.09 \times 10^{14}$ Hz) in ethyl alcohol?

(1) 4.53×10^{-9} m/s
(2) 2.43×10^{2} m/s
(3) 1.24×10^{8} m/s
(4) 2.21×10^{8} m/s

20 ____

21 In the diagram below, an ideal pendulum released from position *A* swings freely to position *B*.

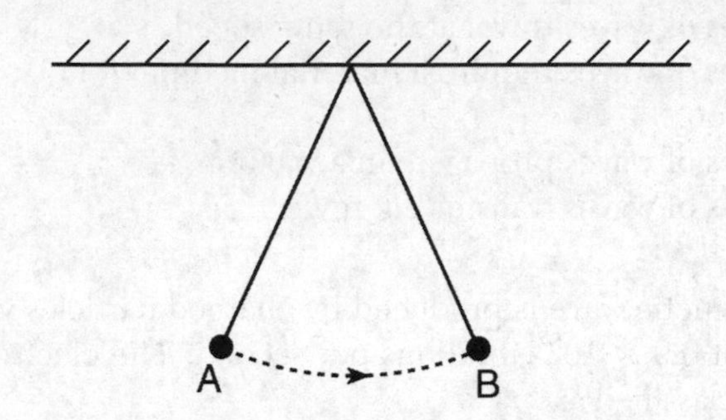

As the pendulum swings from *A* to *B*, its total mechanical energy

(1) decreases, then increases
(2) increases, only
(3) increases, then decreases
(4) remains the same

21 ____

22 The diagram below represents a periodic wave.

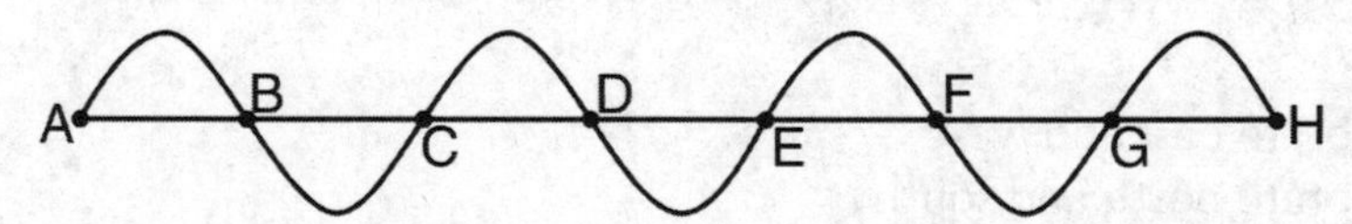

Which two points on the wave are out of phase?

(1) *A* and *C*
(2) *B* and *F*
(3) *C* and *E*
(4) *D* and *G*

22 ____

23 A dry plastic rod is rubbed with wool cloth and then held near a thin stream of water from a faucet. The path of the stream of water is changed, as represented in the diagram below.

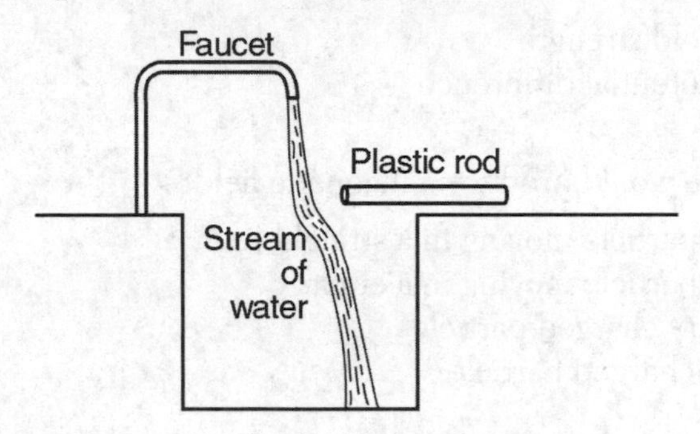

Which force causes the path of the stream of water to change due to the plastic rod?

(1) nuclear
(2) magnetic
(3) electrostatic
(4) gravitational

23 _____

24 A distance of 1.0×10^{-2} meter separates successive crests of a periodic wave produced in a shallow tank of water. If a crest passes a point in the tank every 4.0×10^{-1} second, what is the speed of this wave?

(1) 2.5×10^{-4} m/s
(2) 4.0×10^{-3} m/s
(3) 2.5×10^{-2} m/s
(4) 4.0×10^{-1} m/s

24 _____

25 One vibrating 256-hertz tuning fork transfers energy to another 256-hertz tuning fork, causing the second tuning fork to vibrate. This phenomenon is an example of

(1) diffraction
(2) reflection
(3) refraction
(4) resonance

25 _____

26 Sound waves are produced by the horn of a truck that is approaching a stationary observer. Compared to the sound waves detected by the driver of the truck, the sound waves detected by the observer have a greater

(1) wavelength
(2) frequency
(3) period
(4) speed

26 _____

27 The electronvolt is a unit of

(1) energy
(2) charge
(3) electric field strength
(4) electric potential difference

27 ____

28 Which particle would produce a magnetic field?

(1) a neutral particle moving in a straight line
(2) a neutral particle moving in a circle
(3) a stationary charged particle
(4) a moving charged particle

28 ____

29 A physics student takes her pulse and determines that her heart beats periodically 60 times in 60 seconds. The period of her heartbeat is

(1) 1 Hz
(2) 60 Hz
(3) 1 s
(4) 60 s

29 ____

30 Moving 4.0 coulombs of charge through a circuit requires 48 joules of electric energy. What is the potential difference across this circuit?

(1) 190 V
(2) 48 V
(3) 12 V
(4) 4.0 V

30 ____

31 The diagram below shows currents in a segment of an electric circuit.

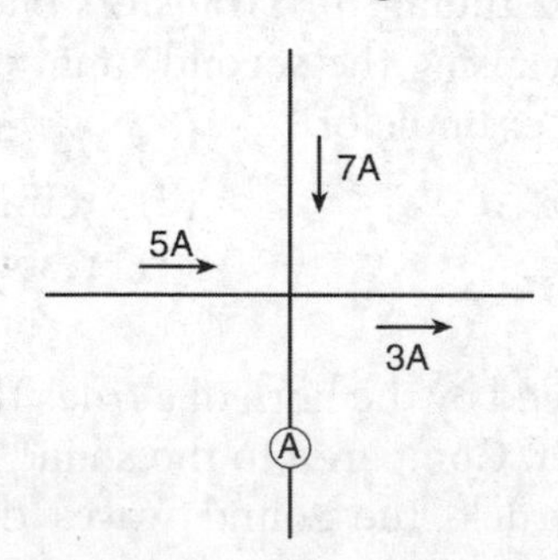

What is the reading of ammeter *A*?

(1) 1 A
(2) 5 A
(3) 9 A
(4) 15 A

31 ____

32 An electric dryer consumes 6.0×10^6 joules of electrical energy when operating at 220 volts for 1.8×10^3 seconds. During operation, the dryer draws a current of

(1) 10 A
(2) 15 A
(3) 9.0×10^2 A
(4) 3.3×10^3 A

32 ____

33 Which net charge could be found on an object?

(1) $+4.80 \times 10^{-19}$ C
(2) $+2.40 \times 10^{-19}$ C
(3) -2.40×10^{-19} C
(4) -5.60×10^{-19} C

33 ____

34 A photon is emitted as the electron in a hydrogen atom drops from the $n = 5$ energy level directly to the $n = 3$ energy level. What is the energy of the emitted photon?

(1) 0.85 eV
(2) 0.97 eV
(3) 1.51 eV
(4) 2.05 eV

34 ____

35 In a process called pair production, an energetic gamma ray is converted into an electron and a positron. It is *not* possible for a gamma ray to be converted into two electrons because

(1) charge must be conserved
(2) momentum must be conserved
(3) mass-energy must be conserved
(4) baryon number must be conserved

35 ____

PART B–1

Answer all questions in this part.

Directions (36–50): For *each* statement or question, select the *number* of the word or expression that, of those given, best completes the statement or answers the question. Some questions may require the use of the *2006 Edition Reference Tables for Physical Setting/Physics*. Record your answers in the spaces provided.

36 The approximate length of an unsharpened No. 2 pencil is

(1) 2.0×10^{-2} m
(2) 2.0×10^{-1} m
(3) 2.0×10^{0} m
(4) 2.0×10^{1} m

36 ____

37 The diagram below shows an 8.0-kilogram cart moving to the right at 4.0 meters per second about to make a head-on collision with a 4.0-kilogram cart moving to the left at 6.0 meters per second.

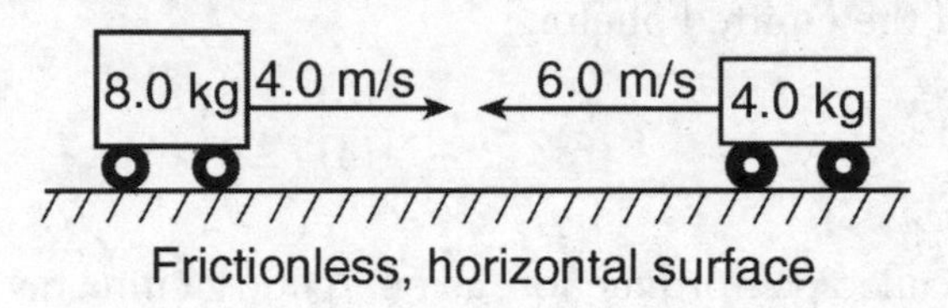

After the collision, the 4.0-kilogram cart moves to the right at 3.0 meters per second. What is the velocity of the 8.0-kilogram cart after the collision?

(1) 0.50 m/s left
(2) 0.50 m/s right
(3) 5.5 m/s left
(4) 5.5 m/s right

37 ____

38 Four forces act concurrently on a block on a horizontal surface as shown in the diagram below.

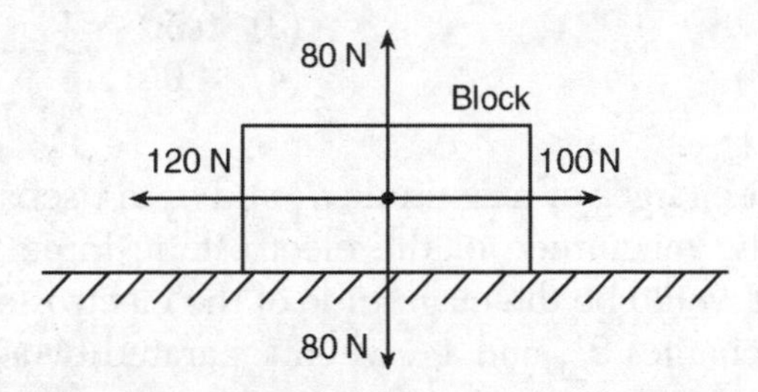

As a result of these forces, the block

(1) moves at constant speed to the right
(2) moves at constant speed to the left
(3) accelerates to the right
(4) accelerates to the left

38 ____

39 If a motor lifts a 400-kilogram mass a vertical distance of 10 meters in 8.0 seconds, the *minimum* power generated by the motor is

(1) 3.2×10^2 W
(2) 5.0×10^2 W
(3) 4.9×10^3 W
(4) 3.2×10^4 W

39 ____

40 A 4.0-kilogram object is accelerated at 3.0 meters per second^2 north by an unbalanced force. The same unbalanced force acting on a 2.0-kilogram object will accelerate this object toward the north at

(1) 12 m/s^2
(2) 6.0 m/s^2
(3) 3.0 m/s^2
(4) 1.5 m/s^2

40 ____

41 An electron is located in an electric field of magnitude 600 newtons per coulomb. What is the magnitude of the electrostatic force acting on the electron?

(1) 3.75×10^{21} N
(2) 6.00×10^2 N
(3) 9.60×10^{-17} N
(4) 2.67×10^{-22} N

41 ____

42 The current in a wire is 4.0 amperes. The time required for 2.5×10^{19} electrons to pass a certain point in the wire is

(1) 1.0 s (3) 0.50 s
(2) 0.25 s (4) 4.0 s

42 _____

43 When two point charges of magnitude q_1 and q_2 are separated by a distance, r, the magnitude of the electrostatic force between them is F. What would be the magnitude of the electrostatic force between point charges 2_{q1} and 4_{q2} when separated by a distance of $2r$?

(1) F (3) $16F$
(2) $2F$ (4) $4F$

43 _____

44 The composition of a meson with a charge of -1 elementary charge could be

(1) $s\bar{c}$ (3) $u\bar{b}$
(2) $d\,s\,s$ (4) $\bar{u}\bar{c}\bar{d}$

44 _____

45 Which graph represents the relationship between the kinetic energy and the speed of a freely falling object?

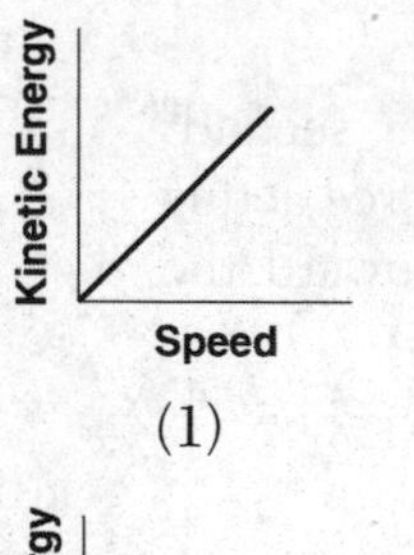

(1)

Kinetic Energy
Speed
(3)

Kinetic Energy
Speed
(2)

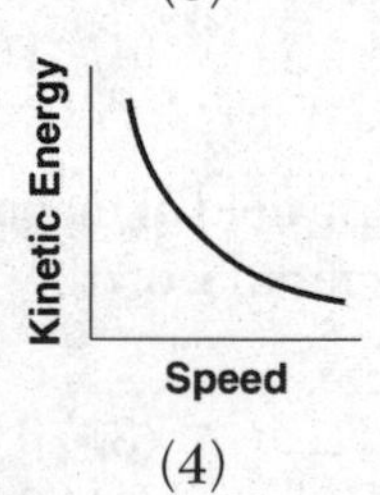

(4)

45 _____

46 Which diagram represents the electric field between two oppositely charged conducting spheres?

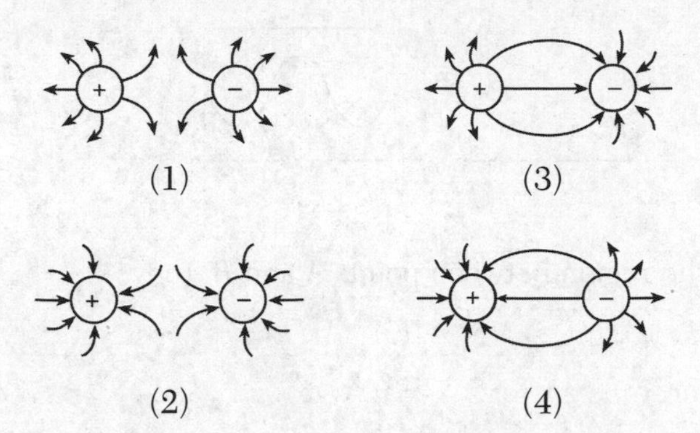

46 _____

47 Which graph represents the relationship between the magnitude of the gravitational force, F_g, between two masses and the distance, r, between the centers of the masses?

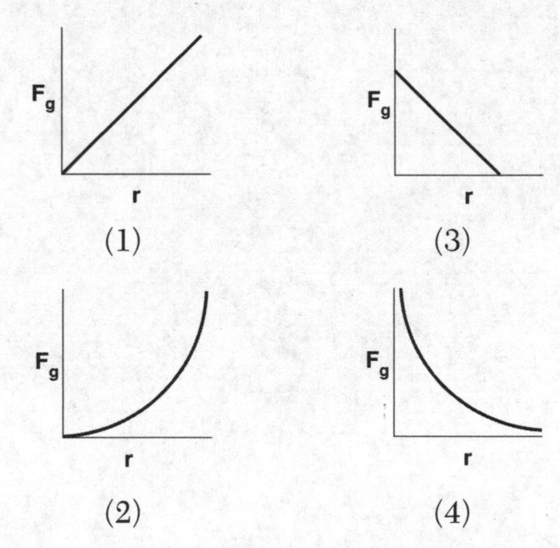

47 _____

48 The diagram below shows two waves traveling toward each other at equal speed in a uniform medium.

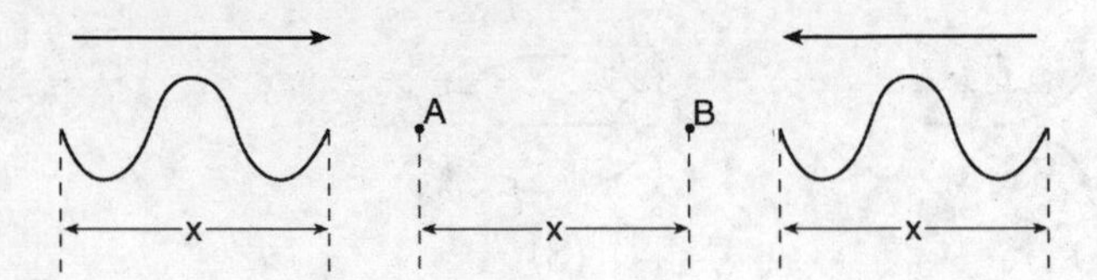

When both waves are in the region between points *A* and *B*, they will undergo

(1) diffraction
(2) the Doppler effect
(3) destructive interference
(4) constructive interference 48 ____

49 The diagram below shows a series of straight wave fronts produced in a shallow tank of water approaching a small opening in a barrier.

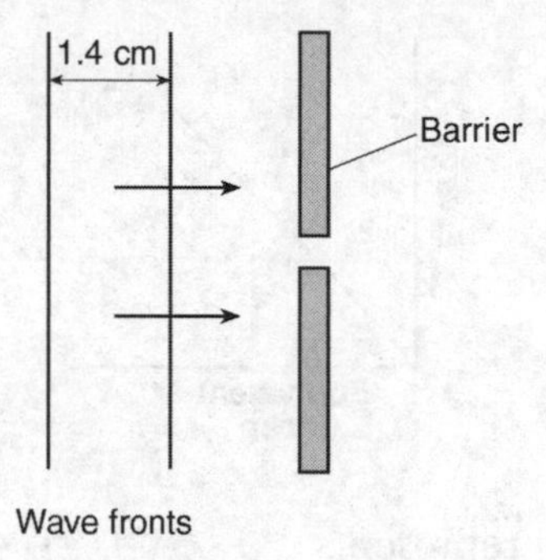

Which diagram represents the appearance of the wave fronts after passing through the opening in the barrier?

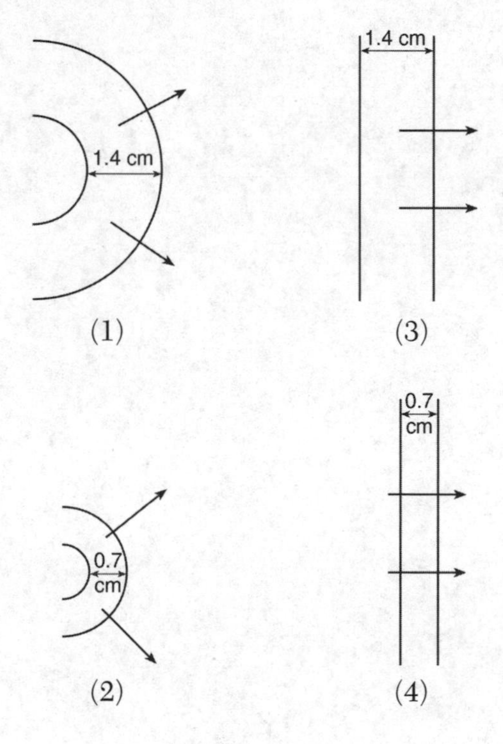

49 _____

50 The graph below represents the relationship between energy and the equivalent mass from which it can be converted.

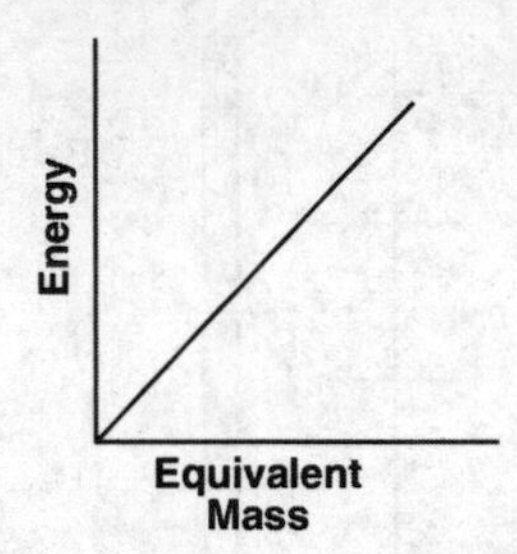

The slope of this graph represents

(1) c
(2) c^2
(3) g
(4) g^2

50 _____

PART B–2

Answer all questions in this part.

Directions (51–65): Record your answers on the answer sheet provided after the questions. Some questions may require the use of the *2006 Edition Reference Tables for Physical Setting/Physics*.

IMPORTANT NOTE:

- 2-point calculation problems appear as two linked questions in the test, for example, 51–52.
- In the first question, 1 point is allowed for showing the equation *and* for substituting the correct values with their units.
- In the second question, 1 point is allowed for a correct answer (number *and* unit) or for an answer that is consistent with your work in the first question.

51–52 A 25.0-meter length of platinum wire with a cross-sectional area of 3.50×10^{-6} meter^2 has a resistance of 0.757 ohm at 20°C. Calculate the resistivity of the wire. [Show all work, including the equation and substitution with units.] [2]

53 The diagram below represents a periodic wave moving along a rope.

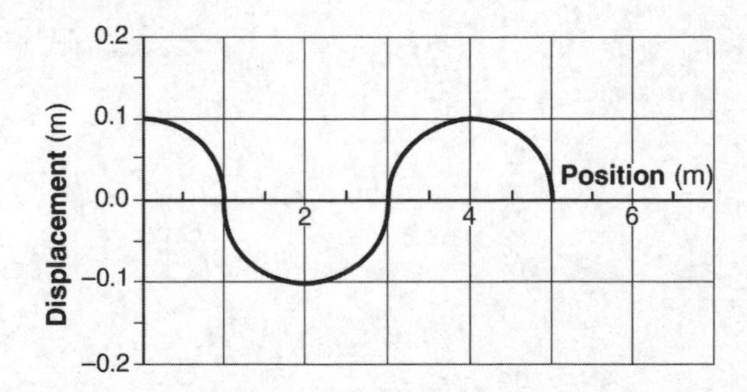

On the grid *on the answer sheet*, draw *at least one* full wave with the same amplitude and half the wavelength of the given wave. [1]

54–55 A baseball bat exerts an average force of 600 newtons east on a ball, imparting an impulse of 3.6 newton • seconds east to the ball. Calculate the amount of time the baseball bat is in contact with the ball. [Show all work, including the equation and substitution with units.] [2]

56 The diagram below shows the north pole of one bar magnet located near the south pole of another bar magnet.

N S

On the diagram *on the answer sheet*, draw *three* magnetic field lines in the region between the magnets. [1]

Base your answers to questions 57 through 59 on the information and graph below.

The graph below shows the relationship between speed and elapsed time for a car moving in a straight line.

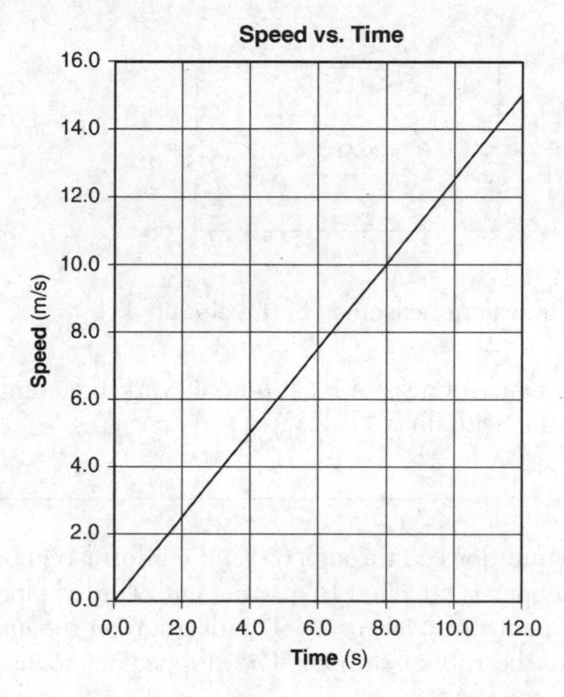

57 Determine the magnitude of the acceleration of the car. [1]

58–59 Calculate the total distance the car traveled during the time interval 4.0 seconds to 8.0 seconds. [Show all work, including the equation and substitution with units.] [2]

Base your answers to questions 60 through 62 on the information below.

A 20-ohm resistor, R_1, and a resistor of unknown resistance, R_2, are connected in parallel to a 30-volt source, as shown in the circuit diagram below. An ammeter in the circuit reads 2.0 amperes.

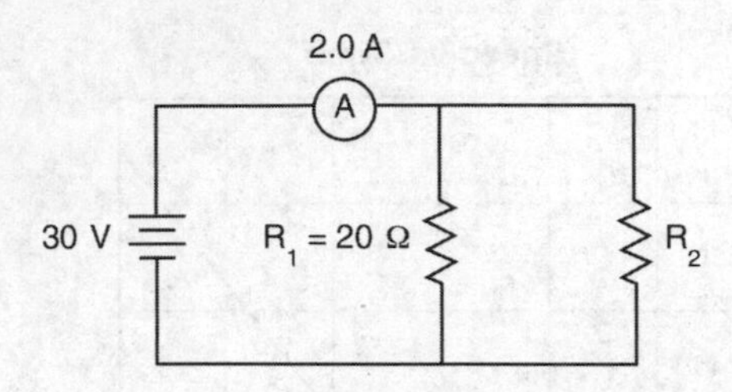

60 Determine the equivalent resistance of the circuit. [1]

61–62 Calculate the resistance of resistor R_2. [Show all work, including the equation and substitution with units.] [2]

Base your answers to questions 63 through 65 on the information below.

A 28-gram rubber stopper is attached to a string and whirled clockwise in a horizontal circle with a radius of 0.80 meter. The diagram *on the answer sheet* represents the motion of the rubber stopper. The stopper maintains a constant speed of 2.5 meters per second.

63–64 Calculate the magnitude of the centripetal acceleration of the stopper. [Show all work, including the equation and substitution with units.] [2]

65 On the diagram *on the answer sheet*, draw an arrow showing the direction of the centripetal force acting on the stopper when it is at the position shown. [1]

PART C

Answer all questions in this part.

Directions (66–85): Record your answers on the answer sheet provided after the questions. Some questions may require the use of the *2006 Edition Reference Tables for Physical Setting/Physics.*

Base your answers to questions 66 through 69 on the information below.

Auroras over the polar regions of Earth are caused by collisions between charged particles from the Sun and atoms in Earth's atmosphere. The charged particles give energy to the atoms, exciting them from their lowest available energy level, the ground state, to higher energy levels, excited states. Most atoms return to their ground state within 10 nanoseconds.

In the higher regions of Earth's atmosphere, where there are fewer interatom collisions, a few of the atoms remain in excited states for longer times. For example, oxygen atoms remain in an excited state for up to 1.0 second. These atoms account for the greenish and red glows of the auroras. As these oxygen atoms return to their ground state, they emit green photons ($f = 5.38 \times 10^{14}$ Hz) and red photons ($f = 4.76 \times 10^{14}$ Hz). These emissions last long enough to produce the changing aurora phenomenon.

IMPORTANT NOTE:

- 2-point calculation problems appear as two linked questions in the test, for example, 67–68.
- In the first question, 1 point is allowed for showing the equation *and* for substituting the correct values with their units.
- In the second question, 1 point is allowed for a correct answer (number *and* unit) or for an answer that is consistent with your work in the first question.

66 What is the order of magnitude of the time, in seconds, that most atoms spend in an excited state? [1]

67–68 Calculate the energy of a photon, in joules, that accounts for the red glow of the aurora. [Show all work, including the equation and substitution with units.] [2]

69 Explain what is meant by an atom being in its ground state. [1]

Base your answers to questions 70 through 75 on the information below.

A girl rides her bicycle 1.40 kilometers west, 0.70 kilometer south, and 0.30 kilometer east in 12 minutes. The vector diagram *on the answer sheet* represents the girl's first two displacements in sequence from point *P*. The scale used in the diagram is 1.0 centimeter = 0.20 kilometer.

70–71 On the vector diagram *on the answer sheet*, using a ruler and a protractor, construct the following vectors:

- Starting at the arrowhead of the second displacement vector, draw a vector to represent the 0.30 kilometer east displacement. Label the vector with its magnitude. [1]
- Draw the vector representing the resultant displacement of the girl for the entire bicycle trip and label the vector *R*. [1]

72–73 Calculate the girl's average speed for the entire bicycle trip. [Show all work, including the equation and substitution with units.] [2]

74 Determine the magnitude of the girl's resultant displacement for the entire bicycle trip, in kilometers. [1]

75 Determine the measure of the angle, in degrees, between the resultant and the 1.40-kilometer displacement vector. [1]

Base your answers to questions 76 through 80 on the information below.

A light ray with a frequency of 5.09×10^{14} hertz traveling in water has an angle of incidence of 35° on a water-air interface. At the interface, part of the ray is reflected from the interface and part of the ray is refracted as it enters the air.

76 What is the angle of reflection of the light ray at the interface? [1]

77 On the diagram *on the answer sheet*, using a protractor and a straightedge, draw the reflected ray. [1]

78–79 Calculate the angle of refraction of the light ray as it enters the air. [Show all work, including the equation and substitution with units.] [2]

80 Identify *one* characteristic of this light ray that is the same in *both* the water and the air. [1]

Base your answers to questions 81 through 85 on the information and diagram below.

A 30.4-newton force is used to slide a 40.0-newton crate a distance of 6.00 meters at constant speed along an incline to a vertical height of 3.00 meters.

81 Determine the total work done by the 30.4-newton force in sliding the crate along the incline. [1]

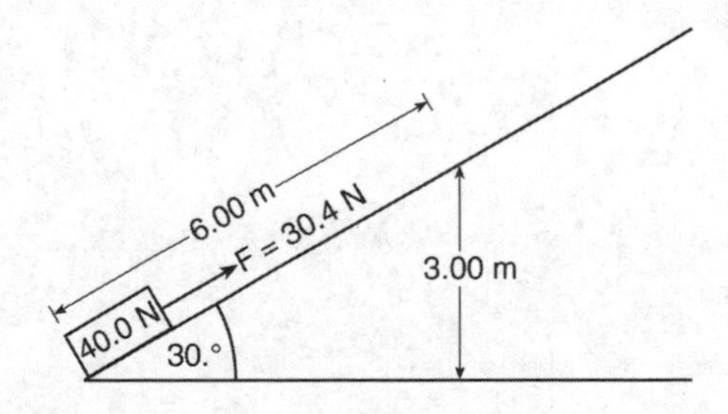

82–83 Calculate the total increase in the gravitational potential energy of the crate after it has slid 6.00 meters along the incline. [Show all work, including the equation and substitution with units.] [2]

84 State what happens to the kinetic energy of the crate as it slides along the incline. [1]

85 State what happens to the internal energy of the crate as it slides along the incline. [1]

Answer Sheet June 2013

Physics: The Physical Setting

PART B–2

51–52

53

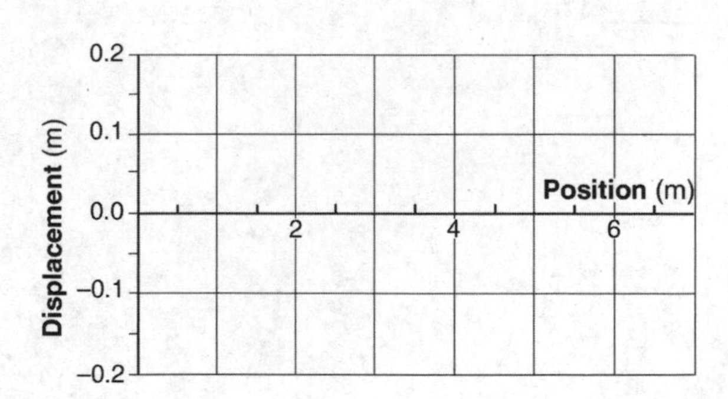

54–55

56

N S

57 ________________ m/s^2

58–59

60 ________________ Ω

61–62

63–64

65

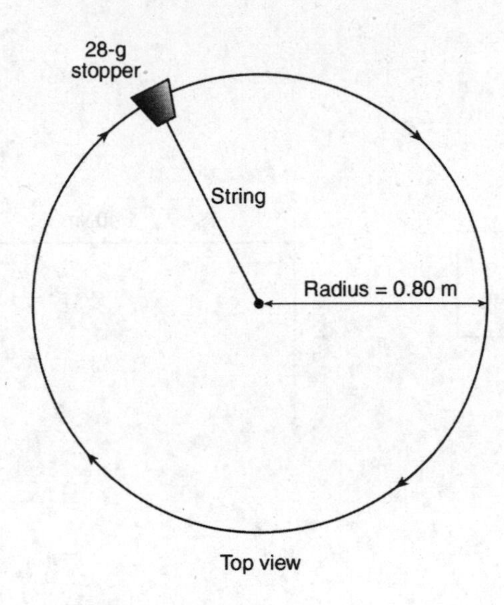

PART C

66 ____________________

67–68

69 ____

____

70–71

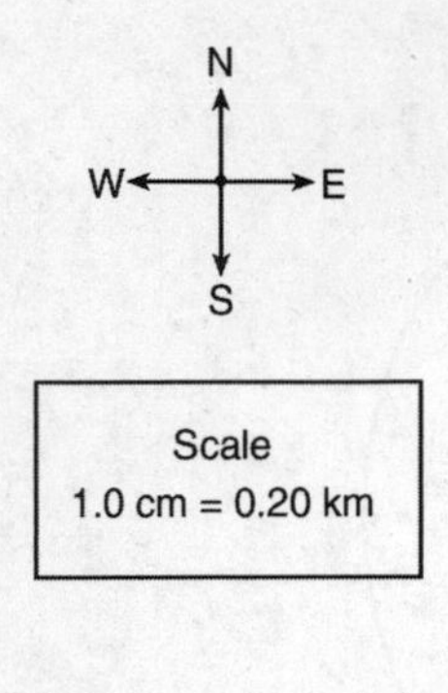

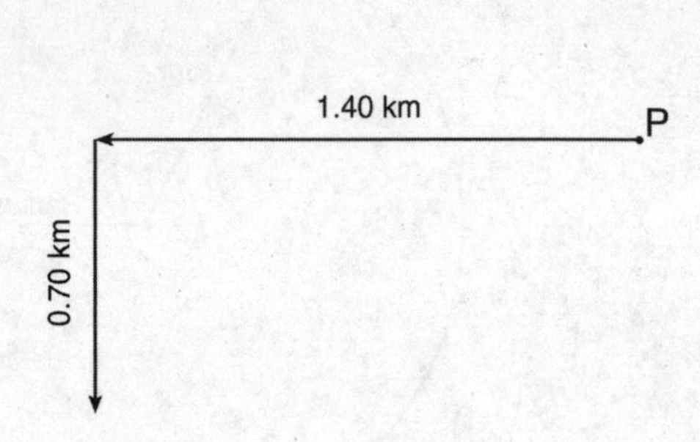

72–73

74 ________________ **km**

75 ________________ °

76 ________________ °

77

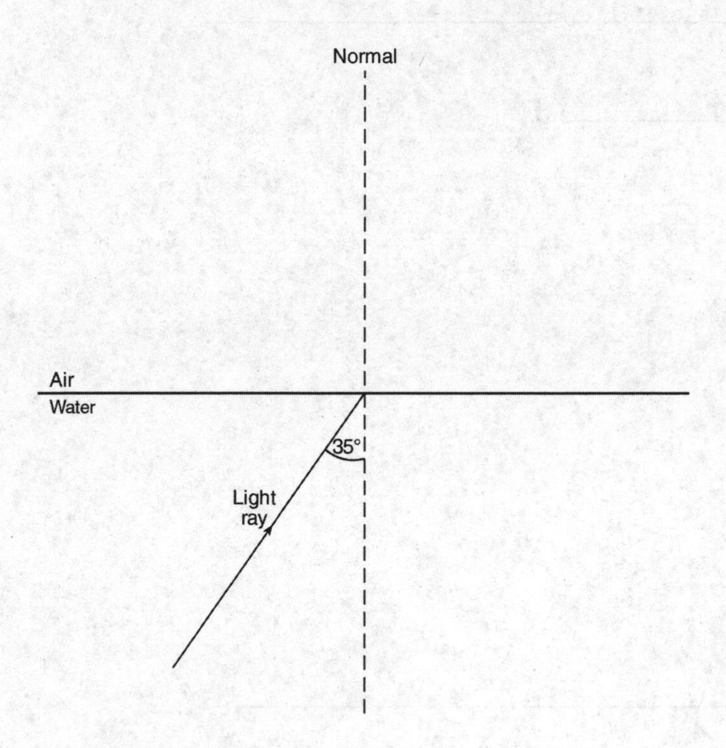

78–79

80 ____

81 __________________ J

82–83

84 ____

85 ____

Answers June 2013

Physics: The Physical Setting

Answer Key

PART A

1. 4	**8.** 1	**15.** 3	**22.** 4	**29.** 3
2. 1	**9.** 3	**16.** 4	**23.** 3	**30.** 3
3. 1	**10.** 3	**17.** 1	**24.** 3	**31.** 3
4. 2	**11.** 1	**18.** 2	**25.** 4	**32.** 2
5. 1	**12.** 2	**19.** 1	**26.** 2	**33.** 1
6. 4	**13.** 2	**20.** 4	**27.** 1	**34.** 2
7. 4	**14.** 4	**21.** 4	**28.** 4	**35.** 1

PART B–1

36. 2	**39.** 3	**42.** 1	**45.** 3	**48.** 4
37. 1	**40.** 2	**43.** 2	**46.** 3	**49.** 1
38. 4	**41.** 3	**44.** 1	**47.** 4	**50.** 2

PART B–2 and **PART C.** *See* **Answers Explained**

Answers Explained

PART A

1 A scalar quantity has magnitude but not direction. Time is never associated with a direction and is therefore a scalar quantity.

WRONG CHOICES EXPLAINED:

(1), (2), (3) Displacement, momentum, and velocity are all associated with both magnitude and direction. They are vector quantities.

2 When two forces act at the same point, they produce the maximum resultant force (40.0 newtons) when the angle between them is 0°.

WRONG CHOICE EXPLAINED:

(4) When the angle between the two forces is 180°, they will produce the minimum resultant force (0.0 newton).

3 Assume that east is positive and west is negative. Use *Equation ME2* on *Reference Table K*:

Solution:
$$a = \frac{\Delta v}{t}$$
$$= \frac{-23.0 \text{ m/s} - (-30.0 \text{ m/s})}{2.00 \text{ s}}$$
$$= +3.5 \text{ m/s}^2 \text{ or } 3.5 \text{ m/s}^2 \text{ east}$$

4 Use *Equation ME4* on *Reference Table K*:

Solution:
$$d = v_i t + \frac{1}{2}at^2$$
$$12 \text{ m} = (0.0 \text{ m/s})(4.0 \text{ s}) + \frac{1}{2}a(4.0 \text{ s})^2$$
$$a = 1.5 \text{ m/s}^2$$

5 When a projectile is in motion, the horizontal component of its velocity, v_x, does not change. So v_x remains 40 m/s. The vertical component, however, does change because of the influence of gravity. In order to find the final value of v_y, use *Equation ME3* on *Reference Table K*:

Solution: $$\begin{aligned} v_f &= v_i + at \\ &= 30 \text{ m/s} + \left(-9.8 \text{ m/s}^2\right)(2.0 \text{ s}) \\ &\approx 10 \text{ s} \end{aligned}$$

6 The vertical component of the ball's velocity determines the time that the ball will remain in the air and, therefore, the height to which it will rise. The vertical component is greatest when the angle with the horizontal is 90°.

WRONG CHOICE EXPLAINED:

(3) If thrown at 45°, the ball will travel the greatest horizontal distance.

7 Inertia is determined solely by an object's mass. The 1500-kilogram car has the greatest inertia.

8 In order to calculate the force of kinetic friction between the wooden block and the wood floor, we must know the coefficient of kinetic friction and the normal force on the block. According to *Reference Table C*, the coefficient of kinetic friction between wood and wood is 0.30. Since the wood is on a horizontal surface, its weight (8.0 newtons) is equal to the normal force on the block. Use *Equation ME9* on *Reference Table K*:

Solution: $$\begin{aligned} F_f &= \mu F_N \\ &= (0.30)(8.0 \text{ N}) \\ &= 2.4 \text{ N} \end{aligned}$$

9 An object in equilibrium has no net force acting on it. Therefore, according to Newton's second law, the object must either be at rest or be traveling at constant speed in a straight line. Of the choices given, only choice (3), a man standing still on a bathroom scale, meets these criteria.

10 At its highest point, the rock is at rest. The net force acting on it is the gravitational force, which is the same as the object's weight.

11 The spring is the means by which cart *A* interacts with cart *B*. Therefore, according to Newton's third law, the force exerted by the spring on cart *A* must be equal in magnitude (and opposite in direction) to the force exerted by the spring on cart *B*.

12 The force acting on the block parallel to the ramp (F_{II}) is the component of the block's weight parallel to the incline of the ramp. This component is equal to the product of the block's weight (*mg*) and the sine of the angle of the ramp (sin θ). Use the following equation:

Solution:
$$\begin{aligned} F_{\parallel} &= mg\sin\theta \\ &= (8.0\ \text{N})\sin 15^\circ \\ &= 2.1\ \text{N} \end{aligned}$$

13 Use *Equation ME11* on *Reference Table K*:

Solution:
$$\begin{aligned} \text{g} &= \frac{F_g}{m} \\ &= \frac{350\ \text{N}}{70\ \text{kg}} \\ &= 5.0\ \text{N/kg} \end{aligned}$$

14 Use *Equation ME16* on *Reference Table K*:

Solution:
$$\begin{aligned} PE_s &= \frac{1}{2}kx^2 \\ k &= \frac{2\,PE_s}{x^2} \\ &= \frac{2(2.34\ \text{J})}{(0.250\ \text{m})^2} \\ &= 74.9\ \text{N/m} \end{aligned}$$

15 A photocell changes light energy into electrical energy and produces an electric current in the circuit containing the fan. The motor in the fan changes the electrical energy into the mechanical energy of the rotating fan blades.

16 All waves transfer energy without transferring mass.

WRONG CHOICES EXPLAINED:

(1) The speed of an electromagnetic wave is much faster than the speed of a mechanical wave.

(2) Only mechanical waves require a material medium for propagation.

(3) Only electromagnetic waves propagate in a vacuum.

17 The vibrating particles produce an electromagnetic wave whose frequency is 3.9×10^8 hertz. See *Reference Table D*. The electromagnetic wave is in the portion of the electromagnetic spectrum known as radio waves.

18 The energy of any mechanical wave depends on the wave's amplitude.

19 Sound is a longitudinal wave in which the air particles vibrate parallel to the wave's direction of motion. Therefore, the air molecules will vibrate east and west.

WRONG CHOICE EXPLAINED:

(2) The particles in this wave are moving perpendicular to the wave's direction of motion. This is the characteristic of a transverse wave.

20 See *Reference Tables A* and *E*. The speed of light (c) is 3.00×10^8 meters per second and the absolute index of refraction of ethyl alcohol is 1.36. Now use *Equation W4* on *Reference Table K*:

Solution:
$$n = \frac{c}{v}$$
$$v = \frac{c}{n}$$
$$= \frac{3.00 \times 10^8 \text{ m/s}}{1.36}$$
$$= 2.21 \times 10^8 \text{ m/s}$$

21 The total mechanical energy of the pendulum is the sum of its kinetic and potential energies. In an ideal pendulum (no friction), kinetic energy is converted to potential energy and vice versa. However, the total mechanical energy does *not* change.

22 When two points on a periodic wave are a whole number of wavelengths apart (0λ, 1λ, 2λ, 3λ, . . .), the two points are in phase with each other. When two points on a periodic wave are an odd number of half-wavelengths apart $\left(\frac{1}{2}\lambda, \frac{3}{2}\lambda, \frac{5}{2}\lambda, \frac{7}{2}, \ldots\right)$, the two points are out of phase with each other. Points D and G are $\frac{3}{2}\lambda$ apart and are, therefore, out of phase.

WRONG CHOICES EXPLAINED:

(1), (3) Points A and C and points C and E are each 1 wavelength apart and are therefore in phase.

(2) Points B and F are 2 wavelengths apart and are therefore in phase.

23 When the plastic rod is rubbed with wool, the rod acquires a negative electric charge. The charged rod causes the stream of water to deviate because it places an electrostatic force on the positive ends of the water molecules.

24 The wave crest travels a distance of 1.0×10^{-2} meter in 4.0×10^{-1} second. Use *Equation ME1* on *Reference Table K*:

Solution:
$$v = \frac{d}{t}$$
$$= \frac{1.0 \times 10^{-2}\ \text{m}}{4.0 \times 10^{-1}\text{s}}$$
$$= 2.5 \times 10^{-2}\ \text{m/s}$$

An alternate way of solving this problem is to recognize that the wavelength (λ) is 1.0×10^{-2} meter and that the frequency (f) is $1/(4.0 \times 10^{-1})$ hertz. Then use *Equation W1* on *Reference Table K*.

25 Sound resonance is the phenomenon by which a vibrating tuning fork can cause another tuning fork with the same frequency to vibrate. This phenomenon can also be observed in vibrating air columns and vibrating strings.

WRONG CHOICES EXPLAINED:

(1) Diffraction is the bending of a wave around a barrier.

(2) Reflection is the bouncing back of a wave when the wave strikes the boundary of two different media.

(3) Refraction is the bending of a wave entering a second medium because the wave speed is different in each of the two media.

26 When a sound source and an observer are in relative motion, the apparent frequency heard by the observer will be different from the frequency emitted by the source. If the source and the observer are moving toward each other, the apparent frequency perceived by the observer will be greater than the emitted frequency. This phenomenon is known as the Doppler effect.

27 The electronvolt is a unit of energy. As shown on *Reference Table A*, 1 electronvolt is equivalent to 1.60×10^{-19} joule.

28 Magnetic fields are produced by charged particles in motion.

29 The period is the time for 1 heartbeat. Since 60 heartbeats take 60 seconds, 1 heartbeat will take 1 second.

30 See *Equation EL3* on *Reference Table K*:

$$\text{Solution:} \quad V = \frac{W}{q}$$

$$= \frac{48\ \text{J}}{4.0\ \text{C}}$$

$$= 12\text{V}$$

31 Current is the flow of charge in a circuit. Since charge is conserved in the circuit, the sum of the currents entering the junction (the place where the two wires cross) must equal the sum of the currents leaving the junction. A total of 12 amperes enter. So a total of 12 amperes must leave the junction. Therefore, the amount of current passing through the ammeter must be 9 amperes.

32 Use *Equation EL8* on *Reference Table K*:

$$\text{Solution:} \quad W = VIt$$

$$I = \frac{W}{Vt}$$

$$= \frac{6.0 \times 10^{6}\ \text{J}}{(220\ \text{V})\left(1.8 \times 10^{3}\ \text{s}\right)}$$

$$= 15\ \text{A}$$

33 All electric charges must be whole-number multiples of the elementary charge. See *Reference Table* A; $e = 1.60 \times 10^{-19}$ C. This relationship is given by the equation $q = ne$, where n is the number of elementary charges comprising the charge q. Of the choices given, only choice (1), $+4.80 \times 10^{-19}$ C, is a whole-number multiple of the elementary charge. (In this case, $q = +3.00e$.)

WRONG CHOICES EXPLAINED:

(2) $q = +1.500e$
(3) $q = -1.500e$
(4) $q = -3.5e$

34 Use the hydrogen energy level diagram on *Reference Table F*:

$$\begin{aligned} E_{photon} &= E_3 - E_5 \\ &= (-1.51 \text{ eV}) - (-0.54 \text{ eV}) \\ &= -0.97 \text{ eV} \end{aligned}$$

The negative sign means that the photon was emitted.

35 A gamma ray photon has no electric charge. Since electric charge must be conserved, the production of an electron, which has a negative charge, must be accompanied by the production of a positive particle.

PART B–1

36 An unsharpened pencil is about 10 inches long. Since 1 inch is approximately equal to 2 centimeters, the length of the pencil is approximately 20 centimeters or 2×10^{-1} meter.

WRONG CHOICES EXPLAINED:

(1) 2×10^{-2} meter is approximately equal to 1 inch.
(3) 2 meters is approximately equal to 80 inches.
(4) 20 meters is approximately equal to 60 feet.

37 This question involves the conservation of (linear) momentum. Use *Equations ME12* and *ME13* on *Reference Table K*:

Solution:

$$p = mv$$

$$p_{\text{before}} = p_{\text{after}}$$

$$(8.0\text{ kg})(+4.0\text{ m/s}) + (4.0\text{ kg})(-6.0\text{ m/s}) = (8.0\text{ kg})v + (4.0\text{ kg})(3.0\text{ m/s})$$

$$v = \frac{(8.0\text{ kg})(+4.0\text{ m/s}) + (4.0\text{ kg})(-6.0\text{ m/s}) - (4.0\text{ kg})(3.0\text{ m/s})}{8.0\text{ kg}}$$

$$= -0.5\text{ m/s}$$

This means the 8.0-kilogram cart is moving to the left at 0.50 meter per second.

38 We need only consider the forces in the x-direction since the vertical forces add to 0 newton. The two horizontal forces produce a net force of 20 newtons to the left. Therefore, the block will accelerate to the left.

39 The force needed to lift the mass is equal to its weight (mg). The power generated by the motor is the work done in lifting the mass (mgh) divided by the time needed to lift it. Use *Equation ME23* on *Reference Table K*:

Solution:

$$P = \frac{W}{t}$$

$$= \frac{Fd}{t}$$

$$= \frac{mgh}{t}$$

$$= \frac{(400\text{ kg})(9.81\text{ m/s}^2)(10\text{ m})}{8.00\text{ s}}$$

$$= 4.9 \times 10^3\text{ W}$$

40 The net force needed to accelerate the 4.0-kilogram object at 3.0 meters per second^2 is determined by using Newton's second law (*Equation ME8* on *Reference Table K*):

$$F_{net} = ma$$
$$= (4.0 \text{ kg})(3.0 \text{ m/s}^2)$$
$$= 12 \text{ N}$$

Now apply Newton's second law to the 2.0-kilogram mass:

Solution: $$a = \frac{F_{net}}{m}$$
$$= \frac{12 \text{ N}}{2.0 \text{ kg}}$$
$$= 6.0 \text{ m/s}^2$$

41 The electron has a charge whose magnitude is equal to the magnitude of the elementary charge. See *Reference Table A* and use *Equation EL2* on *Reference Table K*:

Solution: $$E = \frac{F_e}{q}$$
$$F_e = qE$$
$$= (1.60 \times 10^{-19} \text{ C})(600 \text{ N/C})$$
$$= 9.60 \times 10^{-17} \text{ N}$$

42 See *Reference Table A* for the magnitude of the charge on the electron, and use *Equation EL4* on *Reference Table K*:

Solution: $$I = \frac{\Delta q}{t}$$
$$t = \frac{\Delta q}{I}$$
$$= \frac{(2.5 \times 10^{19} \text{ electrons})\left(\frac{1.6 \times 10^{-19} \text{C}}{1 \text{ electron}}\right)}{4.0 \text{ A}}$$
$$= 1.0 \text{ s}$$

43 Use Coulomb's law (*Equation EL1* on *Reference Table K*):

$$\text{Solution:} \quad F_e = \frac{kq_1q_2}{r^2}$$

$$F_e' = \frac{k(2q_1)(4q_2)}{(2r)^2}$$

$$= \frac{8}{4} \bullet \frac{kq_1q_2}{r^2}$$

$$= 2F$$

44 See *Reference Tables G* and *H*. A meson is composed of a quark and an antiquark. An antiquark always has a charge opposite to that of its companion quark. The combination $s\bar{c}$ is a meson that has a charge of $\left(-\frac{1}{3}e\right) + \left(-\frac{2}{3}e\right) = -1e$ as required.

WRONG CHOICES EXPLAINED:

(2), (4) These particles are not mesons.

(3) The meson $u\bar{b}$ has a charge of $\left(+\frac{2}{3}e\right) + \left(+\frac{1}{3}e\right) = +1e$

45 Refer to *Equation ME20* on *Reference Table K*. Kinetic increases with the square of the speed. A graph of kinetic energy versus speed will produce an upward curve, known as a parabola.

46 Electric field lines always point away from positive charges and toward negative charges. Of the choices given, only the diagram shown in choice (3) meets these criteria.

47 Refer to *Equation ME10* on *Reference Table K.* The gravitational force between two masses varies inversely as the square of the distance between the masses. As the distance decreases, the force increases rapidly. As the distance increases, the force decreases rapidly. Neither the force nor the distance can ever equal zero. Only the graph shown in choice (4) meets all of these criteria.

48 When the waves meet in region *AB*, they will coincide at every point. In other words, the waves will be completely in phase. This is the condition for constructive interference.

49 When the straight wave fronts reach the barrier, they diffract. In other words, they bend around the barrier, causing circular wave fronts to emerge. The circular wave fronts have the same wavelength (1.4 centimeters) as the straight wave fronts incident on the barrier. This is illustrated in choice (1).

50 Refer to *Equation MP3* on *Reference Table K.* According to this relationship, energy is directly proportional to the equivalent mass. A graph of energy versus equivalent mass will yield a straight line that passes through the origin and has a positive slope equal to c^2.

PART B–2

51–52 Use *Equation EL6* on *Reference Table K*:

Solution:

$$R = \frac{\rho L}{A}$$

$$\rho = \frac{RA}{L}$$

$$= \frac{(0.757\ \Omega)\left(3.50 \times 10^{-6}\,\text{m}^2\right)}{25.0\ \text{m}}$$

[1 point]

52 Solution: $\rho = 1.06 \times 10^{-7}\ \Omega \bullet \text{m}$

[1 point]

53 The original periodic wave illustrated in the question has a wavelength of 4 meters. It consists of the right half of a crest (from 0 m to 1 m), a full trough (from 1 m to 3 m), and the left half of a trough (from 3 m to 4 m). The amplitude of the wave is 0.1 meter.

In order to draw a wave with the same amplitude (0.1 m) and half the wavelength (2 m), you must draw the right half of a crest from 0 m to 0.5 m, a full trough from 0.5 m to 1.5 m, and the left half of a crest from 1.5 m to 2 m. See the diagram below.

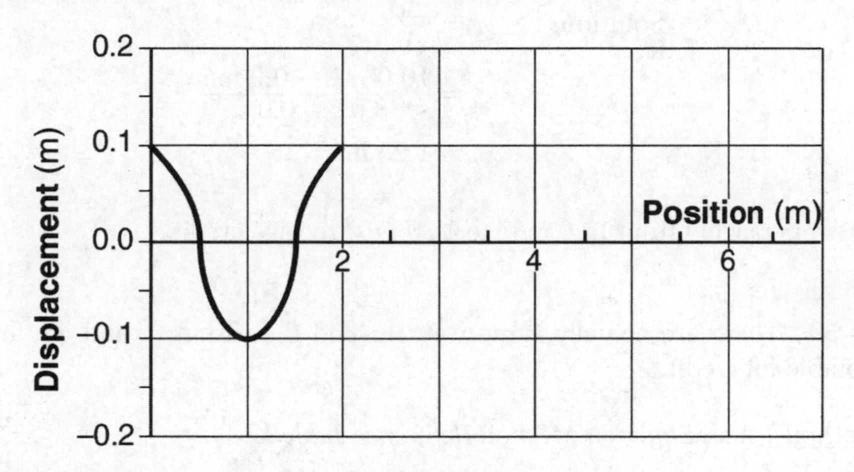

[1 point]

54–55 Use *Equation ME14* on *Reference Table K*:

Solution: $J = F_{net}t$

$$t = \frac{J}{F_{net}}$$

$$= \frac{3.6 \text{ N} \bullet \text{s}}{600 \text{ N}}$$

[1 point]

55 Solution: $t = 0.0060$ s or 6.0×10^{-3} s

[1 point]

56 Magnetic field lines are directed from the north pole to the south pole. Refer to the diagrams below. (Either one is acceptable for credit.)

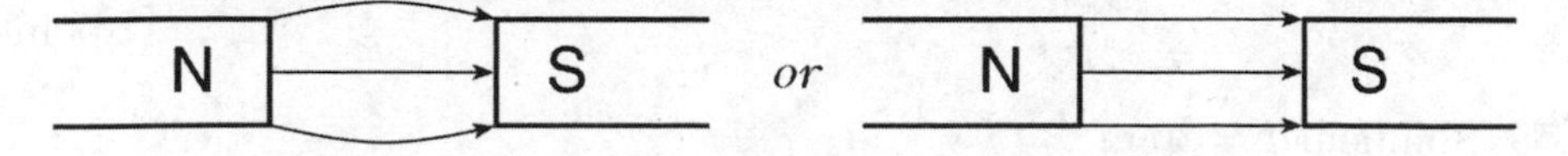

[1 point]

57 Refer to the graph given with this question. The acceleration of the car is determined from the slope of the graph. For example, choose the coordinates (0.0 s, 0.0 m/s) and (8.0 s, 10.0 m/s):

Solution:
$$a = \frac{\Delta v}{\Delta t}$$
$$= \frac{10.0 \text{ m/s} - 0.0 \text{ m/s}}{8.0 \text{ s} - 0.0 \text{ s}}$$
$$= 1.25 \text{ m/s}^2$$

Answers ranging from 1.20 m/s^2 to 1.30 m/s^2 receive credit.

[1 point]

58–59 There are actually three ways to find the distance, and any one is acceptable for credit.

Method 1: Use *Equation ME4* on *Reference Table K*:

Solution:
$$d = v_i t + \frac{1}{2}at^2$$
$$= (5.0 \text{ m/s})(4.0 \text{ s}) + \frac{1}{2}(1.25 \text{ m/s}^2)(4.0 \text{ s})^2$$

Method 2: Use *Equation ME1* on *Reference Table K*. The average speed between 4.0 s and 8.0 s is found by adding the speeds at these two times (5.0 m/s and 10.0 m/s) and then dividing by 2, which is 7.5 m/s:

Solution:
$$\bar{v} = \frac{d}{t}$$
$$d = \bar{v}t$$
$$= (7.5 \text{ m/s})(4.0 \text{ s})$$

Method 3: Find the area under the graph between 4.0 s and 8.0 s by breaking the resultant trapezoid into a right triangle and a rectangle. The area of the rectangle is 20 m and the area of the right triangle is 10 m.

[1 point]

59 Solution: $d = 30$ m

[1 point]

60 Use *Equation EL5* on *Reference Table K*:

Solution:
$$R_{eq} = \frac{V}{I}$$
$$= \frac{30\text{ V}}{2.0\text{ A}}$$
$$= 15\ \Omega$$

[1 point]

61–62 Use *Equation EL10C* on *Reference Table K*:

Solution:
$$\frac{1}{R_{eq}} = \frac{1}{R_1} + \frac{1}{R_2}$$
$$\frac{1}{R_2} = \frac{1}{R_{eq}} - \frac{1}{R_1}$$
$$= \frac{1}{15\ \Omega} - \frac{1}{20\ \Omega}$$
$$= \frac{1}{60\ \Omega}$$

[1 point]

62 Solution: $R_{eq} = 60\ \Omega$

[1 point]

63–64 Refer to the diagram on the answer sheet. The radius of the circular path is 0.80 meter, and the speed of the stopper is 2.5 meters per second. Use *Equation ME18* on *Reference Table K*:

Solution:
$$a_c = \frac{v^2}{r}$$
$$= \frac{(2.5\text{ m/s})^2}{0.80\text{ m}}$$

[1 point]

64 Solution: $a_c = 7.8\text{ m/s}^2$

[1 point]

65 Refer to the diagram below, which shows an appropriate drawing that is acceptable for credit. The arrow pointing to the center of the circle represents the centripetal force vector.

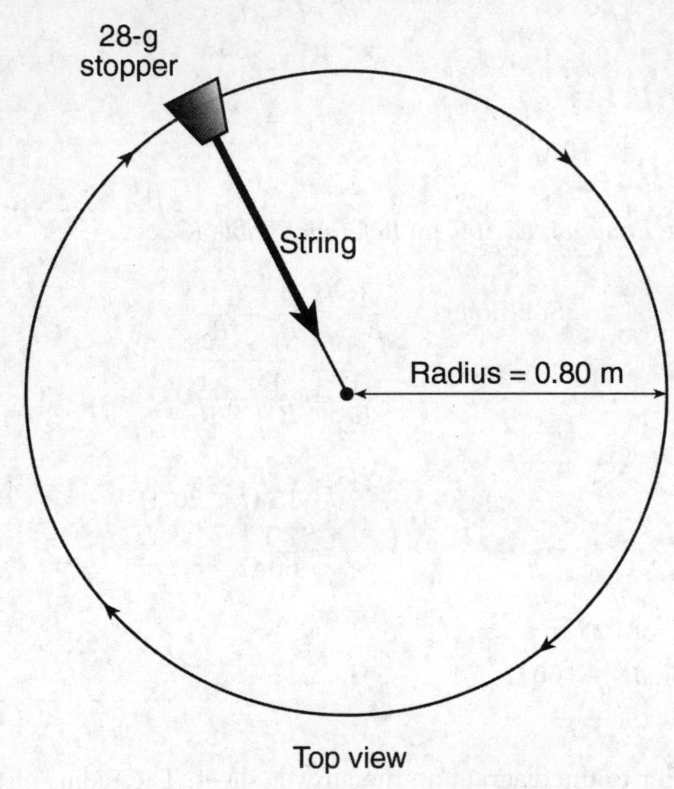

[1 point]

PART C

66 The order of magnitude of a measurement is the measurement stated to the nearest power of 10 or the power itself. For example, the order of magnitude for 10 nanoseconds is 10^{-8} or –8.

This particular question, however, stirred a great deal of controversy due to its wording. Therefore, any of the following answers were also acceptable:

- ≤ -8
- $\leq 10^{-9}$
- < -8
- $< 10^{-9}$
- any whole number less than -8
- any exponent less than -8

[1 point]

67–68 Use *Reference Table A* and *Equation MP1* on *Reference Table K*:

Solution:

$$E_{photon} = hf$$
$$= \left(6.63 \times 10^{-34}\ \text{J} \cdot \text{s}\right)\left(4.76 \times 10^{14}\ \text{Hz}\right)$$

[1 point]

68 Solution: $E_{photon} = 3.16 \times 10^{-19}\ \text{J}$

[1 point]

69 The ground state is the lowest available energy level that an atom can have. The ground state is the most stable energy state of an atom. (Either statement is acceptable for credit.)

[1 point]

70–71 Refer to the diagram below, which applies to questions 70 and 71.

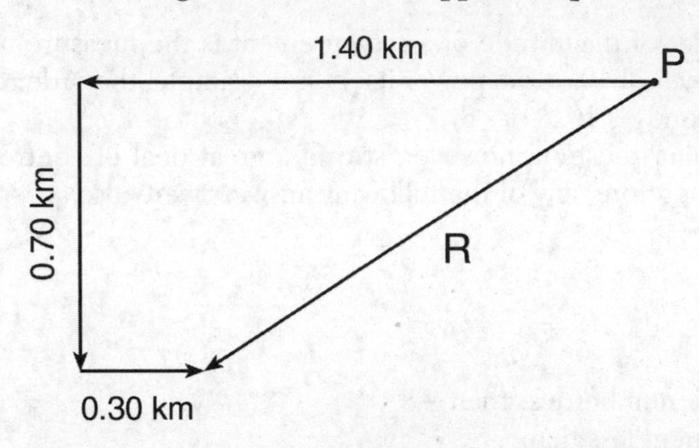

In order to draw the 0.30-kilometer displacement, use the conversion factor provided in the question and employ the factor-label method:

$$0.30 \text{ km}\left(\frac{1.0 \text{ cm}}{0.20 \text{ km}}\right) = 1.5 \text{ cm}$$

Now draw a 1.5-centimeter arrow pointing to the right. Label it "0.30 km," as shown in the diagram.

The range of arrow lengths acceptable for credit is 1.3 cm to 1.7 cm.

[1 point]

71 Draw the resultant vector from point P to the end of the 0.3-kilometer vector. Label the resultant as R, as shown in the diagram above.

[1 point]

72–73 The girl's average speed for the entire trip is the total distance she covers divided by the total elapsed time. Use *Equation ME1* on *Reference Table K*:

Solution:
$$\begin{aligned}\overline{v} &= \frac{d}{t}\\ &= \frac{1.40\text{ km} + 0.70\text{ km} + 0.30\text{ km}}{12\text{ min}}\\ &= \frac{2.4\text{ km}}{12\text{ min}}\end{aligned}$$

or

$$\begin{aligned}\overline{v} &= \frac{d}{t}\\ &= \frac{1400\text{ m} + 700\text{ m} + 300\text{ m}}{12\text{min}\left(\frac{60\text{ s}}{1\text{ min}}\right)}\\ &= \frac{2400\text{ m}}{720\text{ s}}\end{aligned}$$

[1 point]

73 Solution: $\overline{v} = 0.20$ km/min or $\overline{v} = 3.3$ m/s [1 point]

74 In order to calculate the magnitude of the resultant displacement vector, use a ruler to measure the length of the arrow marked *R*. The length is 6.6 centimeters. Now apply the factor-label method using the conversion factor given:

$$\text{Solution: } 6.6\text{ cm}\left(\frac{0.20\text{ km}}{1\text{ cm}}\right) = 1.3\text{ km}$$

Answers ranging from 1.1 km to 1.5 km received credit. [1 point]

75 Measure the angle between the resultant vector and the 1.4-kilometer displacement with a protractor. The correct angle is 32°.

Answers ranging from 30° to 34° receive credit. [1 point]

76 The angle of reflection of any wave at a boundary separating two media is always equal to the angle of incidence. Therefore, the angle of reflection of the light ray at the air-water interface is 35°.

[1 point]

77 Refer to the diagram below:

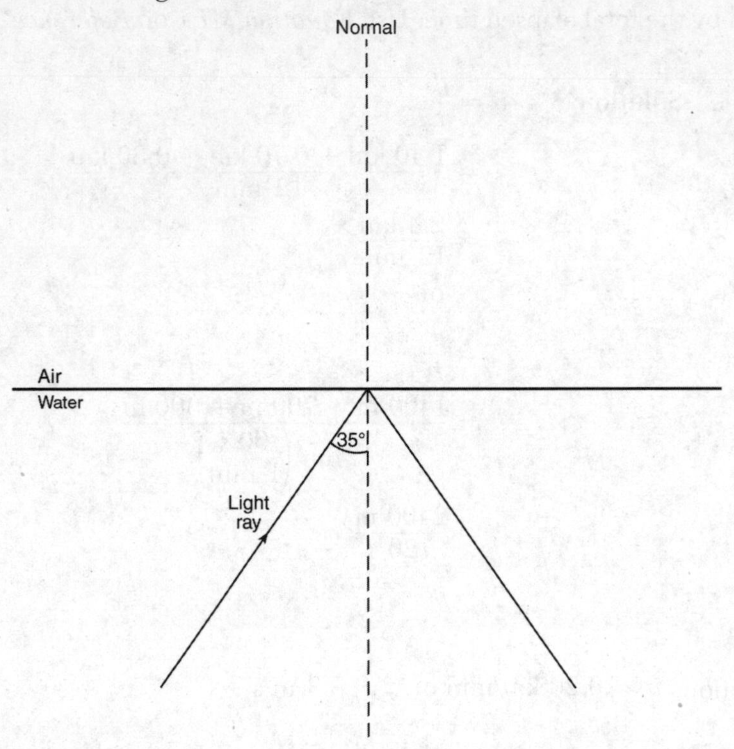

Use a protractor and a straight edge to draw the reflected ray in the water at an angle of 35° to the normal.

[1 point]

78–79 Use *Reference Table E* and *Equation W5* on *Reference Table K*:

Solution:

$$n_{\text{water}} \sin\theta_{\text{water}} = n_{\text{air}} \sin\theta_{\text{air}}$$

$$\sin\theta_{\text{air}} = \frac{n_{\text{water}} \sin\theta_{\text{water}}}{n_{\text{air}}}$$

$$= \frac{(1.33)(\sin 35°)}{1.00}$$

$$= 0.763$$

$$\theta_{\text{air}} = \sin^{-1}(0.763)$$

[1 point]

79 Solution: $\theta_{air} = 49.7°$

Either 49° or 50° is acceptable for credit. [1 point]

80 Any one of the following characteristics is the same in both water and air and is acceptable for credit:

- frequency
- period
- phase
- color
- the transverse nature of the light wave

[1 point]

81 Use *Equation ME21* on *Reference Table K*:

$$\text{Solution:} \quad W = Fd$$
$$= (30.4\ N)(6.00\ m)$$
$$= 182\ \text{J}$$

[1 point]

82–83 Use *Equation ME19* on *Reference Table K*:

$$\text{Solution:} \quad \Delta PE = mgh$$
$$= (40.0\ \text{N})(3.00\ \text{m})$$

Remember:

- The value 40.0 newtons is the weight of the object (mg) and not its mass.
- The vertical distance must be used in the calculation.

[1 point]

83 Solution: $\Delta PE = 120$ J [1 point]

84 Since the speed remains constant, the kinetic energy of the crate remains constant.

[1 point]

85 The work done on the crate is used to counteract the friction between the crate and the incline. As a result, the internal energy of the crate increases.

[1 point]

Topic	Question Numbers (total)	Wrong Answers (x)	Grade
Math Skills	57–59, 66, 70: (5)		$\frac{100(5-x)}{5} = \%$
Mechanics	1–13, 36–38, 40, 47, 54, 55, 63–65, 71–75: (28)		$\frac{100(28-x)}{28} = \%$
Energy	14, 15, 21, 27, 39, 45, 81–85: (11)		$\frac{100(11-x)}{11} = \%$
Electricity/ Magnetism	23, 28, 30–33, 41–43, 46, 51, 52, 56, 60–62: (16)		$\frac{100(16-x)}{16} = \%$
Waves	16–20, 22, 24–26, 29, 48, 49, 53, 76–80: (18)		$\frac{100(18-x)}{18} = \%$
Modern Physics	34, 35, 44, 50, 67–69: (7)		$\frac{100(7-x)}{7} = \%$

Examination June 2014

Physics: The Physical Setting

PART A

Answer all questions in this part.

Directions (1–35): For *each* statement or question, select the *number* of the word or expression that, of those given, best completes the statement or answers the question. Some questions may require the use of the *2006 Edition Reference Tables for Physical Setting/Physics*. Record your answers in the spaces provided.

1 Which quantity is scalar?

(1) mass
(2) force
(3) momentum
(4) acceleration

1 _____

2 What is the final speed of an object that starts from rest and accelerates uniformly at 4.0 meters per second^2 over a distance of 8.0 meters?

(1) 8.0 m/s
(2) 16 m/s
(3) 32 m/s
(4) 64 m/s

2 _____

3 The components of a 15-meters-per-second velocity at an angle of 60° above the horizontal are

(1) 7.5 m/s vertical and 13 m/s horizontal
(2) 13 m/s vertical and 7.5 m/s horizontal
(3) 6.0 m/s vertical and 9.0 m/s horizontal
(4) 9.0 m/s vertical and 6.0 m/s horizontal

3 _____

4 What is the time required for an object starting from rest to fall freely 500 meters near Earth's surface?

(1) 51.0 s
(2) 25.5 s
(3) 10.1 s
(4) 7.14 s

4 ____

5 A baseball bat exerts a force of magnitude F on a ball. If the mass of the bat is three times the mass of the ball, the magnitude of the force of the ball on the bat is

(1) F
(2) $2F$
(3) $3F$
(4) $F/3$

5 ____

6 A 2.0-kilogram mass is located 3.0 meters above the surface of Earth. What is the magnitude of Earth's gravitational field strength at this location?

(1) 4.9 N/kg
(2) 2.0 N/kg
(3) 9.8 N/kg
(4) 20 N/kg

6 ____

7 A truck, initially traveling at a speed of 22 meters per second, increases speed at a constant rate of 2.4 meters per second^2 for 3.2 seconds. What is the total distance traveled by the truck during this 3.2-second time interval?

(1) 12 m
(2) 58 m
(3) 70 m
(4) 83 m

7 ____

8 A 750-newton person stands in an elevator that is accelerating downward. The upward force of the elevator floor on the person must be

(1) equal to 0 N
(2) less than 750 N
(3) equal to 750 N
(4) greater than 750 N

8 ____

9 A 3.0-kilogram object is acted upon by an impulse having a magnitude of 15 newton • seconds. What is the magnitude of the object's change in momentum due to this impulse?

(1) 5.0 kg • m/s
(2) 15 kg • m/s
(3) 3.0 kg • m/s
(4) 45 kg • m/s

9 ____

10 An air bag is used to safely decrease the momentum of a driver in a car accident. The air bag reduces the magnitude of the force acting on the driver by

(1) increasing the length of time the force acts on the driver
(2) decreasing the distance over which the force acts on the driver
(3) increasing the rate of acceleration of the driver
(4) decreasing the mass of the driver

10 _____

11 An electron moving at constant speed produces

(1) a magnetic field, only
(2) an electric field, only
(3) both a magnetic and an electric field
(4) neither a magnetic nor an electric field

11 _____

12 A beam of electrons passes through an electric field where the magnitude of the electric field strength is 3.00×10^3 newtons per coulomb. What is the magnitude of the electrostatic force exerted by the electric field on each electron in the beam?

(1) 5.33×10^{-23} N
(2) 4.80×10^{-16} N
(3) 3.00×10^3 N
(4) 1.88×10^{22} N

12 _____

13 How much work is required to move 3.0 coulombs of electric charge a distance of 0.010 meter through a potential difference of 9.0 volts?

(1) 2.7×10^3 J
(2) 27 J
(3) 3.0 J
(4) 3.0×10^{-2} J

13 _____

14 What is the resistance of a 20.0-meter-long tungsten rod with a cross-sectional area of 1.00×10^{-4} meter^2 at 20°C?

(1) 2.80×10^{-5} Ω
(2) 1.12×10^{-2} Ω
(3) 89.3 Ω
(4) 112 Ω

14 _____

15 Two pieces of flint rock produce a visible spark when they are struck together. During this process, mechanical energy is converted into

(1) nuclear energy and electromagnetic energy
(2) internal energy and nuclear energy
(3) electromagnetic energy and internal energy
(4) elastic potential energy and nuclear energy

15 ____

16 A 15-kilogram cart is at rest on a horizontal surface. A 5-kilogram box is placed in the cart. Compared to the mass and inertia of the cart, the cart-box system has

(1) more mass and more inertia
(2) more mass and the same inertia
(3) the same mass and more inertia
(4) less mass and more inertia

16 ____

17 Transverse waves are to radio waves as longitudinal waves are to

(1) light waves
(2) microwaves
(3) ultraviolet waves
(4) sound waves

17 ____

18 As a monochromatic light ray passes from air into water, two characteristics of the ray that will *not* change are

(1) wavelength and period
(2) frequency and period
(3) wavelength and speed
(4) frequency and speed

18 ____

19 When a mass is placed on a spring with a spring constant of 60.0 newtons per meter, the spring is compressed 0.500 meter. How much energy is stored in the spring?

(1) 60.0 J
(2) 30.0 J
(3) 15.0 J
(4) 7.50 J

19 ____

20 A boy pushes his sister on a swing. What is the frequency of oscillation of his sister on the swing if the boy counts 90 complete swings in 300 seconds?

(1) 0.30 Hz
(2) 2.0 Hz
(3) 1.5 Hz
(4) 18 Hz

20 ____

21 What is the period of a sound wave having a frequency of 340 hertz?

(1) 3.40×10^{2} s
(2) 1.02×10^{0} s
(3) 9.73×10^{-1} s
(4) 2.94×10^{-3} s

21 ____

22 An MP3 player draws a current of 0.120 ampere from a 3.00-volt battery. What is the total charge that passes through the player in 900 seconds?

(1) 324 C
(2) 108 C
(3) 5.40 C
(4) 1.80 C

22 ____

23 A beam of light has a wavelength of 4.5×10^{-7} meter in a vacuum. The frequency of this light is

(1) 1.5×10^{-15} Hz
(2) 4.5×10^{-7} Hz
(3) 1.4×10^{2} Hz
(4) 6.7×10^{14} Hz

23 ____

24 When x-ray radiation and infrared radiation are traveling in a vacuum, they have the same

(1) speed
(2) frequency
(3) wavelength
(4) energy per photon

24 ____

25 The diagram below represents two identical pulses approaching each other in a uniform medium.

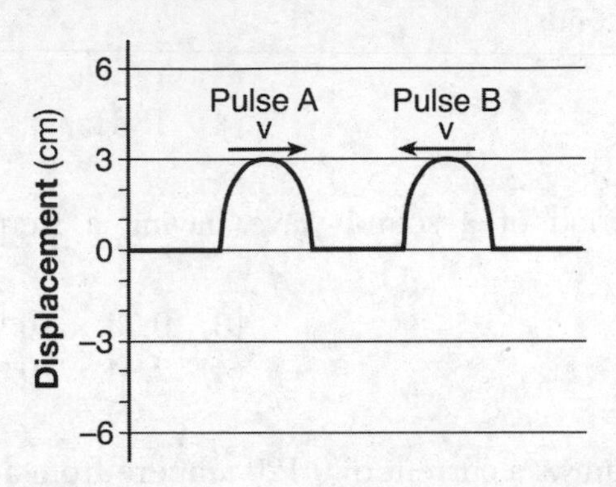

As the pulses meet and are superposed, the maximum displacement of the medium is

(1) –6 cm
(2) 0 cm
(3) 3 cm
(4) 6 cm

25 ____

26 As a car approaches a pedestrian crossing the road, the driver blows the horn. Compared to the sound wave emitted by the horn, the sound wave detected by the pedestrian has a

(1) higher frequency and a lower pitch
(2) higher frequency and a higher pitch
(3) lower frequency and a higher pitch
(4) lower frequency and a lower pitch

26 ____

27 When air is blown across the top of an open water bottle, air molecules in the bottle vibrate at a particular frequency and sound is produced. This phenomenon is called

(1) diffraction
(2) refraction
(3) resonance
(4) the Doppler effect

27 ____

28 An antibaryon composed of two antiup quarks and one antidown quark would have a charge of

(1) +1e
(2) −1e
(3) 0e
(4) −3e

28 ____

29 Which force is responsible for producing a stable nucleus by opposing the electrostatic force of repulsion between protons?

(1) strong
(2) weak
(3) frictional
(4) gravitational

29 ____

30 What is the total energy released when 9.11×10^{-31} kilogram of mass is converted into energy?

(1) 2.73×10^{-22} J
(2) 8.20×10^{-14} J
(3) 9.11×10^{-31} J
(4) 1.01×10^{-47} J

30 ____

31 A shopping cart slows as it moves along a level floor. Which statement describes the energies of the cart?

(1) The kinetic energy increases and the gravitational potential energy remains the same.
(2) The kinetic energy increases and the gravitational potential energy decreases.
(3) The kinetic energy decreases and the gravitational potential energy remains the same.
(4) The kinetic energy decreases and the gravitational potential energy increases.

31 ____

32 Two identically sized metal spheres, *A* and *B*, are on insulating stands, as shown in the diagram below. Sphere *A* possesses an excess of 6.3×10^{10} electrons and sphere *B* is neutral.

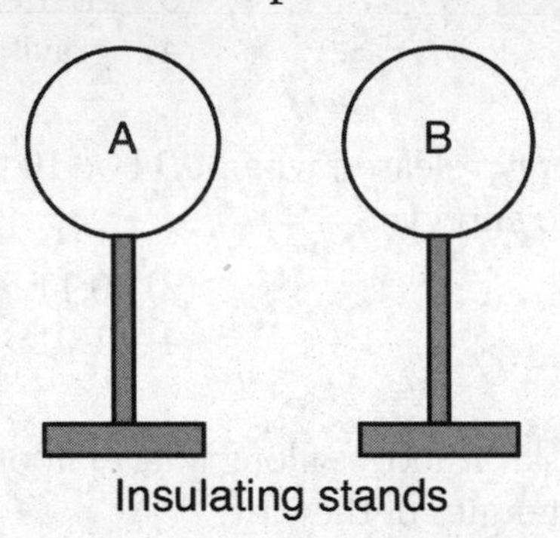

Which diagram best represents the charge distribution on sphere *B*?

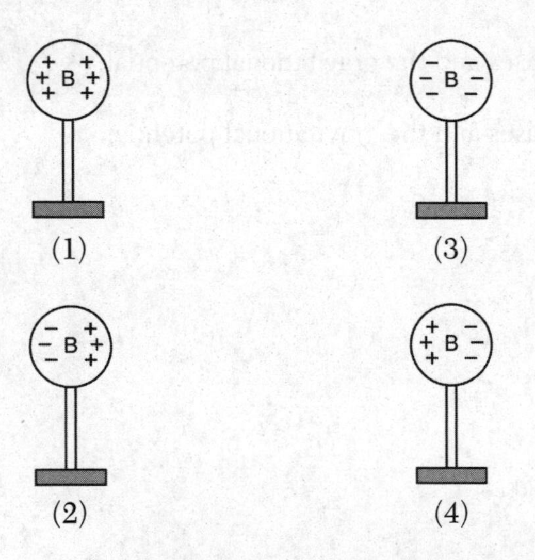

32 ____

33 Two points, *A* and *B*, are located within the electric field produced by a −3.0 nanocoulomb charge. Point *A* is 0.10 meter to the left of the charge and point *B* is 0.20 meter to the right of the charge, as shown in the diagram below.

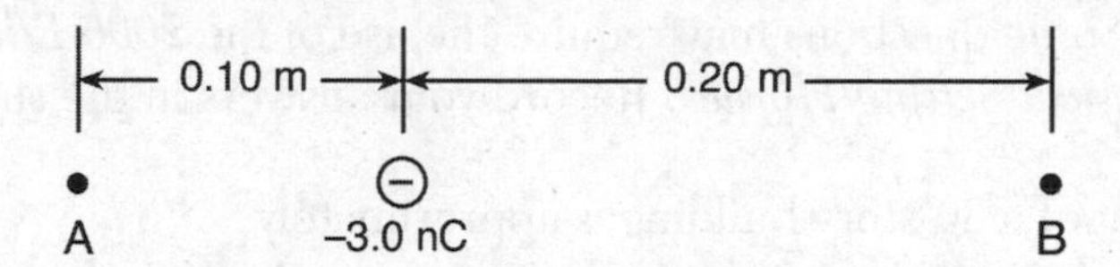

Compared to the magnitude of the electric field strength at point *A*, the magnitude of the electric field strength at point *B* is

(1) half as great
(2) twice as great
(3) one-fourth as great
(4) four times as great

33 ____

34 The diagram below represents two waves, *A* and *B*, traveling through the same uniform medium.

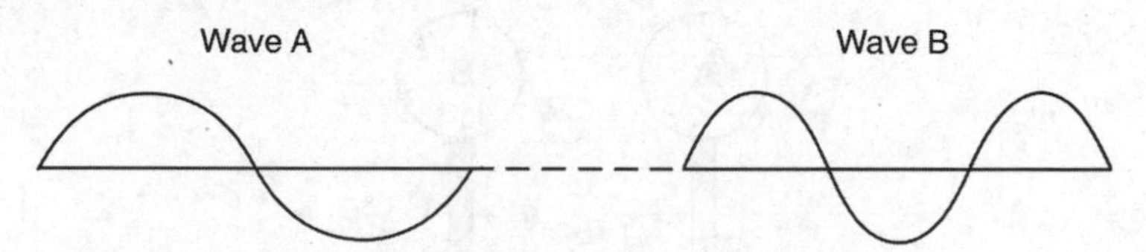

Which characteristic is the same for both waves?

(1) amplitude
(2) frequency
(3) period
(4) wavelength

34 ____

35 The diagram below shows a periodic wave.

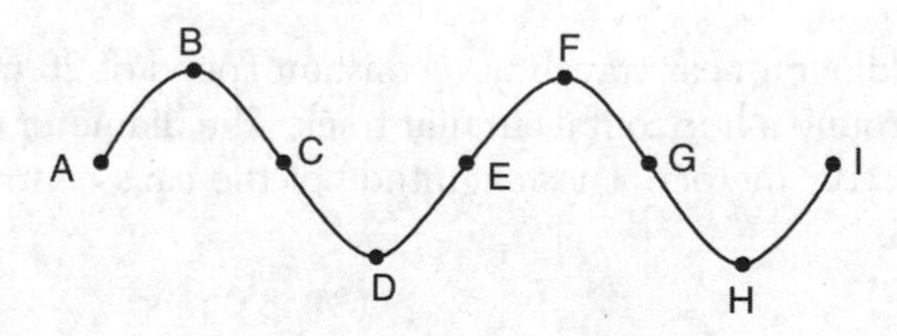

Which two points on the wave are 180° out of phase?

(1) *A* and *C*
(2) *B* and *E*
(3) *F* and *G*
(4) *D* and *H*

35 ____

PART B–1

Answer all questions in this part.

Directions (36–50): For *each* statement or question, select the *number* of the word or expression that, of those given, best completes the statement or answers the question. Some questions may require the use of the *2006 Edition Reference Tables for Physical Setting/Physics*. Record your answers in the spaces provided.

36 The height of a 30-story building is approximately

(1) 10^0 m
(2) 10^1 m
(3) 10^2 m
(4) 10^3 m

36 ____

37 Two identically sized metal spheres on insulating stands are positioned as shown below. The charge on sphere *A* is -4.0×10^{-6} coulomb and the charge on sphere *B* is -8.0×10^{-6} coulomb.

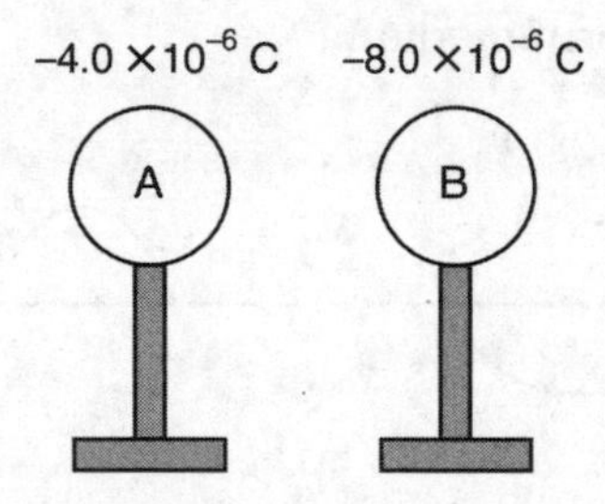

The two spheres are touched together and then separated. The total number of excess electrons on sphere A after the separation is

(1) 2.5×10^{13}
(2) 3.8×10^{13}
(3) 5.0×10^{13}
(4) 7.5×10^{13}

37 ____

38 A 1.0×10^3-kilogram car travels at a constant speed of 20 meters per second around a horizontal circular track. The diameter of the track is 1.0×10^2 meters. The magnitude of the car's centripetal acceleration is

(1) 0.20 m/s^2
(2) 2.0 m/s^2
(3) 8.0 m/s^2
(4) 4.0 m/s^2

38 ____

39 Which combination of units can be used to express electrical energy?

(1) $\frac{\text{volt}}{\text{coulomb}}$

(2) $\frac{\text{coulomb}}{\text{volt}}$

(3) volt • coulomb

(4) volt • coulomb • second

39 _____

40 The total amount of electrical energy used by a 315-watt television during 30.0 minutes of operation is

(1) 5.67×10^5 J

(2) 9.45×10^3 J

(3) 1.05×10^1 J

(4) 1.75×10^{-1} J

40 _____

41 Which graph best represents the relationship between the absolute index of refraction and the speed of light ($f = 5.09 \times 10^{14}$ Hz) in various media?

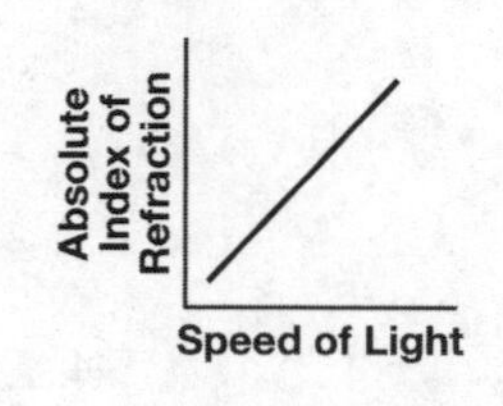

(1)

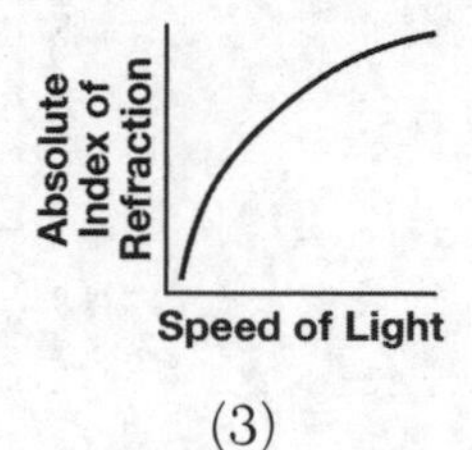

(3)

Absolute Index of Refraction
Speed of Light

(2)

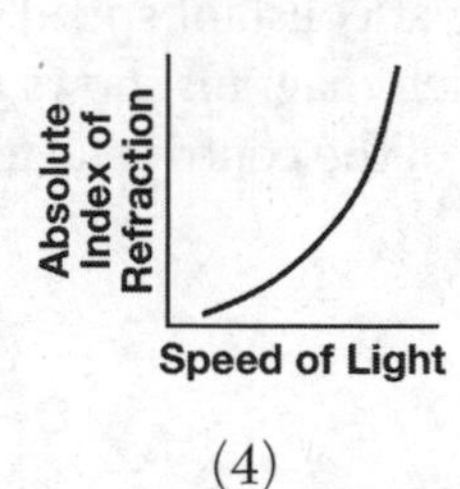

(4)

41 _____

42 A 25-gram paper cup falls from rest off the edge of a tabletop 0.90 meter above the floor. If the cup has 0.20 joule of kinetic energy when it hits the floor, what is the total amount of energy converted into internal (thermal) energy during the cup's fall?

(1) 0.02 J

(2) 0.22 J

(3) 2.2 J

(4) 220 J

42 _____

43 Which electron transition between the energy levels of hydrogen causes the emission of a photon of visible light?

(1) $n = 6$ to $n = 5$
(2) $n = 5$ to $n = 6$
(3) $n = 5$ to $n = 2$
(4) $n = 2$ to $n = 5$

43 ____

44 Which graph best represents an object in equilibrium moving in a straight line?

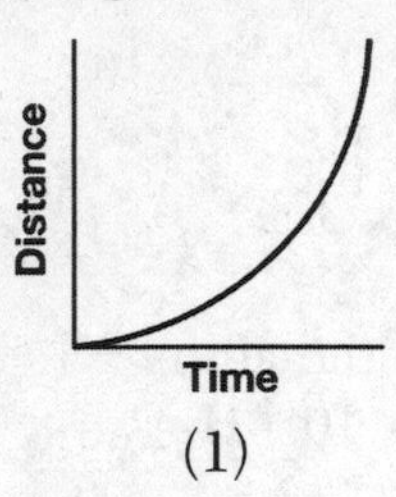

(1)

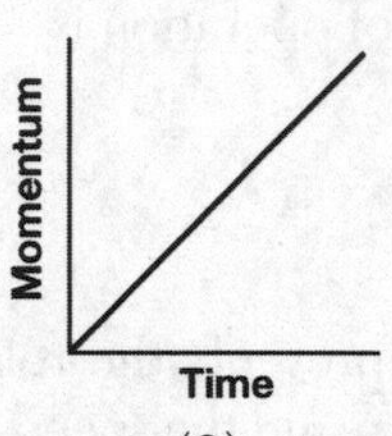

(3)

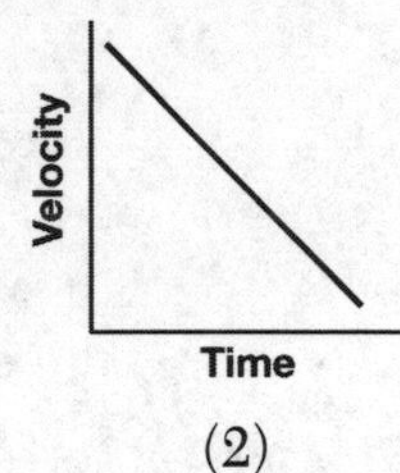

(2)

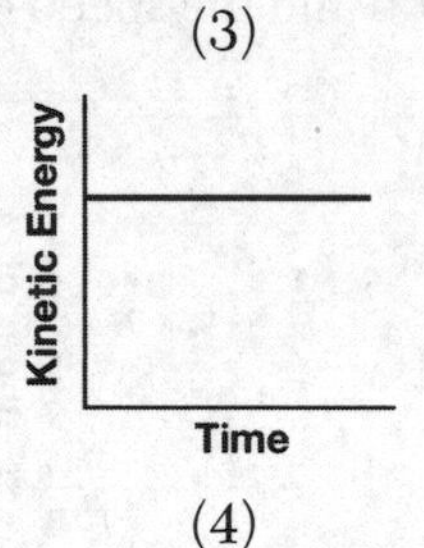

(4)

44 ____

45 A body, B, is moving at constant speed in a horizontal circular path around point P. Which diagram shows the direction of the velocity (v) and the direction of the centripetal force (F_c) acting on the body?

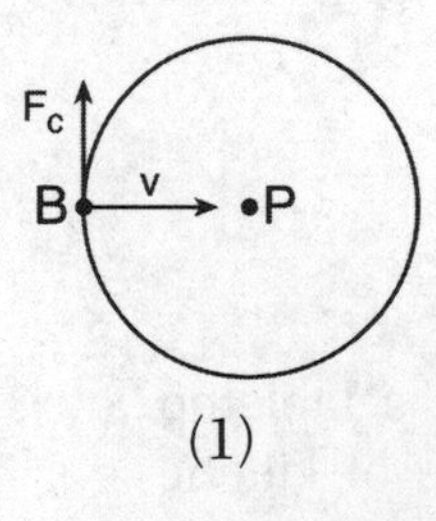

(1)

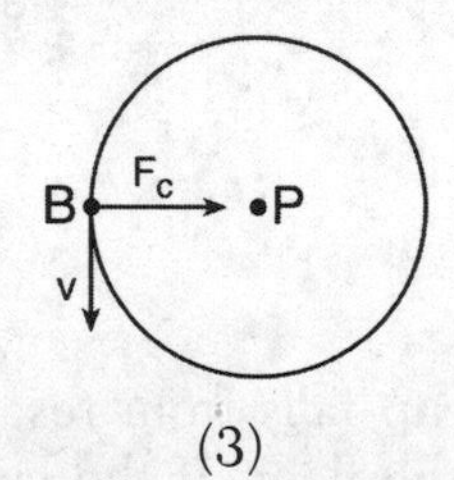

(3)

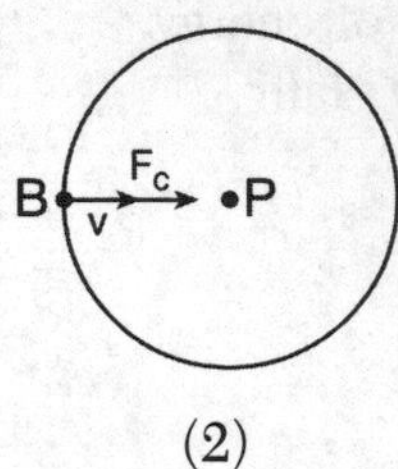

(2)

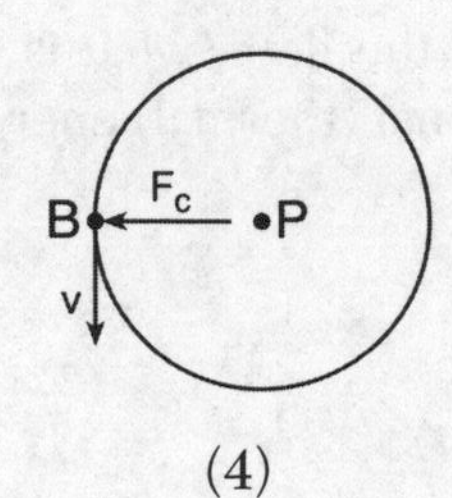

(4)

45 ____

46 Which graph best represents the relationship between photon energy and photon wavelength?

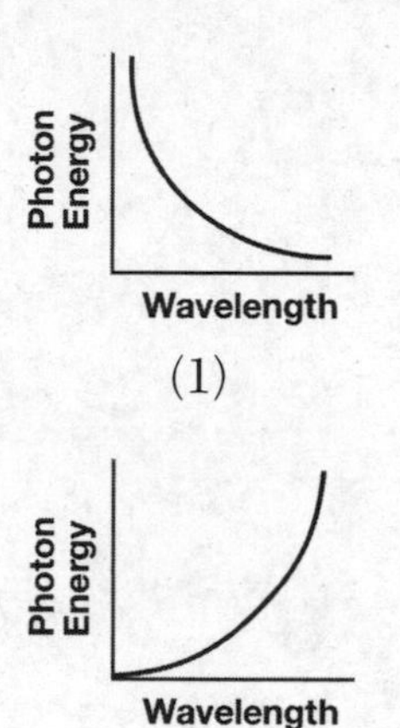

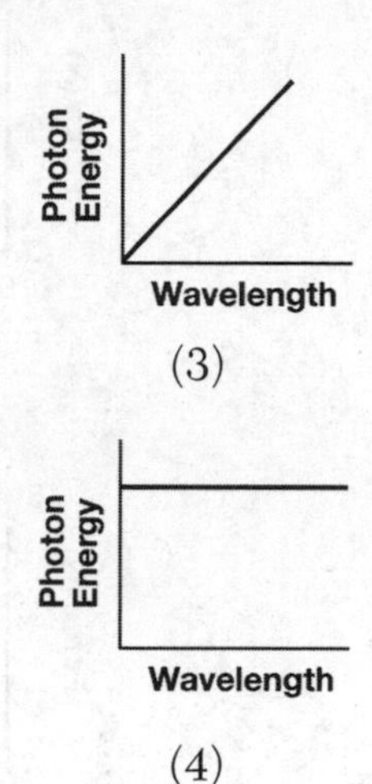

46 ____

47 Which combination of initial horizontal velocity (v_H) and initial vertical velocity (v_v) results in the greatest horizontal range for a projectile over level ground? [Neglect friction.]

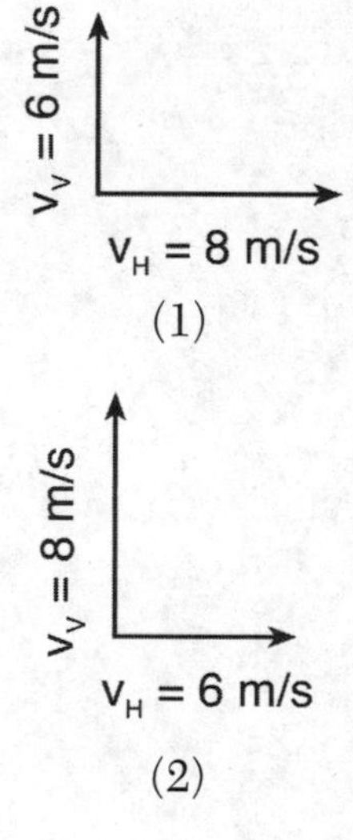

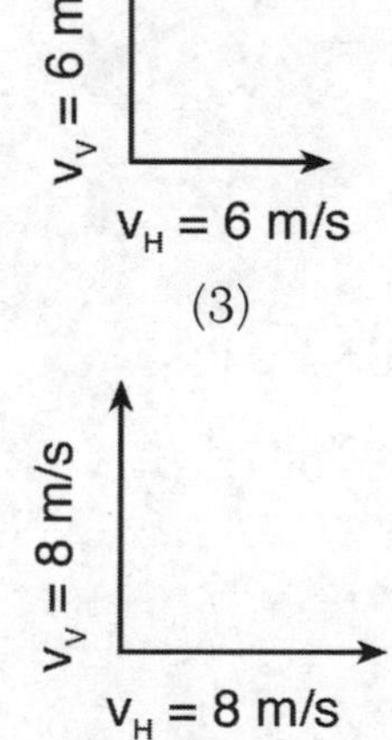

47 ____

48 Which graph represents the greatest amount of work?

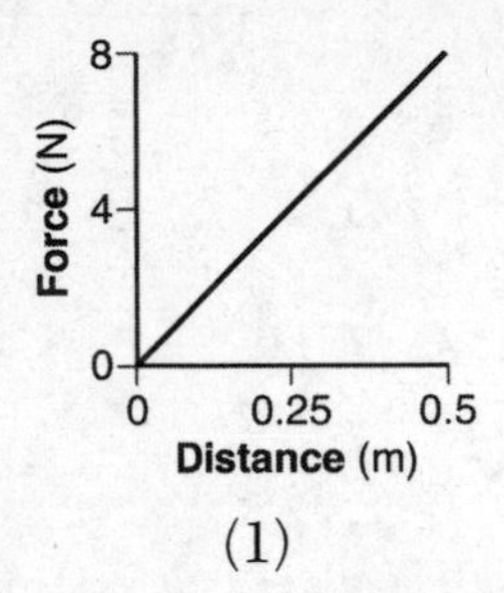

(1)

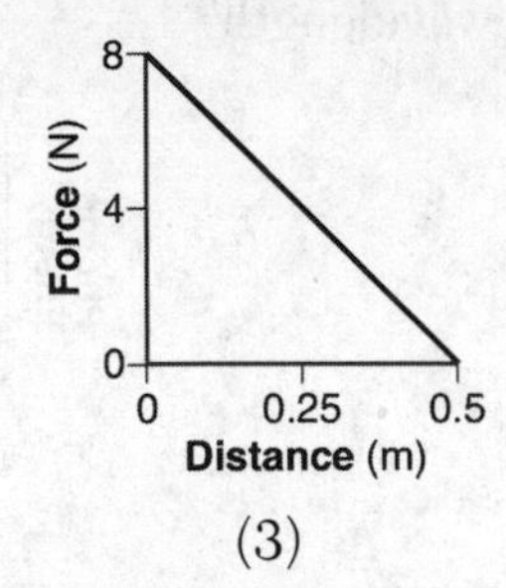

(3)

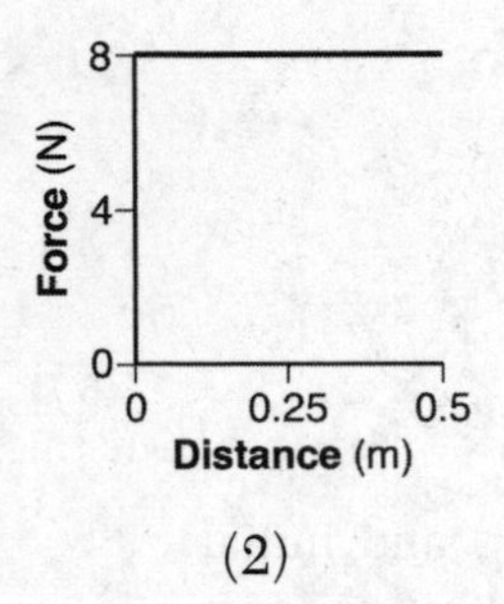

(2)

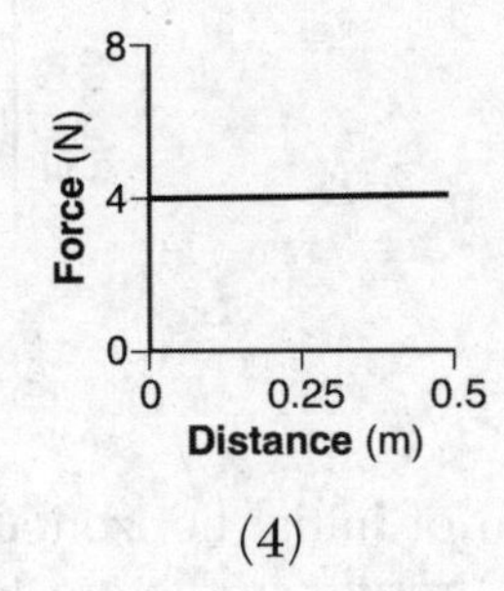

(4)

48 ____

49 When a ray of light traveling in water reaches a boundary with air, part of the light ray is reflected and part is refracted. Which ray diagram best represents the paths of the reflected and refracted light rays?

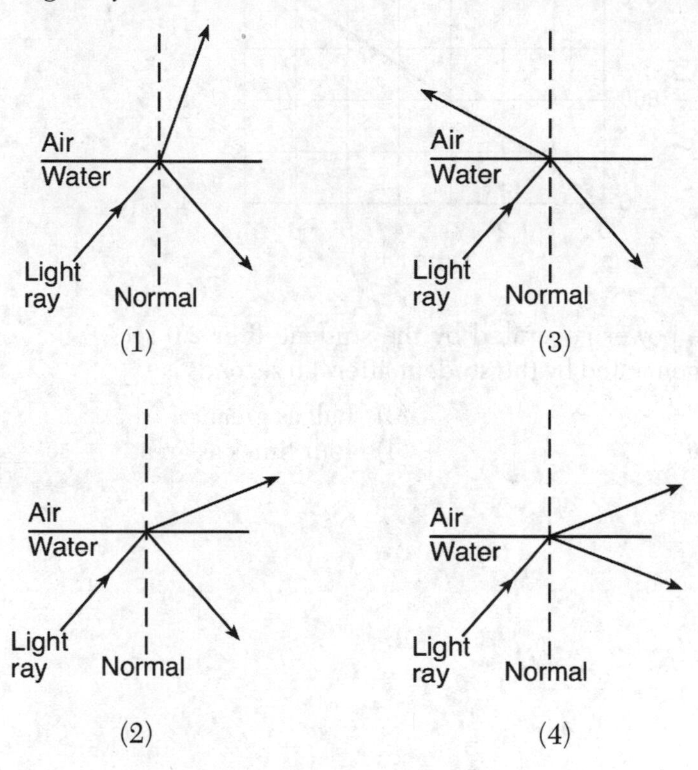

49 _____

50 The graph below represents the work done against gravity by a student as she walks up a flight of stairs at constant speed.

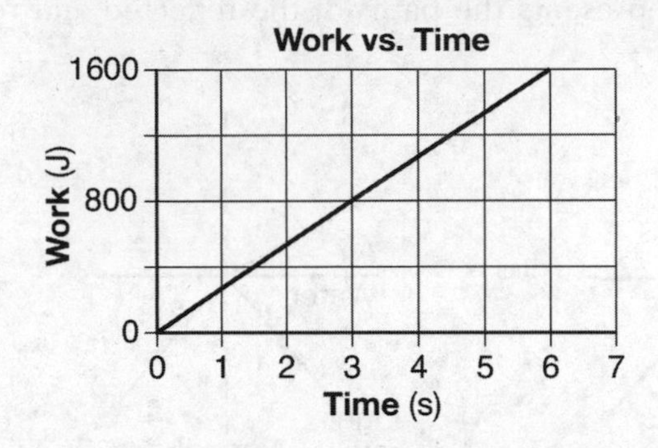

Compared to the power generated by the student after 2.0 seconds, the power generated by the student after 4.0 seconds is

(1) the same
(2) twice as great
(3) half as great
(4) four times as great

50 _____

PART B–2

Answer all questions in this part.

Directions (51–65): Record your answers on the answer sheet provided after the questions. Some questions may require the use of the *2006 Edition Reference Tables for Physical Setting/Physics*.

IMPORTANT NOTE:

- 2-point calculation problems appear as two linked questions in the test, for example, 55–56.
- In the first question, 1 point is allowed for showing the equation *and* for substituting the correct values with their units.
- In the second question, 1 point is allowed for a correct answer (number *and* unit) or for an answer that is consistent with your work in the first question.

Base your answers to questions 51 through 54 on the information below and the scaled vector diagram on the answer sheet and on your knowledge of physics.

Two forces, a 60-newton force east and an 80-newton force north, act concurrently on an object located at point P, as shown.

51 Using a ruler, determine the scale used in the vector diagram. [1]

52 Draw the resultant force vector to scale on the diagram *on the answer sheet*. Label the vector "*R*." [1]

53 Determine the magnitude of the resultant force, *R*. [1]

54 Determine the measure of the angle, in degrees, between north and the resultant force, *R*. [1]

55–56 A 3.00-newton force causes a spring to stretch 60.0 centimeters. Calculate the spring constant of this spring. [Show all work, including the equation and substitution with units.] [2]

57 A 7.28-kilogram bowling ball traveling 8.50 meters per second east collides head-on with a 5.45 kilogram bowling ball traveling 10.0 meters per second west. Determine the magnitude of the total momentum of the two-ball system after the collision. [1]

58–59 Calculate the average power required to lift a 490-newton object a vertical distance of 2.0 meters in 10 seconds. [Show all work, including the equation and substitution with units.] [2]

60 The diagram *on the answer sheet* shows wave fronts approaching an opening in a barrier. The size of the opening is approximately equal to one-half the wavelength of the waves. On the diagram *on the answer sheet*, draw the shape of *at least three* of the wave fronts after they have passed through this opening. [1]

61 The diagram *on the answer sheet* shows a mechanical transverse wave traveling to the right in a medium. Point *A* represents a particle in the medium. Draw an arrow originating at point *A* to indicate the initial direction that the particle will move as the wave continues to travel to the right in the medium. [1]

62 Regardless of the method used to generate electrical energy, the amount of energy provided by the source is always greater than the amount of electrical energy produced. Explain why there is a difference between the amount of energy provided by the source and the amount of electrical energy produced. [1]

[1 point]

Base your answers to questions 63 through 65 on the graph below, which represents the relationship between velocity and time for a car moving along a straight line, and your knowledge of physics.

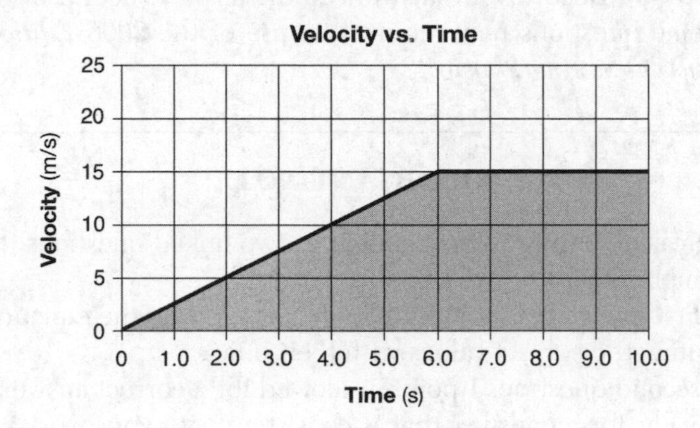

63 Determine the magnitude of the average velocity of the car from $t = 6.0$ seconds to $t = 10$ seconds. [1]

64 Determine the magnitude of the car's acceleration during the first 6.0 seconds. [1]

65 Identify the physical quantity represented by the shaded area on the graph. [1]

PART C

Answer all questions in this part.

Directions (66–85): Record your answers on the answer sheet provided after the questions. Some questions may require the use of the *2006 Edition Reference Tables for Physical Setting/Physics*.

IMPORTANT NOTE:

- 2-point calculation problems appear as two linked questions in the test, for example, 69–70.
- In the first question, 1 point is allowed for showing the equation *and* for substituting the correct values with their units.
- In the second question, 1 point is allowed for a correct answer (number and unit) or for an answer that is consistent with your work in the first question.

Base your answers to questions 66 through 70 on the information below and on your knowledge of physics.

A student constructed a series circuit consisting of a 12.0-volt battery, a 10.0-ohm lamp, and a resistor. The circuit does not contain a voltmeter or an ammeter. When the circuit is operating, the total current through the circuit is 0.50 ampere.

66 In the space *on the answer sheet*, draw a diagram of the series circuit constructed to operate the lamp, using symbols from the *Reference Tables for Physical Setting/Physics*. [1]

67 Determine the equivalent resistance of the circuit. [1]

68 Determine the resistance of the resistor. [1]

69–70 Calculate the power consumed by the lamp. [Show all work, including the equation and substitution with the units.] [2]

Base your answers to questions 71 through 75 on the information below and on your knowledge of physics.

Pluto orbits the Sun at an average distance of 5.91×10^{12} meters. Pluto's diameter is 2.30×10^{6} meters and its mass is 1.31×10^{22} kilograms.

Charon orbits Pluto with their centers separated by a distance of 1.96×10^{7} meters. Charon has a diameter of 1.21×10^{6} meters and a mass of 1.55×10^{21} kilograms.

71–72 Calculate the magnitude of the gravitational force of attraction that Pluto exerts on Charon. [Show all work, including the equation and substitution with units.] [2]

73–74 Calculate the magnitude of the acceleration of Charon toward Pluto. [Show all work, including the equation and substitution with units.] [2]

75 State the reason why the magnitude of the Sun's gravitational force on Pluto is greater than the magnitude of the Sun's gravitational force on Charon. [1]

Base your answers to questions 76 through 80 on the information below and on your knowledge of physics.

A horizontal 20-newton force is applied to a 5.0-kilogram box to push it across a rough, horizontal floor at a constant velocity of 3.0 meters per second to the right.

76 Determine the magnitude of the force of friction acting on the box. [1]

77–78 Calculate the weight of the box. [Show all work, including the equation and substitution with units.] [2]

79–80 Calculate the coefficient of kinetic friction between the box and the floor. [Show all work, including the equation and substitution with units] [2]

Base your answers to questions 81 through 85 on the information below and on your knowledge of physics.

An electron traveling with a speed of 2.50×10^6 meters per second collides with a photon having a frequency of 1.00×10^{16} hertz. After the collision, the photon has 3.18×10^{-18} joule of energy.

81–82 Calculate the original kinetic energy of the electron. [Show all work, including the equation and substitution with units.] [2]

83 Determine the energy in joules of the photon before the collision. [1]

84 Determine the energy lost by the photon during the collision. [1]

85 Name *two* physical quantities conserved in the collision. [1]

Answer Sheet June 2014

Physics: The Physical Setting

PART B–2

51 1.0 cm = ________________ **N**

52

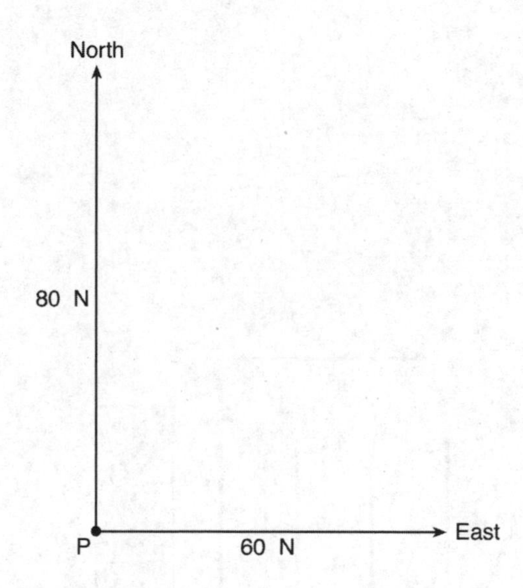

53 ______________ **N**

54 ______________ °

55–56

57 ________________ **kg • m/s**

58–59

60

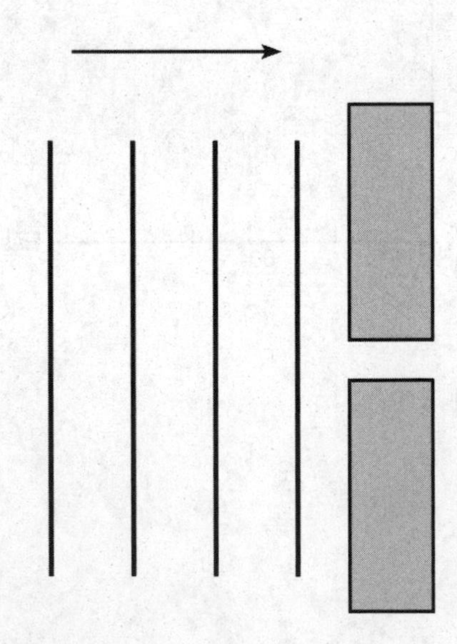

61

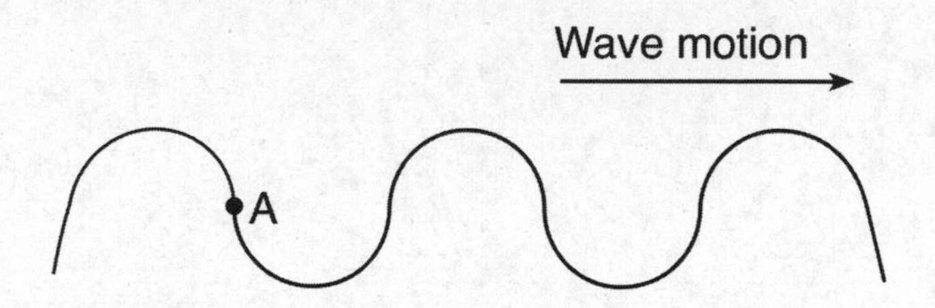

62 ____

____

63 ________________________________ **m/s**

64 ________________________________ **m/s^2**

65 ________________________________

PART C

66

67 ________________ Ω

68 ________________ Ω

69–70

71–72

73–74

75 ____

____

76 ____________________ **N**

77–78

79–80

81–82

83 ____________________ J

84 ____________________ J

85 ____________________ and ____________________

Answers
June 2014

Physics: The Physical Setting

Answer Key

PART A

1. 1	**8.** 2	**15.** 3	**22.** 2	**29.** 1
2. 1	**9.** 2	**16.** 1	**23.** 4	**30.** 2
3. 2	**10.** 1	**17.** 4	**24.** 1	**31.** 3
4. 3	**11.** 3	**18.** 2	**25.** 4	**32.** 4
5. 1	**12.** 2	**19.** 4	**26.** 2	**33.** 3
6. 3	**13.** 2	**20.** 1	**27.** 3	**34.** 1
7. 4	**14.** 2	**21.** 4	**28.** 2	**35.** 1

PART B–1

36. 3	**39.** 3	**42.** 1	**45.** 3	**48.** 2
37. 2	**40.** 1	**43.** 3	**46.** 1	**49.** 2
38. 3	**41.** 2	**44.** 4	**47.** 4	**50.** 1

PART B–2 and **PART C**. *See* **Answers Explained**.

Answers Explained

PART A

1 A scalar quantity has magnitude (a measured value) but not direction. The mass of an object has magnitude only (for example, 50 kilograms) but not direction. Therefore, mass is a scalar quantity.

WRONG CHOICES EXPLAINED:

(2), (3), (4) Force, momentum, and acceleration all have direction associated with them. For example, a force can be 20. newtons north. Since these quantities have both magnitude and direction, they are vector quantities.

2 Use *Equation ME5* on *Reference Table K*:

Solution:

$$\begin{aligned} v^2{}_f &= v_i^2 + 2ad \\ &= (0.0\ \text{m/s})^2 + 2\left(4.0\ \text{m/s}^2\right)(8.0\ \text{m}) \\ &= 64\ \text{m}^2/\text{s}^2 \\ v_f &= 8.0\ \text{m/s} \end{aligned}$$

3 Use *Equations ME6* and *ME7* on *Reference Table K*:

Solution:

$$\begin{aligned} A_y &= A\ \sin\theta \\ &= (15\ \text{m/s})\sin\ 60^\circ \\ &= (15\ \text{m/s})(0.8660) \\ &= 13\ \text{m/s} \\ &\quad (\text{vertical component}) \end{aligned}$$

$$\begin{aligned} A_x &= A\cos\theta \\ &= (15\ \text{m/s})\cos\ 60^\circ \\ &= (15\ \text{m/s})(0.5) \\ &= 7.5\ \text{m/s} \\ &\quad (\text{horizontal component}) \end{aligned}$$

4 Since it is released from rest, the initial speed, v_i, is 0.0 meters per second. The acceleration due to gravity is 9.81 meters per second squared (see *Reference Table A*). Use *Equation ME4* on *Reference Table K*:

Solution:

$$d = v_i t + \frac{1}{2}at^2$$

$$d = \frac{1}{2}at^2$$

$$t = \sqrt{\frac{2d}{a}}$$

$$= \sqrt{\frac{2(500\ \text{m})}{9.81\ \text{m/s}^2}}$$

$$= 10.1\ \text{s}$$

5 Newton's third law states that if object *A* exerts a force on object *B*, then object *B* exerts an equal and opposite force on object *A*. An alternate way of stating this is that for every action, there is an equal and opposite reaction. Hence these forces are called an "action-reaction pair." So the magnitude of the force that the ball exerts on the bat is equal to the force that the bat exerts on the ball.

WRONG CHOICES EXPLAINED:

(2), (3), (4) The relative masses of the objects do not affect the forces described.

6 Earth's gravitational field strength is symbolized by the letter *g*. See the definition of symbols on *Reference Table K*. The letter g is also the symbol for the acceleration due to gravity, which near the surface of Earth is 9.81 m/s^2 (see *Reference Table A*). Since the question states that the mass is 3.0 meters above the surface of Earth, this location is considered to be near the surface. To 2 significant digits, the gravitational field strength can be rounded to 9.8. Note that N/kg are equivalent units to m/s^2.

WRONG CHOICES EXPLAINED:

(1), (4) In the absence of air resistance, all objects accelerate at the same rate when they fall. So the gravitational field strength does not depend on the mass of the falling object.

7 Use *Equation ME4* on *Reference Table K*:

Solution:

$$d = v_i t + \frac{1}{2}at^2$$

$$= (22\ \text{m/s})(3.2\ \text{s}) + \frac{1}{2}(2.4\ \text{m/s}^2)(3.2\ \text{s})^2$$

$$= 83\ \text{m}$$

8 According to Newton's second law, when the forces acting on an object are unbalanced, the object will accelerate in the direction of the greater force. To accelerate downward, the downward force must be greater than the upward force. In this case, the downward force is the gravitational force (the person's weight) of 750 newtons. So the upward force (the normal force coming from the floor of the elevator) must be less than 750 newtons.

9 An impulse is a force acting for a certain amount of time. An impulse acting on an object causes the object to change its momentum. Use *Equation ME14* on *Reference Table K*. The impulse is numerically equal to the object's change in momentum ($J = \Delta p$). If the impulse is 15 newton • seconds, then the change in momentum is 15 kg • m/s. Note that kg • m/s are equivalent units to newton • seconds.

WRONG CHOICES EXPLAINED:

(1), (4) Since impulse and change in momentum are equal, the mass of the object is irrelevant.

10 Use *Equation ME14* on *Reference Table K*. The product of the force acting on an object and the time that force acts equals the object's change in momentum. The change in momentum of a driver in a car accident will be the same whether he or she hits an air bag or the steering wheel. In either case, the driver goes from moving very fast to being stopped. If the driver hits the steering wheel, he or she stops very quickly since the wheel is hard. So a large force must be applied. If the driver hits an air bag, he or she stops more slowly since the air bag compresses. Thus a smaller force is applied, doing less damage to the driver. Look at the equation to see this mathematically:

$$\Delta_{P\text{hitting steering wheel}} = \Delta_{P\text{hitting air bag}}$$

$$F_{net} \times t = F_{net} \times t$$

$$(\text{large force}) \times (\text{small time}) = (\text{small force}) \times (\text{large time})$$

11 All charged particles produce an electric field. Magnetic fields are produced by electric charges that are moving. Thus, a moving electron produces both types of fields.

12 Use *Equation EL2* on *Reference Table K*. Find the elementary charge (the charge on an electron or a proton) from *Reference Table A*:

Solution:
$$E = \frac{F_e}{q}$$
$$F_e = Eq$$
$$= (3.00 \times 10^3 \text{ N/C})(1.60 \times 10^{-19} \text{ C})$$
$$= 4.80 \times 10^{-16} \text{ N}$$

13 The distance is irrelevant. Use *Equation EL3* on *Reference Table K*:

Solution:
$$V = \frac{W}{q}$$
$$W = qV$$
$$= (3.0\text{C})(9.0 \text{ V})$$
$$= 27 \text{ J}$$

14 Use *Equation EL6* on *Reference Table K*. The resistivity (ρ) of tungsten can be found on *Reference Table J*:

Solution:
$$R = \frac{\rho L}{A}$$
$$= \frac{(5.60 \times 10^{-8} \ \Omega \bullet \text{m})(20.0 \text{ m})}{(1.00 \times 10^{-4} \text{ m}^2)}$$
$$= 1.12 \times 10^{-2} \ \Omega$$

15 Mechanical energy (kinetic and potential energy) is converted into electromagnetic energy and internal energy when the two pieces of rock are struck together since a visible spark is produced. Remember that visible light is a type of electromagnetic radiation. (See *Reference Table D.*) In addition, the rock pieces get warmer (increase in internal energy) as a result of the collision.

WRONG CHOICES EXPLAINED:

(1), (2), (4) Nuclear energy is not released since no atom has had its nucleus split. Visible light results instead from electrons transitioning between energy levels. Electrons are not in the nucleus.

16 Inertia is the property of an object that resists a change in motion. The measure of an object's inertia is its mass. Therefore, the more mass an object has, the more inertia it has.

17 All electromagnetic waves are transverse. As shown on *Reference Table D*, radio waves are a type of electromagnetic wave. So radio waves are transverse. Sound waves are longitudinal and mechanical.

WRONG CHOICES EXPLAINED:

(1), (2), (3) Light waves, microwaves, and ultraviolet waves are all on the electromagnetic spectrum as shown on *Reference Table D*. So they are all classified as transverse waves.

18 As a light wave passes from air into water, it slows down. However, its frequency remains the same. Since frequency and period are reciprocals of each other, as shown on *Equation W2* on *Reference Table K*, the period also remains the same.

WRONG CHOICES EXPLAINED:

(1), (3) Speed and wavelength are directly proportional, as shown by *Equation W1* on *Reference Table K*. Since the speed decreases, the wavelength decreases as well.

19 Use *Equation ME16* on *Reference Table K*:

Solution:

$$\begin{aligned} PE_s &= \frac{1}{2}kx^2 \\ &= \frac{1}{2}(60.0 \text{ N/m})(0.500 \text{ m})^2 \\ &= 7.50 \text{ J} \end{aligned}$$

WRONG CHOICE EXPLAINED:

(3) Did you forget to square the 0.500 m? Beware the square!

20 Frequency is a measure of the number of cycles (oscillations) per second:

Solution:
$$f = \frac{\text{number of swings}}{\text{number of seconds}}$$
$$= \frac{90 \text{ swings}}{300 \text{ seconds}}$$
$$= 0.30 \text{ Hz}$$

21 Use *Equation W2* on *Reference Table K*:

Solution:
$$T = \frac{1}{f}$$
$$= \frac{1}{340 \text{ Hz}}$$
$$= 2.94 \times 10^{-3} \text{ s}$$

22 The voltage is irrelevant. Use *Equation EL4* on *Reference Table K*:

Solution:
$$I = \frac{\Delta q}{t}$$
$$\Delta q = I \cdot t$$
$$= (0.120 \text{ A})(900 \text{ s})$$
$$= 108 \text{ C}$$

23 Use *Equation W1* on *Reference Table K* and see *Reference Table A*. The speed of all electromagnetic waves in a vacuum, including light, is 3.00×10^8 meters per second:

Solution:
$$v = f\lambda$$
$$f = \frac{v}{\lambda}$$
$$= \frac{3.00 \times 10^8 \text{ m/s}}{4.5 \times 10^{-7} \text{ m}}$$
$$= 6.7 \times 10^{14} \text{ Hz}$$

24 See *Reference Table A* and *Reference Table D*. Since the speed of all electromagnetic waves in a vacuum is 3.00×10^8 meters per second, x-ray and infrared radiation (waves) have the same speed. They are both types of electromagnetic waves.

WRONG CHOICES EXPLAINED:

(2), (3) You can also see on *Reference Table D* that x-ray radiation and infrared radiation have different frequencies and wavelengths.

(4) The energy per photon depends on the frequency (or wavelength) of the photon. See *Equation MP1* on *Reference Table K*.

25 Each pulse has an amplitude (height or maximum displacement) of 3 centimeters. When the two pulses meet, the amplitude of the resultant pulse will be the sum of the two individual amplitudes, which is 6 centimeters. In other words, the medium will be displaced twice as much when the waves overlap. The waves will be superposed.

26 When a source of sound (the horn) and an observer (the pedestrian) are moving relative to each other, the apparent frequency heard by the observer will be different than the actual frequency emitted by the source. This is known as the Doppler effect. When the source and the observer are moving toward each other, the apparent frequency heard by the observer will be greater than the actual frequency emitted by the source. Additionally, the term *pitch* is a common term that means frequency. Hence, a high pitch sound has a high frequency.

27 All systems have one or more frequencies at which they naturally tend to vibrate. Think of a guitar string or a child on a swing. If they are forced to vibrate at one of these natural frequencies by some external driving force, the amount they vibrate is large. In this case, blowing across the top of the bottle forces the air molecules in the bottle to vibrate enough to produce a sound. This phenomenon is known as *resonance*.

WRONG CHOICES EXPLAINED:

(1) Diffraction is the bending of a wave as it passes through a small opening or around a barrier.

(2) Refraction is the change in direction of a wave when it crosses a boundary between two different media at an angle if the speed of the wave is different in the two media.

(4) The Doppler effect is the apparent change in frequency of a wave due to the relative motion between the source of the wave and the observer of the wave.

28 See the note that is written below *Reference Table H*. The charge on any antiparticle is the opposite charge of the associated particle. Each antiup quark has a charge of $-\frac{2}{3}$e, and the antidown quark has a charge of $+\frac{1}{3}$e. Add these together to find the total charge on the antibaryon:

Solution: $\left(-\frac{2}{3}e\right)+\left(-\frac{2}{3}e\right)+\left(+\frac{1}{3}e\right)=-1e$

29 The strong force is an attractive force that governs the interactions of particles in the nucleus.

WRONG CHOICES EXPLAINED:

(2) The weak force governs certain types of radioactive decay.

(3) Frictional force is a type of electromagnetic force.

(4) The gravitational force governs the interactions among particles that have mass.

30 Use *Equation MP3* on *Reference Table K*. Find the speed of light in a vacuum (c) on *Reference Table* A:

Solution:
$$\begin{aligned} E &= mc^2 \\ &= \left(9.11\times10^{-31}\,\text{kg}\right)\left(3.00\times10^{8}\ \text{m/s}\right)^2 \\ &= 8.20\times10^{-14}\ \text{J} \end{aligned}$$

WRONG CHOICE EXPLAINED:

(1) Did you forget to square the speed of light? Beware the square!

31 As shown in *Equation ME20* on *Reference Table K*, kinetic energy is a measure of the energy of motion. This means that the faster an object travels, the more kinetic energy it has. As shown in *Equation ME19* on *Reference Table K*, gravitational potential energy is a measure of the energy of position. This means that the higher an object is above the ground, the more gravitational potential energy it has. Since the shopping cart is slowing down, its kinetic energy decreases. Since the shopping cart is on a level floor, its gravitational potential energy remains the same.

WRONG CHOICE EXPLAINED:

(4) Did you think that whenever an object's kinetic energy decreases, its gravitational potential energy increases (like when a ball is thrown up into the air)? That is not always the case!

32 Since sphere *B* is neutral, it has an equal number of protons and electrons. Because like charges repel each other, the electrons on sphere *A* will repel the electrons on sphere *B* to the side farthest away from sphere *A*.

WRONG CHOICES EXPLAINED:

(1), (3) Neither of these show a neutral sphere.

33 Use *Equations EL1 and EL2* on *Reference Table K*. Let $q_1 = -3.0$ nC, and let q_2 be a test charge placed at points *A* and *B*.

$$F_e = \frac{kq_1q_2}{r^2} \text{ and } E = \frac{F_e}{q_2}$$

$$E = \frac{\frac{kq_1q_2}{r^2}}{q_2}$$

$$= \frac{kq_1}{r^2}$$

Since k and q_1 remain the same and the distance is doubled, proportional reasoning can be used:

Solution:

$$E = \frac{(1)(1)}{(2)^2}$$

$$= \frac{1}{4}$$

34 Amplitude is the maximum displacement of the wave from equilibrium, which is the maximum height of the wave. By looking at the diagram, the height of each wave appears to be the same.

WRONG CHOICES EXPLAINED:

(2), (3) Since frequency and period are reciprocals of each other, see *Equation W2* on *Reference Table K*, these are either both the same or both different.

(4) Wavelength is the length of the wave and can be measured from one crest to the next crest. By looking at the diagram, the wavelength of wave *B* appears to be shorter than the wavelength of wave *A*.

35 Two points on a wave are out of phase by 180° if the horizontal distance between them is an odd number of half-wavelengths. Points *A* and *C* are one half-wavelength (½ λ) apart from each other.

WRONG CHOICES EXPLAINED:

(2) *B* and *E* are ¾ λ apart, so they are out of phase by 270°.
(3) *F* and *G* are ¼ λ apart, so they are out of phase by 90°.
(4) *D* and *H* are a whole wavelength apart (1λ), so they are in phase.

PART B–1

36 Each story of a building is approximately 3 meters. How would you know that? Take a look at the room you're in right now. Imagine how many meter sticks would be needed to go from floor to ceiling—about 3 of them. A 30-story building would then be about 90 meters high. This is approximately 100 meters, which is 10^2 meters or roughly the size of a football field. Of the choices, this seems the most reasonable.

WRONG CHOICES EXPLAINED:

(1) 10^0 m is 1 meter, which is clearly too small.
(2) 10^1 m is 10 meters. This is a better answer than choice (1) but is still too small. This would be a the height of a 3-story building.
(4) 10^3 m is 1,000 m or 1 kilometer, which is about $\frac{2}{3}$ of a mile. This measurement is much too large.

37 Since the spheres are identical, the final charge on each sphere is the average of the two initial charges. The electrons are shared equally between the two spheres after the spheres are touched together. This charge has to be converted by the factor-label method into the number of elementary charges (electrons) it represents. Use the conversion factor found on *Reference Table A*:

Solution:

$$\text{Average charge} = \frac{\left(-4.0 \times 10^{-6}\,\text{C}\right) + \left(-8.0 \times 10^{-6}\,\text{C}\right)}{2}$$

$$= -6.0 \times 10^{-6}\,\text{C}$$

$$\text{Number of electrons} = \left(\frac{-6.0 \times 10^{-6}\,\text{C}}{1}\right)\left(\frac{1\text{e}}{-1.60 \times 10^{-19}\,\text{C}}\right)$$

$$\text{Number of electrons} = 3.8 \times 10^{13}\,\text{e}$$

38 Use *Equation ME18* on *Reference Table K*. The mass of the car is irrelevant. Be sure to divide the diameter by 2 to get the radius:

Solution:
$$a_c = \frac{v^2}{r}$$
$$= \frac{(20 \text{ m/s})^2}{(0.5 \times 10^2 \text{ m})}$$
$$= 8.0 \text{ m/s}^2$$

WRONG CHOICES EXPLAINED:

(1) You probably forgot to square the velocity and also to divide the diameter by 2.

(4) You probably forgot to divide the diameter by 2.

39 Use the equation for electrical energy found at *Equation EL3* on *Reference Table K*. Substitute the appropriate units without including any numbers:

Solution:
$$V = \frac{W}{q}$$
$$W = Vq$$
$$= (\text{V})(C)$$
$$= \text{volt} \bullet \text{coulomb}$$

40 Use *Equation EL8* on *Reference Table K*. Be sure to convert minutes into seconds by using the factor-label method:

Solution:
$$t = \left(\frac{30.0 \text{ minutes}}{1}\right)\left(\frac{60 \text{ seconds}}{1.0 \text{ minute}}\right)$$
$$t = 1{,}800 \text{ seconds}$$
$$W = Pt$$
$$= (315 \text{ W})(1{,}800 \text{ s})$$
$$= 5.67 \times 10^5 \text{ J}$$

WRONG CHOICE EXPLAINED:

(2) You probably forgot to convert minutes into seconds.

41 Use *Equation W4* on *Reference Table K*. The absolute index of refraction of a substance is inversely proportional to the speed of light in that substance. Only choice (2) shows a graph in which the index decreases as the speed increases.

42 According to the conservation of energy principle, the total energy (E_T) of a closed system at one instant is the same as its total energy at another instant. In other words, $E_T = E_T$. Use *Equation ME22* and *Equation ME19* on *Reference Table K* and use *Reference Table A*. Be sure to convert grams to kilograms:

Solution:

$$\begin{aligned} E_T &= E_T \\ E_T &= PE + KE + Q \\ PE + KE + Q &= PE + KE + Q \\ mg\Delta h + 0 + 0 &= 0 + KE + Q \\ (0.025\text{ kg})(9.81\text{ m/s}^2)(0.90\text{ m}) &= (0.20\text{ J}) + Q \\ 0.22\text{ J} &= (0.20\text{ J}) + Q \\ Q &= 0.02\text{ J} \end{aligned}$$

WRONG CHOICE EXPLAINED:

(4) You probably forgot to convert grams into kilograms.

43 Photons are emitted (as opposed to absorbed) only when an electron transitions from a higher energy level to a lower one. This reduces the possible answer choices to only (1) or (3). Check both of these choices to see if the emitted photon is visible, which involves a multistep process:

First determine the energy of the photon in choice (3), when the photon drops from energy level 5 to 2. Use *Equation MP2* on *Reference Table K* and the energy level diagram for hydrogen on *Reference Table F*:

$$\begin{aligned} E_{photon} &= E_i - E_f \\ &= E_5 - E_2 \\ &= (-0.54\text{ eV}) - (-3.40\text{ eV}) \\ &= 2.86\text{ eV} \end{aligned}$$

Then convert this energy from electronvolts into joules by the factor-label method. Use the conversion factor found in *Reference Table A*:

$$E_{photon} = \left(\frac{2.86 \text{ eV}}{1}\right)\left(\frac{1.60 \times 10^{-19} \text{ J}}{1 \text{ eV}}\right)$$
$$= 4.58 \times 10^{-19} \text{ J}$$

Next determine the frequency of this photon. Use *Equation MP1* on *Reference Table K* and Planck's constant (h) found on *Reference Table A*:

$$E_{photon} = hf$$
$$f = \frac{E_{photon}}{h}$$
$$= \frac{4.58 \times 10^{-19} \text{ J}}{6.63 \times 10^{-34} \text{ J} \cdot \text{s}}$$
$$= 6.91 \times 10^{14} \text{ Hz}$$

Finally check to see if this frequency is in the range of visible light by looking at the electromagnetic spectrum found in *Reference Table D*. Light with this frequency is in the range of violet light. Whew! If you had determined the energy of the photon in choice (1) instead, you would have found that its frequency was not in the visible light range. At that point, you would not have had to redo the equations using choice (3). You could simply have chosen choice (3) because you would have eliminated all other answer choices.

44 An object in equilibrium is either at rest or moving with a constant velocity (constant speed in a straight line). Since this object is moving in a straight line, it must have a constant speed. *Equation ME20* on *Reference Table K* shows that if the speed is constant, so is the kinetic energy.

WRONG CHOICES EXPLAINED:

(1) This graph shows an object that is accelerating, not moving at a constant speed.

(2) This graph shows an object whose velocity (speed) is decreasing.

(3) This graph shows an object whose velocity (speed) is increasing. As shown in *Equation ME12* on *Reference Table K*, momentum is the product of mass and velocity.

45 When an object travels at constant speed in a horizontal circle, the force causing it to move in a circle (centripetal force) is always pointed toward the center of the circle. The word "centripetal" means *center-seeking*. If it were not for this force, according to Newton's first law, the object would continue traveling in a straight line in the direction it was already moving, that is, tangent to the circle.

46 Use *Equation MP1* on *Reference Table K*. The energy of a photon is inversely proportional to its wavelength. Only choice (1) shows a graph in which the photon energy decreases as the wavelength increases.

47 The greatest horizontal range for a projectile, in the absence of air resistance, is when the angle of launch is at 45° above the horizontal. See *Equations ME6 and ME7* on *Reference Table K*. This means that both the horizontal and vertical components of the initial velocity are equal.

48 Use *Equation ME21* on *Reference Table K*. The total work done is the area underneath a force-distance graph. By looking at the graphs, you can see that if you shade in the area underneath each graph for the first 0.5 meter, the greatest shading would be under graph (2). Alternatively, you could calculate the area underneath each graph.

49 When a ray of light travels from a substance with a higher index of refraction to a substance with a lower index of refraction, see *Reference Table E*, the light ray bends away from the normal. Additionally, the reflected ray must reflect at the same angle that the incident ray makes with the normal. This is called the law of reflection. See *Equation W3* on *Reference Table K*. Choice (2) shows both of these behaviors.

WRONG CHOICES EXPLAINED:

(1) This diagram shows the refracted ray bending toward the normal, not away from it.

(3) This diagram shows the refracted ray crossing to the other side of the normal.

(4) This diagram shows the angle of reflection not equal to the angle of incidence.

50 Use *Equation ME23* on *Reference Table K*. The power generated by the student is equal to the slope of the graph. Since the slope is the same at all points, the power generated is the same as well.

PART B–2

51 Using a ruler, measure the length of either vector. Then use the factor-label method to convert the answer to newtons. For example, the east vector is marked 60. N and measures 6.0 centimeters:

Solution: $$\left(\frac{1.0\text{ cm}}{1}\right)\left(\frac{60\text{ N}}{6.0\text{ cm}}\right) = 10\text{ N}$$

Answers ranging from 9.0 N to 11 N receive credit.

[1 point]

52 Refer to the diagram below. Be sure to draw an arrowhead at the end of the resultant vector or you will not receive credit.

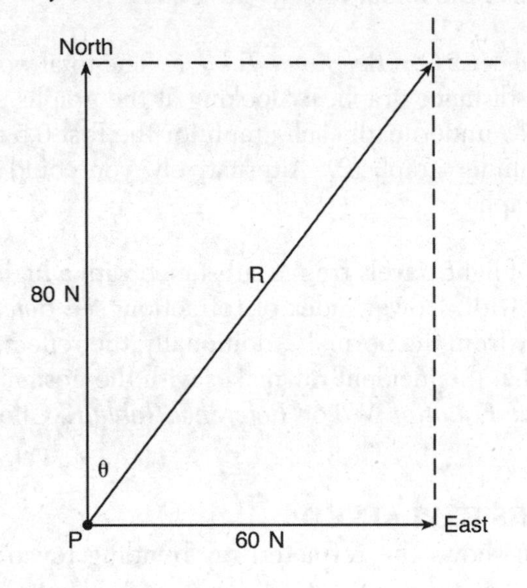

[1 point]

53 This question can be done in one of two ways, either by measuring it or by calculating it.

Method 1: Measure the resultant vector with a ruler, and use your scale to determine its magnitude. The vector measures 10.0 centimeters:

Solution: $$\left(\frac{10.0\text{ cm}}{1}\right)\left(\frac{10\text{ N}}{1.0\text{ cm}}\right) = 100\text{ N}$$

Method 2: Use the Pythagorean theorem found in *Equation GT4A* on *Reference Table K* to calculate the magnitude of the resultant vector:

Solution: $$c^2 = a^2 + b^2$$
$$= (60\text{ N})^2 + (80\text{ N})^2$$
$$c = 100\text{ N}$$

Answers ranging from 97 N to 103 N receive credit.

[1 point]

54 This question can be done in one of two ways, either by measuring it or by calculating it.

Method 1: Measure the angle with a protractor.

Method 2: Use one of the trigonometric formulas found in *Equation GT4B*, *Equation GT4C*, or *Equation GT4D* on *Reference Table K* to calculate the angle:

Solution: $$\sin\theta = \frac{a}{c}$$
$$= \frac{60\text{ N}}{100\text{ N}}$$
$$\theta = 37°$$

Answers ranging from 35° to 39° receive credit.

[1 point]

55–56 Use *Equation ME15* on *Reference Table K*. Converting centimeters into meters is optional:

Solution: $$F_s = kx$$
$$k = \frac{F_s}{x}$$
$$= \frac{3.00\text{ N}}{0.600\text{ m}}$$

or

$$F_s = kx$$
$$k = \frac{F_s}{x}$$
$$= \frac{3.00\text{ N}}{60.0\text{ cm}}$$

[1 point]

56 Use the equation you found in question 55.

Solution: $k = 5.00$ N/m or $k = 0.0500$ N/cm

[1 point]

57 Use *Equation ME12* and *Equation ME13* on *Reference Table K*. Assume east is the positive direction. Be sure to set negative any velocity that is directed west:

Solution:

$$p_{\text{before}} = p_{after}$$
$$p_{1\text{initial}} + p_{2\text{initial}} = p_{\text{after}}$$
$$m_1 v_{1i} + m_2 v_{2i} = p_{\text{after}}$$
$$(7.28 \text{ kg})(8.50 \text{ m/s}) + (5.45 \text{ kg})(-10.0 \text{ m/s}) = p_{\text{after}}$$
$$61.88 \text{ kg} \bullet \text{m/s} + (-54.5\text{kg} \bullet \text{m/s}) = p_{after}$$
$$p_{after} = 7.38\text{kg} \bullet \text{m/s}$$

Credit is allowed for 7.3 kg • m/s or 7.4 kg • m/s.

[1 point]

58–59 Use *Equation ME23* on *Reference Table K*:

Solution:

$$P = \frac{Fd}{t}$$
$$\frac{(490 \text{ N})(2.0 \text{ m})}{10 \text{ s}}$$

[1 point]

59 Use the equation you found in question 57.

Solution: $P = 98$ W

[1 point]

60 Refer to the diagram below, which shows an appropriate drawing that will receive credit.

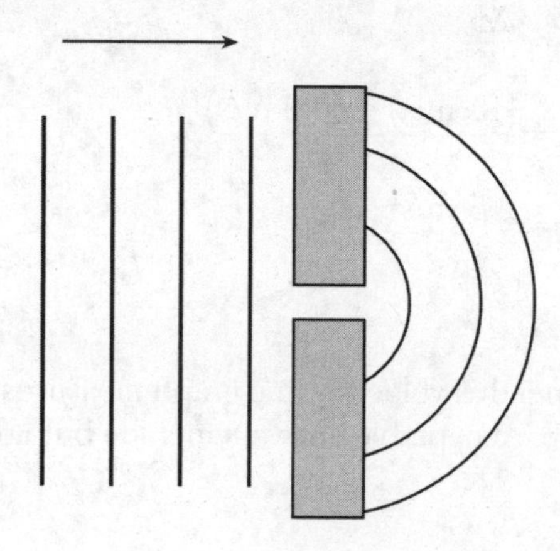

[1 point]

61 Refer to the diagram below, which shows an appropriate drawing that will receive credit.

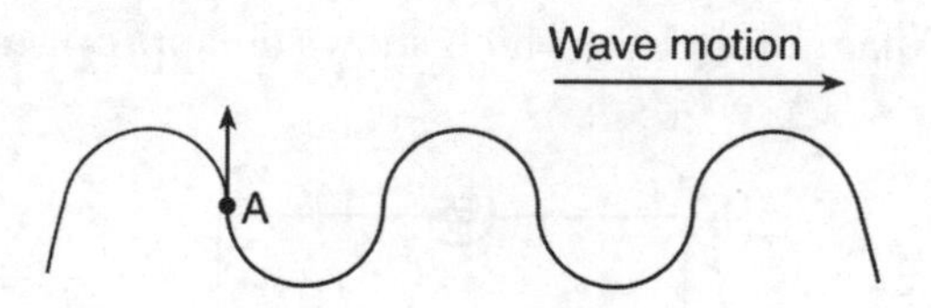

[1 point]

62 Acceptable answers include but are not limited to:

- Energy is needed to overcome friction.
- Energy is converted into internal (thermal) energy in the moving parts.
- Energy is converted into sound.

[1 point]

63 Since the velocity of the car is constant from 6.0 seconds to 10 seconds, the answer may be read directly off the graph.

Solution: 15 m/s

[1 point]

64 Use *Equation ME2* on *Reference Table K*:

Solution:
$$a = \frac{\Delta v}{t}$$
$$= \frac{(15 \text{ m/s}) - (0.0 \text{ m/s})}{6.0 \text{ s}}$$
$$= 2.5 \text{ m/s}^2$$

[1 point]

65 The area underneath a velocity-time graph measures how far an object has traveled in a given time. Acceptable answers include but are not limited to:

- displacement
- distance
- how far the car traveled

[1 point]

PART C

66 Refer to the diagram below, which shows an appropriate drawing that will receive credit.

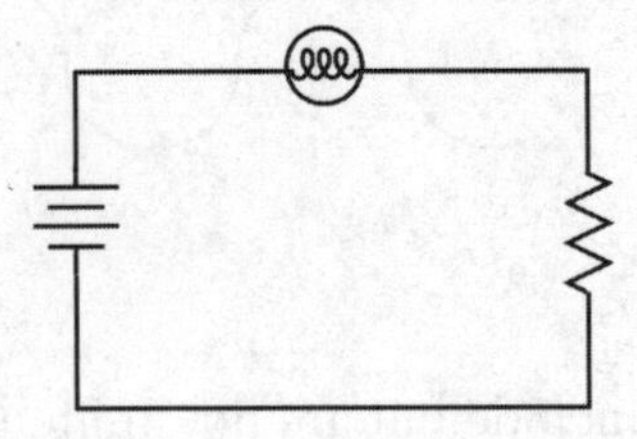

[1 point]

67 Use *Equation EL5* on *Reference Table K*. Since you are calculating the total resistance (equivalent resistance) of the circuit, use the total potential difference and the total current:

Solution:
$$R = \frac{V}{I}$$
$$= \frac{12.0 \text{ V}}{0.50 \text{ A}}$$
$$= 24\ \Omega$$

[1 point]

68 Use *Equation EL9C* on *Reference Table K*:

Solution:
$$R_{eq} = R_1 + R_2$$
$$R_2 = R_{eq} - R_1$$
$$R_2 = 24\ \Omega - 10.0\ \Omega$$
$$R_2 = 14\ \Omega$$

[1 point]

69–70 Use *Equation EL7* on *Reference Table K*:

Solution:
$$P = I^2R$$
$$= (0.50\ \text{A})^2(10.0\ \Omega)$$

[1 point]

70 Use the equation you found in question 69.

Solution: $P = 2.5\ \text{W}$

[1 point]

71–72 Use *Equation ME10* on *Reference Table K*. The universal gravitational constant (G) is found on *Reference Table A*. The diameters of Pluto and Charon are irrelevant:

Solution:
$$F_g = \frac{Gm_1m_2}{r^2}$$
$$= \frac{\left(6.67 \times 10^{-11}\ \text{N} \cdot \text{m}^2/\text{kg}^2\right)\left(1.31 \times 10^{22}\ \text{kg})(1.55 \times 10^{21}\ \text{kg}\right)}{\left(1.96 \times 10^7\ \text{m}\right)^2}$$

[1 point]

72 Use the equation you found in question 71.

Solution: $F_g = 3.53 \times 10^{18}\ \text{N}$

[1 point]

73–74 Use *Equation ME8* on *Reference Table K*. Use the mass of Charon since it is the object that is accelerating:

Solution:

$$a = \frac{F_{net}}{m}$$

$$= \frac{3.53 \times 10^{18}\ \text{N}}{1.55 \times 10^{21}\ \text{kg}}$$

[1 point]

74 Use the equation you found in question 73.

Solution: $a = 2.28 \times 10^{-3}\ \text{m/s}^2$ or $a = 2.27 \times 10^{-3}\ \text{m/s}^2$

Credit is given for either answer as long as it is consistent with the equation you found in question 73.

[1 point]

75 The gravitational force of attraction between two objects depends on the mass of both objects. The gravitational force from the Sun is greater on Pluto than on Charon because the mass of Pluto is greater than the mass of Charon.

[1 point]

76 According to Newton's first law, since the box is moving at a constant velocity, the forces acting on it are balanced. Therefore, the force of friction acting to the left is equal to the applied force pushing the box to the right:

Solution: 20 N

[1 point]

77–78 Use *Equation ME11* on *Reference Table K*. The acceleration due to gravity (g) is found on *Reference Table A*.

Solution:

$$g = \frac{F_g}{m}$$

$$F_g = mg$$

$$F_g = (5.0\ \text{kg})(9.81\ \text{m/s}^2)$$

[1 point]

78 Use the equation you found in question 77.

Solution: $F_g = 49$ N

[1 point]

79–80 Use *Equation ME9* on *Reference Table K*. The normal force, *FN*, which is the force that the floor exerts upward on the box, is equal and opposite to the weight of the box:

Solution:
$$F_f = \mu F_N$$
$$\mu = \frac{F_f}{F_N}$$
$$= \frac{20\text{ N}}{49\text{ N}}$$

[1 point]

80 Use the equation you found in question 79.

Solution: $\mu = 0.41$ or 0.40.

Note that the answer has no units.

[1 point]

81–82 Use *Equation ME20* on *Reference Table K*. The mass of the electron is found on *Reference Table A:*

Solution:
$$KE = \frac{1}{2}mv^2$$
$$= \frac{1}{2}\left(9.11 \times 10^{-31}\text{ kg}\right)\left(2.50 \times 10^6\text{ m/s}\right)^2$$

[1 point]

82 Use the equation you found in question 81.

Solution: $KE = 2.85\ 10^{-18}$ J

[1 point]

83 Use *Equation MP1* on *Reference Table K*. The value of Planck's constant (h) is found on *Reference Table A*:

Solution:
$$\begin{aligned} E_{photon} &= hf \\ &= (6.63 \times 10^{-34}\ \text{J} \bullet \text{s})(1.00 \times 10^{16}\ \text{Hz}) \\ &= 6.63 \times 10^{-18}\ \text{J} \end{aligned}$$

[1 point]

84 The energy lost by the photon during the collision is the difference between its initial and final energies.

Solution:
$$\begin{aligned} E_{\text{lost}} &= E_{\text{i}} - E_{\text{f}} \\ &= (6.63 \times 10^{-18}\ \text{J}) - (3.18 \times 10^{-18}\ \text{J}) \\ &= 3.45 \times 10^{-18}\ \text{J} \end{aligned}$$

[1 point]

85 1 point is given for *two* acceptable quantities. Acceptable answers include but are not limited to:

- mass
- charge
- momentum
- energy

[1 point]

Topic	Question Numbers (total)	Wrong Answers (x)	Grade
Math Skills	41, 46, 48, 51–54, 64, 65: (9)		$\frac{100\ (9-x)}{9} = \%$
Mechanics	1–10, 16, 36, 38, 44, 45, 47, 55–57, 63, 71–80: (30)		$\frac{100\ (30-x)}{30} = \%$
Energy	15, 19, 31, 42, 50, 58, 59, 62, 81, 82, 84: (11)		$\frac{100\ (11-x)}{11} = \%$
Electricity/ Magnetism	11–14, 22, 32, 33, 37, 39, 40, 66–70: (15)		$\frac{100\ (15-x)}{15} = \%$
Waves	17, 18, 20, 21, 23–27, 34, 35, 49, 60, 61: (14)		$\frac{100\ (14-x)}{14} = \%$
Modern Physics	28–30, 43, 83, 85: (6)		$\frac{100\ (6-x)}{6} = \%$

Examination June 2015

Physics: The Physical Setting

PART A

Answer all questions in this part.

Directions (1–35): For *each* statement or question, select the *number* of the word or expression that, of those given, best completes the statement or answers the question. Some questions may require the use of the *2006 Edition Reference Tables for Physical Setting/Physics.* Record your answers in the spaces provided.

1 Which quantities are scalar?

(1) speed and work
(2) velocity and force
(3) distance and acceleration
(4) momentum and power

1 ____

2 A 3.00-kilogram mass is thrown vertically upward with an initial speed of 9.80 meters per second. What is the maximum height this object will reach? [Neglect friction.]

(1) 1.00 m
(2) 4.90 m
(3) 9.80 m
(4) 19.6 m

2 ____

3 An airplane traveling north at 220 meters per second encounters a 50-meters-per-second crosswind from west to east, as represented in the diagram below.

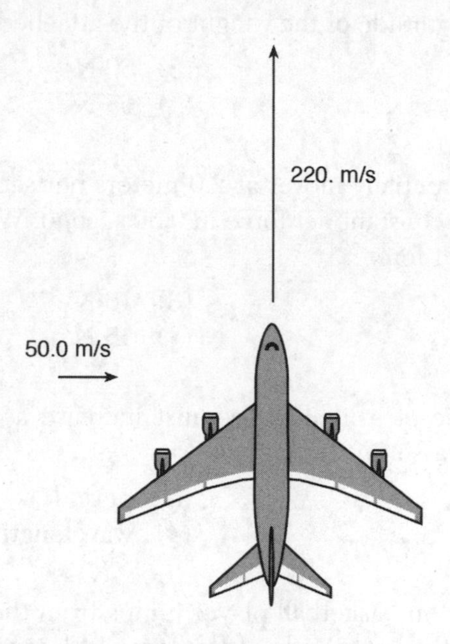

What is the resultant speed of the plane?

(1) 170 m/s
(2) 214 m/s
(3) 226 m/s
(4) 270 m/s

3 ____

4 A 160-kilogram space vehicle is traveling along a straight line at a constant speed of 800 meters per second. The magnitude of the net force on the space vehicle is

(1) 0 N
(2) 1.60×10^2 N
(3) 8.00×10^2 N
(4) 1.28×10^5 N

4 ____

5 A student throws a 5.0-newton ball straight up. What is the net force on the ball at its maximum height?

(1) 0.0 N
(2) 5.0 N, up
(3) 5.0 N, down
(4) 9.8 N, down

5 ____

6 A vertical spring has a spring constant of 100 newtons per meter. When an object is attached to the bottom of the spring, the spring changes from its unstretched length of 0.50 meter to a length of 0.65 meter. The magnitude of the weight of the attached object is

(1) 1.1 N
(2) 15 N
(3) 50 N
(4) 65 N

6 ____

7 A 1.5-kilogram cart initially moves at 2.0 meters per second. It is brought to rest by a constant net force in 0.30 second. What is the magnitude of the net force?

(1) 0.40 N
(2) 0.90 N
(3) 10 N
(4) 15 N

7 ____

8 Which characteristic of a light wave must increase as the light wave passes from glass into air?

(1) amplitude
(2) frequency
(3) period
(4) wavelength

8 ____

9 As a 5.0×10^2-newton basketball player jumps from the floor up toward the basket, the magnitude of the force of her feet on the floor is 1.0×10^3 newtons. As she jumps, the magnitude of the force of the floor on her feet is

(1) 5.0×10^2 N
(2) 1.0×10^3 N
(3) 1.5×10^3 N
(4) 5.0×10^5 N

9 ____

10 A 0.0600-kilogram ball traveling at 60.0 meters per second hits a concrete wall. What speed must a 0.0100-kilogram bullet have in order to hit the wall with the same magnitude of momentum as the ball?

(1) 3.60 m/s
(2) 6.00 m/s
(3) 360 m/s
(4) 600 m/s

10 ____

11 The Hubble telescope's orbit is 5.6×10^5 meters above Earth's surface. The telescope has a mass of 1.1×10^4 kilograms. Earth exerts a gravitational force of 9.1×10^4 newtons on the telescope. The magnitude of Earth's gravitational field strength at this location is

(1) 1.5×10^{-20} N/kg
(2) 0.12 N/kg
(3) 8.3 N/kg
(4) 9.8 N/kg

11 ____

12 When two point charges are a distance d apart, the magnitude of the electrostatic force between them is F. If the distance between the point charges is increased to $3d$, the magnitude of the electrostatic force between the two charges will be

(1) $\frac{1}{9}F$
(2) $\frac{1}{3}F$
(3) $2F$
(4) $4F$

12 ____

13 A radio operating at 3.0 volts and a constant temperature draws a current of 1.8×10^{-4} ampere. What is the resistance of the radio circuit?

(1) 1.7×10^4 Ω
(2) 3.0×10^1 Ω
(3) 5.4×10^{-4} Ω
(4) 6.0×10^{-5} Ω

13 ____

14 Which energy transformation occurs in an operating electric motor?

(1) electrical → mechanical
(2) mechanical → electrical
(3) chemical → electrical
(4) electrical → chemical

14 ____

15 A block slides across a rough, horizontal tabletop. As the block comes to rest, there is an increase in the block-tabletop system's

(1) gravitational potential energy
(2) elastic potential energy
(3) kinetic energy
(4) internal (thermal) energy

15 ____

16 How much work is required to move an electron through a potential difference of 3.00 volts?

(1) 5.33×10^{-20} J
(2) 4.80×10^{-19} J
(3) 3.00 J
(4) 1.88×10^{19} J

16 ____

17 During a laboratory experiment, a student finds that at 20° Celsius, a 6.0-meter length of copper wire has a resistance of 1.3 ohms. The cross-sectional area of this wire is

(1) 7.9×10^{-8} m^2
(2) 1.1×10^{-7} m^2
(3) 4.6×10^{0} m^2
(4) 1.3×10^{7} m^2

17 ____

18 A net charge of 5.0 coulombs passes a point on a conductor in 0.050 second. The average current is

(1) 8.0×10^{-8} A
(2) 1.0×10^{-2} A
(3) 2.5×10^{-1} A
(4) 1.0×10^{2} A

18 ____

19 If several resistors are connected in series in an electric circuit, the potential difference across each resistor

(1) varies directly with its resistance
(2) varies inversely with its resistance
(3) varies inversely with the square of its resistance
(4) is independent of its resistance

19 ____

20 The amplitude of a sound wave is most closely related to the sound's

(1) speed
(2) wavelength
(3) loudness
(4) pitch

20 ____

21 A duck floating on a lake oscillates up and down 5.0 times during a 10.-second interval as a periodic wave passes by. What is the frequency of the duck's oscillations?

(1) 0.10 Hz
(2) 0.50 Hz
(3) 2.0 Hz
(4) 50.0 Hz

21 ____

22 Which diagram best represents the position of a ball, at equal time intervals, as it falls freely from rest near Earth's surface?

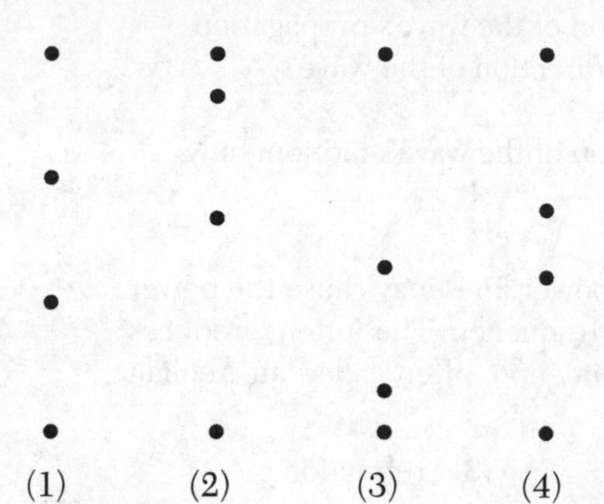

22 ____

23 A gamma ray and a microwave traveling in a vacuum have the same

(1) frequency
(2) period
(3) speed
(4) wavelength

23 ____

24 A student produces a wave in a long spring by vibrating its end. As the frequency of the vibration is doubled, the wavelength in the spring is

(1) quartered
(2) halved
(3) unchanged
(4) doubled

24 ____

25 Which two points on the wave shown in the diagram below are in phase with each other?

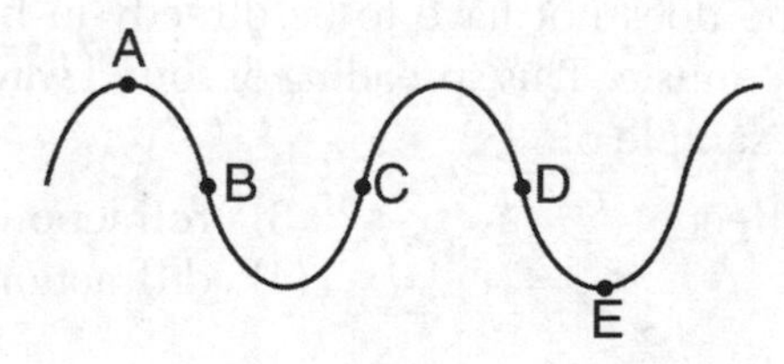

(1) *A* and *B*
(2) *A* and *E*
(3) *B* and *C*
(4) *B* and *D*

25 ____

26 As a longitudinal wave moves through a medium, the particles of the medium

(1) vibrate parallel to the direction of the wave's propagation
(2) vibrate perpendicular to the direction of the wave's propagation
(3) are transferred in the direction of the wave's motion, only
(4) are stationary

26 _____

27 Wind blowing across suspended power lines may cause the power lines to vibrate at their natural frequency. This often produces audible sound waves. This phenomenon, often called an Aeolian harp, is an example of

(1) diffraction
(2) the Doppler effect
(3) refraction
(4) resonance

27 _____

28 A student listens to music from a speaker in an adjoining room, as represented in the diagram below.

She notices that she does not have to be directly in front of the doorway to hear the music. This spreading of sound waves beyond the doorway is an example of

(1) the Doppler effect
(2) resonance
(3) refraction
(4) diffraction

28 _____

29 What is the minimum energy required to ionize a hydrogen atom in the $n = 3$ state?

(1) 0.00 eV
(2) 0.66 eV
(3) 1.51 eV
(4) 12.09 eV

29 ____

Base your answers to questions 30 and 31 on the diagram below and on your knowledge of physics. The diagram represents two small, charged, identical metal spheres, *A* and *B*, that are separated by a distance of 2.0 meters.

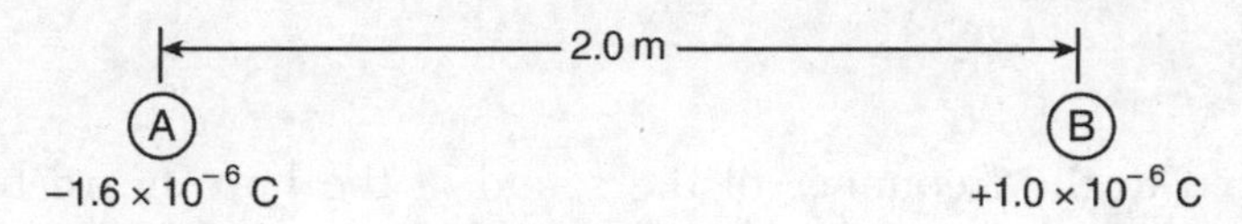

30 What is the magnitude of the electrostatic force exerted by sphere *A* on sphere *B*?

(1) 7.2×10^{-3} N
(2) 3.6×10^{-3} N
(3) 8.0×10^{-13} N
(4) 4.0×10^{-13} N

30 ____

31 If the two spheres were touched together and then separated, the charge on sphere A would be

(1) -3.0×10^{-7} C
(2) -6.0×10^{-7} C
(3) -1.3×10^{-6} C
(4) -2.6×10^{-6} C

31 ____

32 The horn of a moving vehicle produces a sound of constant frequency. Two stationary observers, *A* and *C*, and the vehicle's driver, *B*, positioned as represented in the diagram below, hear the sound of the horn.

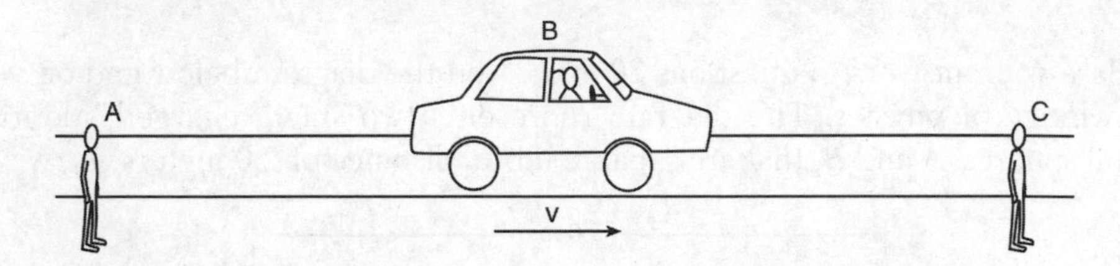

Compared to the frequency of the sound of the horn heard by driver *B*, the frequency heard by observer *A* is

(1) lower and the frequency heard by observer *C* is lower
(2) lower and the frequency heard by observer *C* is higher
(3) higher and the frequency heard by observer *C* is lower
(4) higher and the frequency heard by observer *C* is higher

32 _____

33 A different force is applied to each of four different blocks on a frictionless, horizontal surface. In which diagram does the block have the greatest inertia 2.0 seconds after starting from rest?

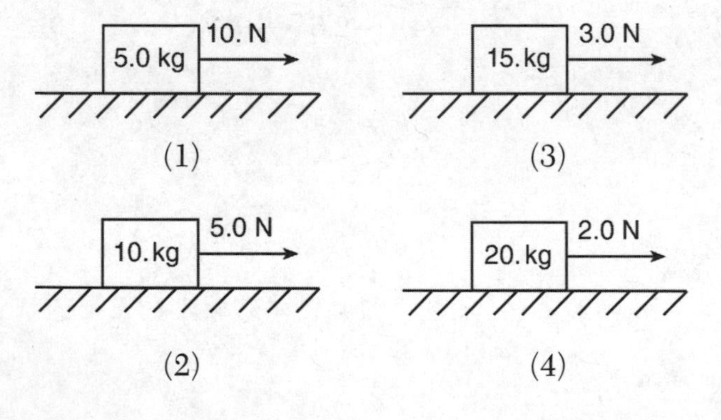

33 _____

34 The diagram below shows a ray of monochromatic light incident on a boundary between air and glass.

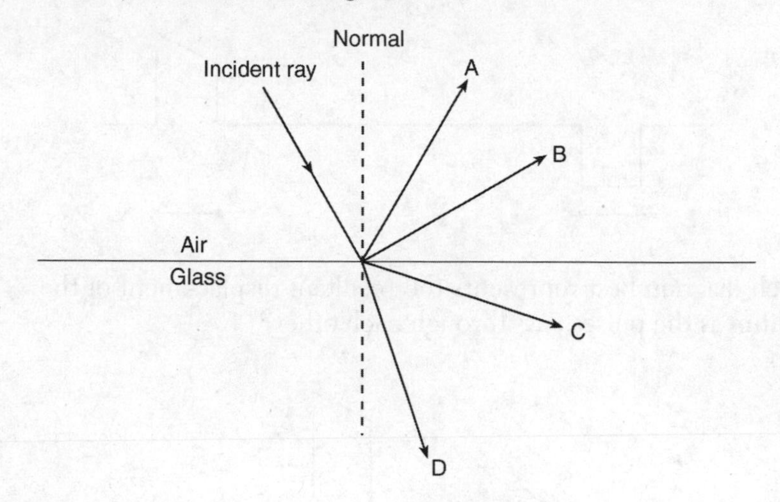

Which ray best represents the path of the reflected light ray?

(1) *A*
(2) *B*
(3) *C*
(4) *D*

34 _____

35 Two pulses approach each other in the same medium. The diagram below represents the displacements caused by each pulse.

Which diagram best represents the resultant displacement of the medium as the pulses pass through each other?

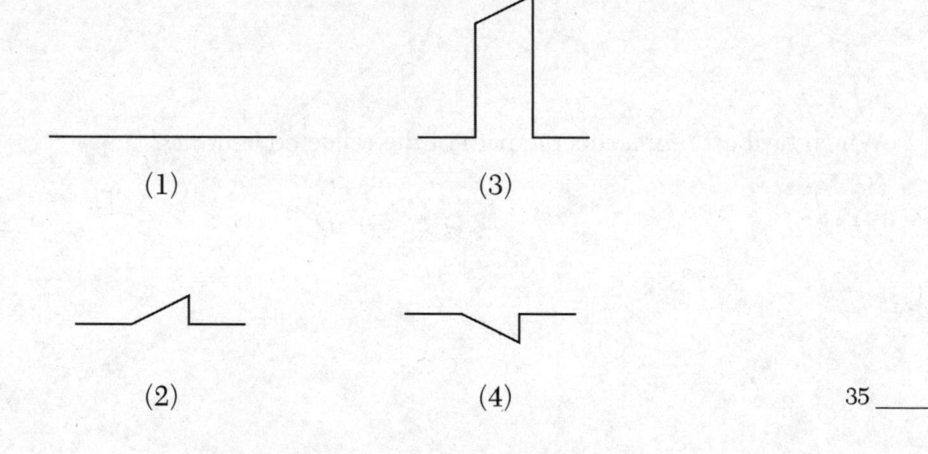

35 _____

PART B–1

Answer all questions in this part.

Directions (36–50): For *each* statement or question, select the *number* of the word or expression that, of those given, best completes the statement or answers the question. Some questions may require the use of the *2006 Edition Reference Tables for Physical Setting/Physics.* Record your answers in the spaces provided.

36 The diameter of an automobile tire is closest to

(1) 10^{-2} m
(2) 10^{0} m
(3) 10^{1} m
(4) 10^{2} m

36 ____

37 The vector diagram below represents the velocity of a car traveling 24 meters per second 35° east of north.

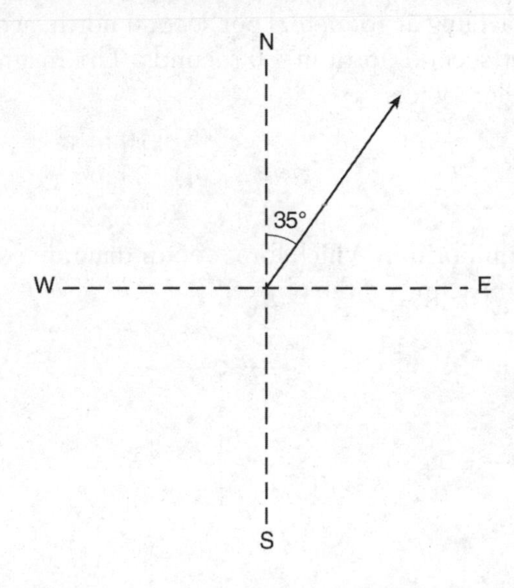

What is the magnitude of the component of the car's velocity that is directed eastward?

(1) 14 m/s
(2) 20. m/s
(3) 29 m/s
(4) 42 m/s

37 ____

38 Without air resistance, a kicked ball would reach a maximum height of 6.7 meters and land 38 meters away. With air resistance, the ball would travel

(1) 6.7 m vertically and more than 38 m horizontally
(2) 38 m horizontally and less than 6.7 m vertically
(3) more than 6.7 m vertically and less than 38 m horizontally
(4) less than 38 m horizontally and less than 6.7 m vertically

38 _____

39 A car is moving with a constant speed of 20 meters per second. What total distance does the car travel in 2.0 minutes?

(1) 10 m
(2) 40 m
(3) 1200 m
(4) 2400 m

39 _____

40 A car, initially traveling at 15 meters per second north, accelerates to 25 meters per second north in 4.0 seconds. The magnitude of the average acceleration is

(1) 2.5 m/s^2
(2) 6.3 m/s^2
(3) 10 m/s^2
(4) 20 m/s^2

40 _____

41 An object is in equilibrium. Which force vector diagram could represent the force(s) acting on the object?

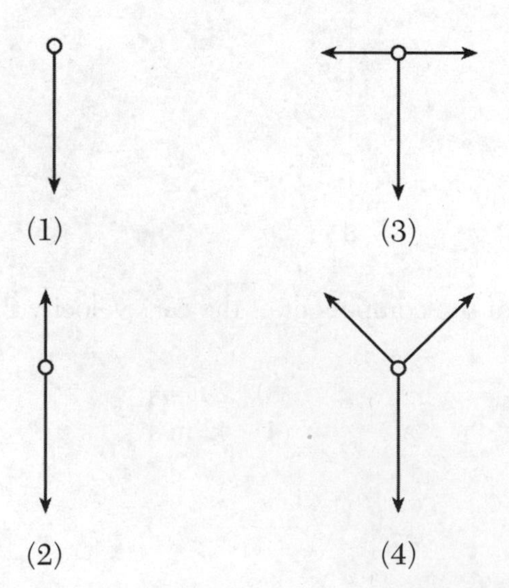

41 _____

42 Which combination of fundamental units can be used to express the amount of work done on an object?

(1) kg • m/s
(2) kg • m/s^2
(3) kg • m^2/s^2
(4) kg • m^2/s^3

42 ____

43 Which graph best represents the relationship between the potential energy stored in a spring and the change in the spring's length from its equilibrium position?

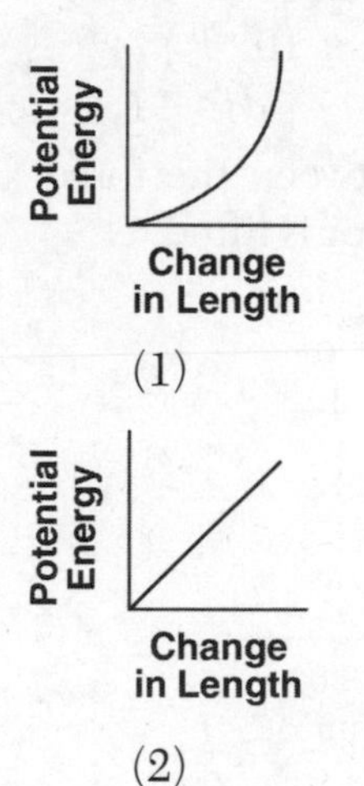

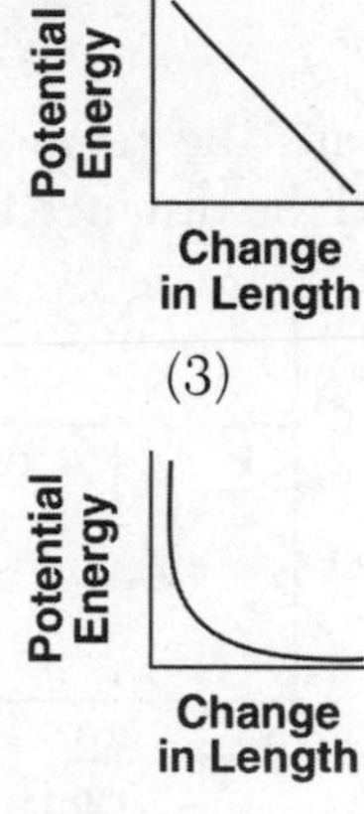

43 ____

44 An electric motor has a rating of 4.0×10^2 watts. How much time will it take for this motor to lift a 50.-kilogram mass a vertical distance of 8.0 meters? [Assume 100% efficiency.]

(1) 0.98 s
(2) 9.8 s
(3) 98 s
(4) 980 s

44 ____

45 A compressed spring in a toy is used to launch a 5.00-gram ball. If the ball leaves the toy with an initial horizontal speed of 5.00 meters per second, the minimum amount of potential energy stored in the compressed spring was

(1) 0.0125 J
(2) 0.0250 J
(3) 0.0625 J
(4) 0.125 J

45 ____

46 A ray of yellow light ($f = 5.09 \times 10^{14}$ Hz) travels at a speed of 2.04×10^8 meters per second in

(1) ethyl alcohol
(2) water
(3) lucite
(4) glycerol

46 ____

47 A blue-light photon has a wavelength of 4.80×10^{-7} meter. What is the energy of the photon?

(1) 1.86×10^{22} J
(2) 1.44×10^2 J
(3) 4.14×10^{-19} J
(4) 3.18×10^{-26} J

47 ____

48 The graph below represents the relationship between the force exerted on an elevator and the distance the elevator is lifted.

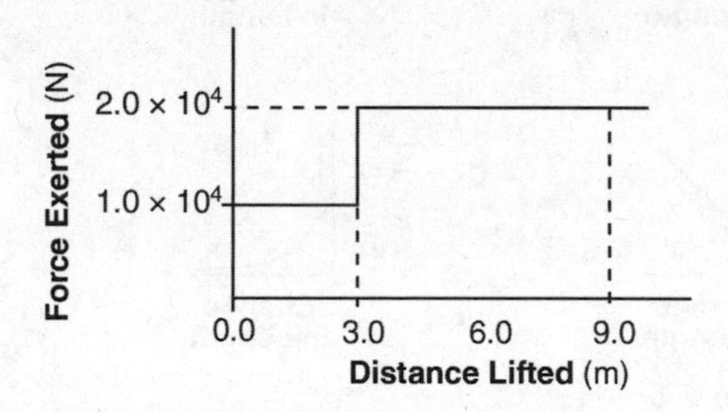

How much total work is done by the force in lifting the elevator from 0.0 m to 9.0 m?

(1) 9.0×10^4 J
(2) 1.2×10^5 J
(3) 1.5×10^5 J
(4) 1.8×10^5 J

48 ____

49 The diagram below shows waves *A* and *B* in the same medium.

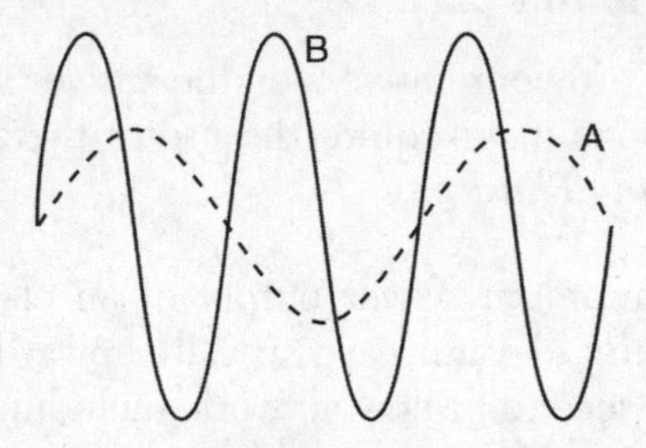

Compared to wave *A*, wave *B* has

(1) twice the amplitude and twice the wavelength
(2) twice the amplitude and half the wavelength
(3) the same amplitude and half the wavelength
(4) half the amplitude and the same wavelength 49 _____

50 What is the quark composition of a proton?

(1) uud
(2) udd
(3) csb
(4) uds 50 _____

PART B–2

Answer all questions in this part.

Directions (51–65): Record your answers on the answer sheet provided after the questions. Some questions may require the use of the *2006 Edition Reference Tables for Physical Setting/Physics*.

51–52 Calculate the minimum power output of an electric motor that lifts a 1.30×10^4-newton elevator car vertically upward at a constant speed of 1.50 meters per second. [Show all work, including the equation and substitution with units.] [2]

53–54 A microwave oven emits a microwave with a wavelength of 2.00×10^{-2} meter in air. Calculate the frequency of the microwave. [Show all work, including the equation and substitution with units.] [2]

55–56 Calculate the energy equivalent in joules of the mass of a proton. [Show all work, including the equation and substitution with units.] [2]

Base your answers to questions 57 through 59 on the information and diagram below and on your knowledge of physics.

A 1.5×10^3-kilogram car is driven at a constant speed of 12 meters per second counterclockwise around a horizontal circular track having a radius of 50. meters, as represented below.

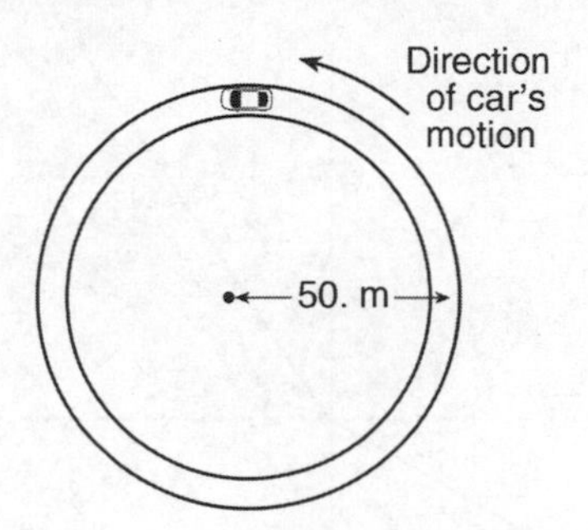

Track, as Viewed from Above

57 On the diagram *on the answer sheet*, draw an arrow to indicate the direction of the velocity of the car when it is at the position shown. Start the arrow on the car. [1]

58–59 Calculate the magnitude of the centripetal acceleration of the car. [Show all work, including the equation and substitution with units.] [2]

Base your answers to questions 60 through 62 on the information below and on your knowledge of physics.

A football is thrown at an angle of 30° above the horizontal. The magnitude of the horizontal component of the ball's initial velocity is 13.0 meters per second. The magnitude of the vertical component of the ball's initial velocity is 7.5 meters per second. [Neglect friction.]

60 On the axes *on the answer sheet*, draw a graph representing the relationship between the horizontal displacement of the football and the time the football is in the air. [1]

61–62 The football is caught at the same height from which it is thrown. Calculate the total time the football was in the air. [Show all work, including the equation and substitution with units.] [2]

Base your answers to questions 63 through 65 on the information and diagram below and on your knowledge of physics.

A ray of light ($f = 5.09 \times 10^{14}$ Hz) traveling through a block of an unknown material, passes at an angle of incidence of 30° into air, as shown in the diagram below.

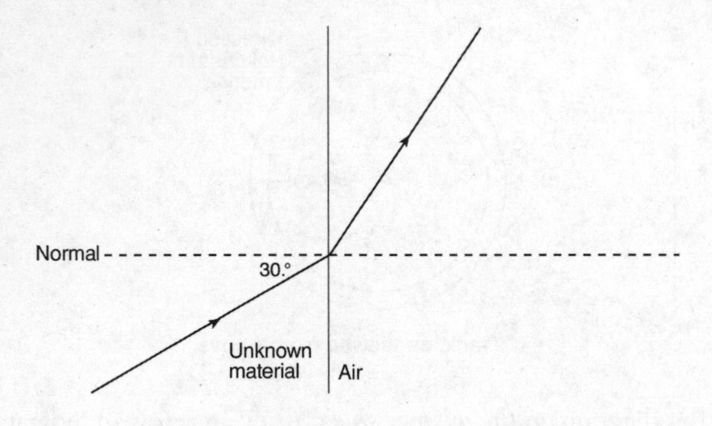

63 Use a protractor to determine the angle of refraction of the light ray as it passes from the unknown material into air. [1]

64–65 Calculate the index of refraction of the unknown material. [Show all work, including the equation and substitution with units.] [2]

PART C

Answer all questions in this part.

Directions (66–85): Record your answers on the answer sheet provided after the questions. Some questions may require the use of the *2006 Edition Reference Tables for Physical Setting/Physics.*

Base your answers to questions 66 through 70 on the information below and on your knowledge of physics.

The diagram below represents a 4.0-newton force applied to a 0.200-kilogram copper block sliding to the right on a horizontal steel table.

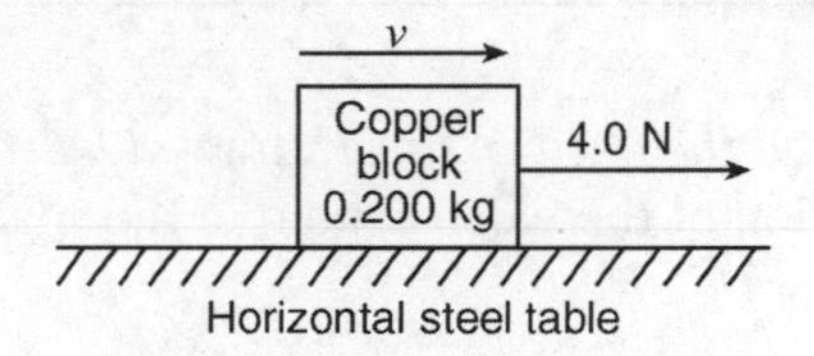

66 Determine the weight of the block. [1]

67–68 Calculate the magnitude of the force of friction acting on the moving block. [Show all work, including the equation and substitution with units.] [2]

69 Determine the magnitude of the net force acting on the moving block. [1]

70 Describe what happens to the magnitude of the velocity of the block as the block slides across the table. [1]

Base your answers to questions 71 through 75 on the information and diagram below and on your knowledge of physics.

Two conducting parallel plates 5.0×10^{-3} meter apart are charged with a 12-volt potential difference. An electron is located midway between the plates. The magnitude of the electrostatic force on the electron is 3.8×10^{-16} newton.

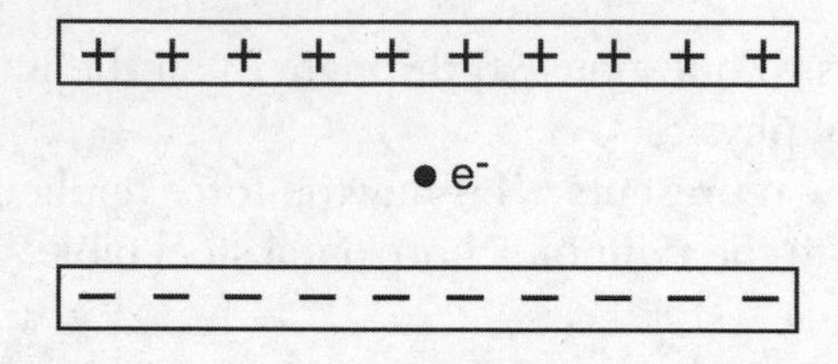

71 On the diagram *on the answer sheet*, draw *at least three* field lines to represent the direction of the electric field in the space between the charged plates. [1]

72 Identify the direction of the electrostatic force that the electric field exerts on the electron. [1]

73–74 Calculate the magnitude of the electric field strength between the plates, in newtons per coulomb. [Show all work, including the equation and substitution with units.] [2]

75 Describe what happens to the magnitude of the net electrostatic force on the electron as the electron is moved toward the positive plate. [1]

Base your answers to questions 76 through 80 on the information below and on your knowledge of physics.

An electron in a mercury atom changes from energy level *b* to a higher energy level when the atom absorbs a single photon with an energy of 3.06 electron-volts.

76 Determine the letter that identifies the energy level to which the electron jumped when the mercury atom absorbed the photon. [1]

77 Determine the energy of the photon, in joules. [1]

78–79 Calculate the frequency of the photon. [Show all work, including the equation and substitution with units.] [2]

80 Classify the photon as one of the types of electromagnetic radiation listed in the electromagnetic spectrum. [1]

Base your answers to questions 81 through 85 on the information and circuit diagram below and on your knowledge of physics.

Three lamps are connected in parallel to a 120-volt source of potential difference, as represented below.

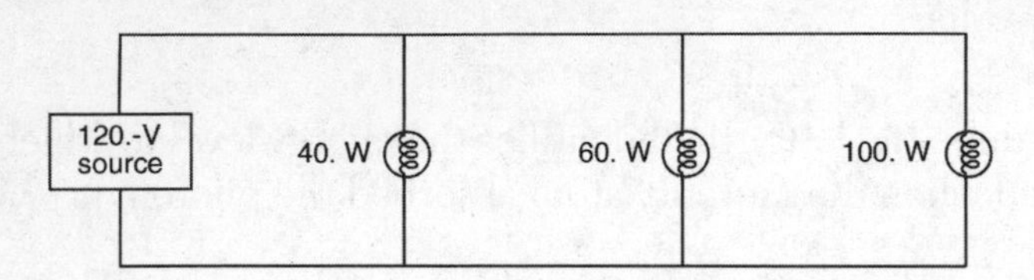

81–82 Calculate the resistance of the 40-watt lamp. [Show all work, including the equation and substitution with units.] [2]

83 Describe what change, if any, would occur in the power dissipated by the 100-watt lamp if the 60-watt lamp were to burn out. [1]

84 Describe what change, if any, would occur in the equivalent resistance of the circuit if the 60-watt lamp were to burn out. [1]

85 The circuit is disassembled. The same three lamps are then connected in series with each other and the source. Compare the equivalent resistance of this series circuit to the equivalent resistance of the parallel circuit. [1]

Answer Sheet
June 2015

Physics: The Physical Setting

PART B–2

51–52

53–54

55–56

57

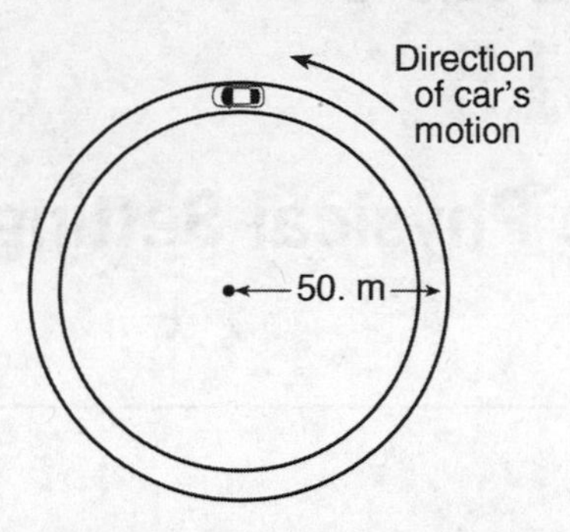

Track, as Viewed from Above

58–59

60

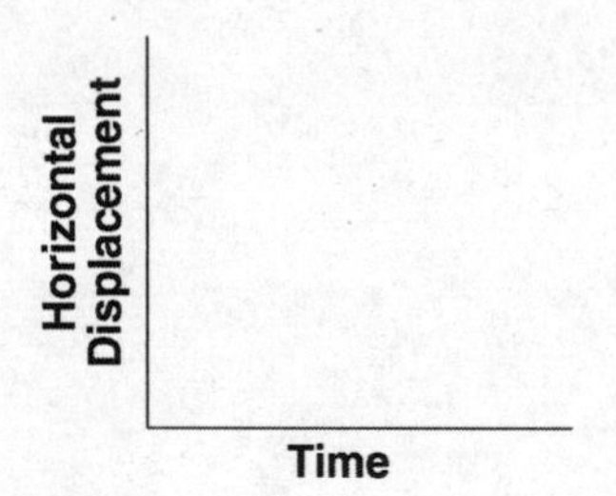

61–62

63 ________________ °

64–65

PART C

66 ______________________ N

67–68

69 ______________________ N

70 ____

____

71

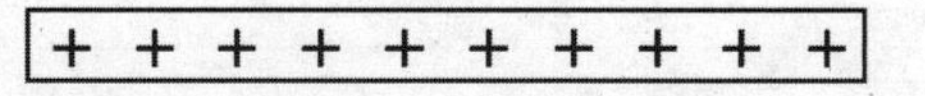

● e^-

72 ____________________________

73–74

75 ____

____

76 ________________

77 ________________ **J**

78–79

80 ___________________

81–82

83 ______________________________________

____

84 ______________________________________

____

85 ______________________________________

____

Answers June 2015

Physics: The Physical Setting

Answer Key

PART A

1. 1	**8.** 4	**15.** 4	**22.** 2	**29.** 3
2. 2	**9.** 2	**16.** 2	**23.** 3	**30.** 2
3. 3	**10.** 3	**17.** 1	**24.** 2	**31.** 1
4. 1	**11.** 3	**18.** 4	**25.** 4	**32.** 2
5. 3	**12.** 1	**19.** 1	**26.** 1	**33.** 4
6. 2	**13.** 1	**20.** 3	**27.** 4	**34.** 1
7. 3	**14.** 1	**21.** 2	**28.** 4	**35.** 2

PART B–1

36. 2	**39.** 4	**42.** 3	**45.** 3	**48.** 3
37. 1	**40.** 1	**43.** 1	**46.** 4	**49.** 2
38. 4	**41.** 4	**44.** 2	**47.** 3	**50.** 1

PART B–2 and **PART C**. *See* **Answers Explained**.

PART A

1 Both speed and work are scalar quantities.

WRONG CHOICES EXPLAINED:

(2) Both velocity and force are vector quantities.
(3) Distance is a scalar quantity while acceleration is a vector quantity.
(4) Momentum is a vector quantity while power is a scalar quantity.

2 When dealing with a word problem requiring a mathematical solution, it is most helpful to translate the words into which quantities/variables are explicitly given and which constants or other variables can be deduced from the situation or words. We also write down the variable that needs to be solved for or found. Then it is much easier to determine which formula is required and then use it to solve the problem.

Given:	Find:	Solution:
$m = 3.00$ kg	$d_{max} = ?$	$v_f^2 = v_i^2 + 2ad$
$v_i = 9.80$ m/s		$d = \dfrac{v_f^2 - v_i^2}{2a}$
$a = -9.81$ m/s^2		$d = \dfrac{(0\text{ m/s})^2 - (9.80\text{ m/s})^2}{2(-9.81\text{ m/s}^2)}$
$v_f = 0$ m/s		$d = 4.90$ m

3 To solve this problem we use the technique of vector addition. We recognize that if we place the vectors head to tail, the resultant vector would be the hypotenuse of a right triangle and we can use the Pythagorean theorem to solve algebraically for the resultant vector:

$$a^2 + b^2 = c^2$$

$$c = \sqrt{a^2 + b^2}$$

$$c = \sqrt{(220\text{ m/s})^2 + (50\text{ m/s})^2}$$

$$c = 226\text{ m/s}$$

4 Because the vehicle is traveling in a straight line at constant speed, according to Newton's First Law, the net force on the vehicle would be equal to 0 N.

5 The only force acting on the ball is the force due to gravity, in this case the weight of the ball, which is 5.0 N.

6

Given:	Find:	Solution:
$k = 100$ N/m	$F = ?$	$F = (100 \text{ N/m})(0.15 \text{ m})$
$x = 0.65$ m – 0.50 m		$F = 15$ N

7

Given:	Find:	Solution:
$m = 1.5$ kg	$F = ?$	$F_{net}\Delta t = \Delta p$
$v_i = 2.0$ m/s		$F_{net}\Delta t = m\Delta v$
$v_f = 0$ m/s		$F_{net}\Delta t = m(v_f - v_i)$
$t = 0.30$ s		$F_{net} = \dfrac{m\left(v_f - v_i\right)}{\Delta t}$
		$F_{net} = \dfrac{(1.5 \text{ kg})(0 \text{ m/s} - 2.0 \text{ m/s})}{(0.30 \text{ s})}$
		$F_{net} = 10$ N

8 When a light wave passes from glass into air, the speed of the light wave increases. Velocity of wave is equal to the product of the frequency of the wave times its wavelength ($v = f\lambda$). Since the frequency does not change, the wavelength changes in proportion to the change in velocity. Since the velocity increases, the wavelength must also increase.

WRONG CHOICES EXPLAINED:

(1) Amplitude is related to the energy carried by the wave.

(2) Frequency of the wave does not change.

(3) Period is the reciprocal of frequency and, since frequency does not change, neither would the period.

9 According to Newton's Third Law, if object A exerts a force on object B, then object B exerts an equal and opposite force back on object A. In this case, the basketball player's feet exert a 1.0×10^3 N force on the floor; therefore, the floor exerts a 1.0×10^3 N force on her feet.

10

Given:	Find:	Solution
$m_{ball} = 0.0600$ kg	$v_{bullet} = ?$	$p_{ball} = p_{bullet}$
$v_{ball} = 60.0$ m/s		$m_{ball} \bullet v_{ball} = m_{bullet} \bullet v_{bullet}$
$m_{bullet} = 0.0100$ kg		$v_{bullet} = \dfrac{m_{ball} \bullet v_{ball}}{m_{bullet}}$
		$v_{bullet} = \dfrac{(0.0600 \text{ kg})(60 \text{ m/s})}{(0.0100 \text{ kg})}$
		$v_{bullet} = 360$ m/s

11

Given:	Find:	Solution:
$m = 1.1 \times 10^4$ kg	$g = ?$	$g = \dfrac{F_g}{m}$
$F_g = 9.1 \times 10^4$ N		$g = \dfrac{9.1 \times 10^4 \text{ N}}{1.1 \times 10^4 \text{ kg}}$
		$g = 8.3$ N/kg

12

$$Fe = \frac{kq_1q_2}{r^2}$$

Electrostatic force is inversely proportional to the square of the distance between the centers of the point charges. If the electrostatic force is equal to F if $r = d$, then:

$$F = \frac{kq_1q_2}{d^2}$$

If $r = 3d$, then:

$$F_e = \frac{kq_1q_2}{(3d)^2}$$

$$F_e = \frac{kq_1q_2}{9d^2}$$

$$F_e = \left(\frac{1}{9}\right)\left(\frac{kq_1q_2}{d^2}\right)$$

$$F_e = \frac{1}{9}F$$

13

Given:	Find:	Solution:
$V = 3.0\text{ V}$	$R = ?$	$V = IR$
$I = 1.8 \times 10^{-4}\text{ A}$		$R = \frac{V}{I}$
		$R = \frac{3.0\text{ V}}{1.8 \times 10^{-4}\text{ A}}$
		$R = 1.7 \times 10^4\ \Omega$

14 An electric motor converts electrical energy into mechanical energy.

WRONG CHOICES EXPLAINED:

(1) A generator converts mechanical into electrical energy.

(2) A battery is an example of converting chemical to electrical energy.

(3) Charging a battery would be an example of converting electrical into chemical energy.

15 The law of conservation of energy states that the total energy of an isolated system remains constant: $E_{Ti} = E_{Tf}$.

$$PE_i + KE_i + Q_i = PE_f + KE_f + Q_f.$$

If the block comes to rest, then the kinetic energy of the block decreases to zero. Neither the block nor the table changes in height; therefore, there is no change in potential energy. The force of friction between the table and the block which causes the block to come to rest results in an increase in the heat (thermal) energy of the system.

WRONG CHOICES EXPLAINED:

(1) No change in height relative to earth surface so no change in gravitational potential energy

(2) No springs in the system, so no change in elastic potential energy

(3) Since the block came to rest and velocity decreased, there is a decrease in kinetic energy, not an increase.

16

Given:	Find:	Solution:
$q = 1.6 \times 10^{-19}$ J	$W = ?$	$V = \dfrac{W}{q}$
$V = 3.00$ V		$W = Vq$
		$W = (3.00 \text{ V})(1.6 \times 10^{-19} \text{ J})$
		$W = 4.80 \times 10^{-19}$ J

17

Given:	Find:	Solution:
$T = 20^{\circ}\text{C}$	$A = ?$	$R = \frac{\rho L}{A}$
Copper wire		$A = \frac{\rho L}{R}$
$\rho = 1.72 \times 10^{-8}\ \Omega\ \text{m}$		$A = \frac{(1.72 \times 10^{-8}\Omega\text{m})(6.0\ \text{m})}{1.3\ \Omega}$
$L = 6.0\ \text{m}$		$A = 7.9 \times 10^{-8}\ \text{m}^2$
$R = 1.3\ \Omega$		

18

Given:	Find:	Solution:
$q = 5.0\ \text{C}$	$I = ?$	$I = \frac{q}{t}$
$t = 0.050\ \text{s}$		$I = \frac{5.0\ \text{C}}{0.050\ \text{s}}$
		$I = 1.0 \times 10^{2}\ \text{A}$

19 Resistors that are connected in series (all in a row) have the identical amount of current flow through each of them. The relationship between potential difference, current, and resistance is calculated using the relationship $V = IR$. If I remains constant, potential difference is directly proportional to resistance.

20 Amplitude is related to the amount of energy in a wave. In a sound wave, amplitude corresponds to loudness.

21 Frequency is defined as the amount of times a motion repeats itself in a given time period. In this instance

$$\frac{5.0}{10\ \text{s}} = 0.50\ \text{Hz}$$

22 As a ball falls freely from rest, it is acted upon by a gravitational force downward, which results in its speed increasing from rest at a rate of 9.81 meters per second. Therefore, with each increasing second, the ball will fall a greater distance during each time second interval.

WRONG CHOICES EXPLAINED:

(1) This diagram represents movement at constant velocity as the distance for each time interval is identical.

(3) This diagram represents movement with decreasing velocity as the distance traveled for each time interval decreases.

(4) This diagram represents movement with decreasing then increasing velocity as the distance traveled for each time interval decreases then increases.

23 Gamma rays and microwaves are both electromagnetic waves and thus travel at the same speed, the speed of light in a vacuum.

WRONG CHOICES EXPLAINED:

(1), (2), (4) Gamma rays and microwaves travel at the same speed but have different frequencies and thus different periods (reciprocal of frequency) and different wavelengths [wavelength and frequency are inversely proportional if velocity is constant ($v = f\lambda$)].

24 If velocity is constant, wavelength and frequency are inversely proportional ($v = f\lambda$). If frequency is doubled, wavelength is halved.

$$f\lambda = (2f)\left(\frac{1}{2}\lambda\right)$$

25 Points on a wave that are at equal displacements from their rest position and are experiencing identical movements, i.e., toward or away from the rest position, are said to be in phase. Points *B* and *D* fit this definition.

WRONG CHOICES EXPLAINED:

(2) *A* and *E* are completely out of phase.

(1), (3) *A* and *B* and *B* and *C* are partially out of phase.

26 As a longitudinal wave travels through a medium, the particles of the medium vibrate in simple harmonic motion in a direction parallel to the direction of motion of the wave.

WRONG CHOICES EXPLAINED:

(2) This motion is characteristic for a transverse wave.

(3) The energy of the wave is transferred in the direction of motion.

27 Resonance is the spontaneous vibration of an object at a frequency equal to that of the wave that initiates that resonant vibration. The resulting standing wave created has amplitudes large enough that sound can be heard.

WRONG CHOICES EXPLAINED:

(1) Diffraction is the bending of a wave around a barrier.

(2) The Doppler effect is the apparent change in frequency that results when a wave source and an observer are in relative motion with respect to each other.

(3) Refraction is the change in direction of a wave when it passes obliquely from one medium into another in which it moves at a different speed.

28 Diffraction is the bending of a wave around a barrier.

WRONG CHOICES EXPLAINED:

(1) The Doppler effect is the apparent change in frequency that results when a wave source and an observer are in relative motion with respect to each other.

(2) Resonance is the spontaneous vibration of an object at a frequency equal to that of the wave that initiates that resonant vibration.

(3) Refraction is the change in direction of a wave when it passes obliquely from one medium into another in which it moves at a different speed.

29 Refer to the *Energy Level Diagram for Hydrogen* in *Reference Table F*. At $n = 3$, the energy level of the hydrogen atom is equal to –1.51 eV. Therefore, 1.51 eV is the minimum amount of energy required to ionize the hydrogen atom, i.e., disassociate the electron from the atom by bringing the energy level to zero.

30

Given:	Find:	Solution:
$q_A = -1.6 \times 10^{-6}$ C	$F_e = ?$	$F_e = \dfrac{kq_1q_2}{r^2}$
$q_B = 1.0 \times 10^{-6}$ C		$F_e =$
		$\dfrac{(8.99 \times 10^9 \text{ N m}^2/\text{C}^2)(-1.6 \times 10^{-6} \text{ C})(1.0 \times 10^{-6} \text{ C})}{(2.0 \text{ m})^2}$
$r = 2.0$ m		$F_e = 3.6 \times 10^{-3}$ N
$k = 8.99 \times 10^9$ Nm^2/C^2		

31 If the two spheres were touched together there would be a redistribution of the total charge of the 2 spheres such that each sphere would have the same net charge. We add the charges on all the spheres and divide by the total number of spheres:

$$\frac{(-1.6 \times 10^{-6} \text{ C}) + (1.0 \times 10^{-6} \text{ C})}{2} = -3.0 \times 10^{-7} \text{C}$$

32 This scenario is an example of the Doppler effect. The vehicle is moving away from observer *A* and moving toward observer *C*. The apparent pitch or frequency is decreased as the sound source moves away from and increased as the sound source moves toward an observer.

WRONG CHOICES EXPLAINED:

(1) The sound source would have to be traveling away from both observers for this to be true.

(3) The vehicle would have to be moving toward observer *A* and away from observer *C* for this to be true.

(4) The sound source would have to be traveling toward both observers for this to be true.

33 Mass is the quantitative measure of inertia. The block in diagram 4 has the greatest mass of all the 4 blocks, 20 kg, and thus has the greatest inertia.

34 When light rays are reflected from a surface, the angle that the incident ray makes with the normal to the surface is equal to the angle that the reflected ray makes with the normal to the surface. Ray *A* meets this definition.

WRONG CHOICES EXPLAINED:

(2) Ray *B* has an angle of reflection greater than the angle of incidence.

(3), (4) Rays *C* and *D* represent refracted rays.

35 These two displacements are in opposite directions and when they meet will result in destructive interference. The resulting displacement is the algebraic sum of the displacements of the individual pulses. The pulse on the right is the virtual opposite to the pulse on the left with a small addition. When they meet, the rectangular portion of each will cancel out and all that will be left is the small top triangular area above the center. The diagram in choice (2) accurately represents the resultant displacement at the instant the pulses pass through each other.

WRONG CHOICES EXPLAINED:

(1) The two pulses would have to be equal in magnitude and opposite in direction for this diagram to be the resultant.

(3) The two pulses would have to have displacements in the same direction for this diagram to be the resultant.

(4) The pulse on the left would have had to have been displaced up instead of down and the pulse on the right displaced down instead of up for this diagram to be the resultant.

PART B–1

36 This is an order of magnitude question. 10 to the zero power is equivalent to 1. One meter is approximately equal to 3.3 feet, which is a reasonable approximation for the diameter of a tire.

WRONG CHOICES EXPLAINED:

(1) 10 centimeters is not a reasonable approximation of the diameter of a tire.

(3) 10 meters (approx. 33 ft) is not a reasonable approximation of the diameter of a tire.

(4) 100 meters (approx. 330 ft) is not a reasonable approximation of the diameter of a tire.

37

Given:	Find:	Solution:
$v = 24$ m/s	$v_{east} = ?$	$v_{east} = v \sin \theta$
$\theta = 35°$ E of N		$v_{east} = (24 \text{ m/s})(\sin 35°)$
		$v_{east} = 14$ m/s

38 Air resistance would have the effect of slowing the motion or velocity of the ball. As a result the horizontal and vertical components of the velocity would be decreased. The ball would not reach as great a height. It would therefore have a shorter amount of time in the air or hang time and thus not travel as far horizontally.

39

Given:	Find:	Solution:
$v = 20$ m/s	$d = ?$	$v = \frac{d}{t}$
$t = 2.0 \text{ min} \bullet (60 \text{ s}/1 \text{ min}) = 120$ s		$d = vt$
		$d = (20 \text{ m/s})(120 \text{ s})$
		$d = 2{,}400$ m

40

Given:	Find:	Solution:
$v_i = 15$ m/s	$a = ?$	$a = \frac{\Delta v}{t}$
$v_f = 25$ m/s		$a = \frac{v_f - v_i}{t}$
$t = 4.0$ s		$a = \frac{(25 \text{ m/s} - 15 \text{ m/s})}{4.0 \text{ s}}$
		$a = 2.5 \text{ m/s}^2$

41 In diagram 4 the horizontal components of the generally upward vector forces could negate each other and the vertical components could negate the downward force vector.

WRONG CHOICES EXPLAINED:

(1) This diagram indicates a net force in the downward direction.

(2) In this diagram, the magnitude of the upward force vector is smaller than the downward vector, which would result in a net force downward.

(3) In this diagram, the 2 horizontal vectors look equal in magnitude so, being that they are opposite in direction, would result in a net force of zero in the horizontal direction. However, there is no force to counteract the downward vector; therefore, there would be a net force in the downward direction.

42

$$W = Fd$$
$$F = ma$$

Therefore $W = mad$

$$W = (\text{kg})\left(\text{m/s}^2\right)(\text{m})$$
$$W = \text{kg} \bullet \text{m}^2/\text{s}^2$$

WRONG CHOICES EXPLAINED:

(1) unit for momentum

(2) equivalent to newtons, unit for force

(4) equivalent to joules per second or watts, unit for power

43 The potential energy stored in a spring is proportional to the square of the displacement of the spring from its equilibrium position.

$$PE_s = \frac{1}{2}kx^2$$

The graph in choice (1) illustrates an exponential relationship.

WRONG CHOICES EXPLAINED:

(2) This graph illustrates a direct proportionality.

(3) This graph illustrates an inverse relationship.

(4) This graph illustrates an inversely proportional relationship.

44

Given:	Find:	Solution:
$P = 4.0 \times 10^2$ W	$t = ?$	$P = \frac{Fd}{t}$
$m = 50$ kg		$t = \frac{Fd}{P}$ $\quad F_g = mg$
$d = 8.0$ m		$t = \frac{mgd}{P}$
$g = 9.81$ m/s^2		$t = \frac{(50.\text{ kg})(9.81\text{ m/s}^2)(8.0\text{ m})}{(4.0 \times 10^2\text{ W})}$
		$t = 9.8$ s

45

Given:	Find:	Solution:
$m_{ball} = 5.00$ g • (1 kg/1,000 g)		
$= 5.00 \times 10^{-3}$ kg	$PE_s = ?$	$PE_s = KE_{ball}$
$v_{ball} = 5.00$ m/s		$PE_s = \frac{1}{2} m_{ball} v_{ball}{}^2$
		$PE_s =$
		$\frac{1}{2}(5.00 \times 10^{-3}\text{ kg})(5.00\text{ m/s})^2$
		$PE_s = 0.0625$ J

46 Refer to the *Absolute Indices of Refraction* table in *Reference Table E*.

Given:	Find:	Solution:
$v = 2.04 \times 10^8$ m/s	$n = ?$	$n = \frac{c}{v}$
$c = 3.00 \times 10^8$ m/s		$n = \frac{3.00 \times 10^8 \text{ m/s}}{2.04 \times 10^8 \text{ m/s}}$
		$n = 1.47$

An absolute index of refraction of 1.47 corresponds to corn oil or glycerol.

WRONG CHOICES EXPLAINED:

(1) has an index of refraction of 1.36.
(2) has an index of refraction of 1.33.
(3) has an index of refraction of 1.50.

47

Given:	Find:	Solution:
$\lambda = 4.80 \times 10^{-7}$ m	$E_{photon} = ?$	$E_{photon} = \frac{hc}{\lambda}$
		$E_{photon} = \frac{(6.63 \times 10^{-34} \text{ J} \bullet \text{s})(3.00 \times 10^8 \text{ m/s})}{(4.80 \times 10^{-7} \text{ m})}$
		$E_{photon} = 4.14 \times 10^{-19}$ J

48

Given:	Find:	Solution:
$F_1 = 1.0 \times 10^4$ N	$W_{total} = ?$	$W_{total} = W_1 + W_2$
$d_1 = 3.0$ m		$W_{total} = F_1d_1 + F_2d_2$
$F_2 = 2.0 \times 10^4$ N		$W_{total} =$
		$(1.0 \times 10^4 \text{ N})(3.0 \text{ m}) + (2.0 \times 10^4 \text{ N})(6.0 \text{ m})$
$d_2 = 6.0$ m		$W_{total} = 3.0 \times 10^4 \text{ Nm} + 1.2 \times 10^5$
Nm		
		$W_{total} = 1.5 \times 10^5$ J

49 Wave B has a greater amplitude than wave A, i.e., a greater displacement from rest position, so we can eliminate choices (3) and (4). Wave B has a smaller, or shorter wavelength than wave A so we can eliminate choice (1).

50 A proton is made up of up and down quarks and has a net charge of +1. Refer to the *Particles of the Standard Model* chart in *Reference Table H*. The correct combination of up and down quarks to yield a net charge of +1 would be uud.

WRONG CHOICES EXPLAINED:

(2) This would yield a net charge of 0 and is the correct combination for a neutron.

(3) This would yield a net charge of 0.

(4) This would yield a net charge of 0.

PART B–2

51–52

Given:	Find:	Solution:
$F = 1.30 \times 10^4$ N	$P = ?$	$P = F\overline{v}$
$\overline{v} = 1.50$ m/s		$P = (1.30 \times 10^4 \text{ N})(1.50 \text{ m/s})$
		$P = 1.95 \times 10^4$ W

51 One credit is awarded for the correct equation and substitution of values with units.

52 One credit is awarded for the correct answer with units or for an answer, with units, that is consistent with the answer to question 51.

Note: Students will not be penalized more than 1 credit for errors in units in questions 51 and 52.

53–54

Given:	Find:	Solution:
$\lambda = 2.00 \times 10^{-2}$ m	$f = ?$	$c = f\lambda$
$c = 3.0 \times 10^{8}$ m/s		$f = \frac{c}{\lambda}$
		$f = \frac{3.0 \times 10^{8}\ \text{m/s}}{2.00 \times 10^{-2}\ \text{m}}$
		$f = 1.50 \times 10^{10}$ Hz

53 One credit is awarded for the correct equation and substitution of values with units.

54 One credit is awarded for the correct answer with units or for an answer, with units, that is consistent with the answer to question 53.

Note: Students will not be penalized more than 1 credit for errors in units in questions 53 and 54.

55–56

Given:	Find:	Solution:
$m = 1.67 \times 10^{-27}$ kg	$E = ?$	$E = mc^2$
$c = 3.0 \times 10^{8}$ m/s		$E = (1.67 \times 10^{-27}\ \text{kg})(3.0 \times 10^{8}\ \text{m/s})^2$
		$E = 1.50 \times 10^{-10}$ J

Or

Given:	Find:	Solution:
$m = 1$ u $= 9.31 \times 10^{2}$ eV)	$E = ?$	$E = (9.31 \times 10^{8}\ \text{eV})(1.60 \times 10^{-19}\ \text{J/eV})$
MeV $= 9.31 \times 10^{8}$ eV		
1 eV $= 1.60 \times 10^{-19}$ J		$E = 1.50 \times 10^{-10}$ J

55 One credit is awarded for the correct equation and substitution of values with units.

56 One credit is awarded for the correct answer with units or for an answer, with units, that is consistent with the answer to question 55.

Note: Students will not be penalized more than 1 credit for errors in units in questions 55 and 56.

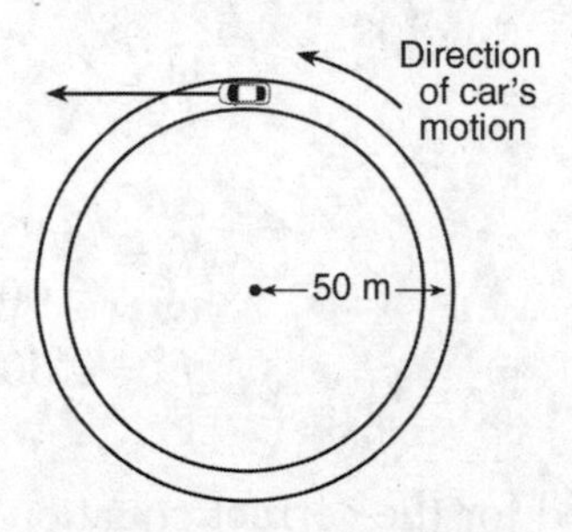

Track, as Viewed from Above

57 One credit is awarded for an arrow drawn from the car with the arrowhead pointed to the left.

58–59

Given:	Find:	Solution:
$m = 1.5 \times 10^3$ kg	$a_c = ?$	$a_c = \frac{v^2}{r}$
$v = 12$ m/s		$a_c = \frac{(12 \text{ m/s})^2}{50 \text{ m}}$
$r = 50$ M		$a_c = 2.9 \text{ m/s}^2$

58 One credit is awarded for the correct equation and substitution of values with units.

59 One credit is awarded for the correct answer with units or for an answer, with units, that is consistent with the answer to question 58.

Note: Students will not be penalized more than 1 credit for errors in units in questions 58 and 59.

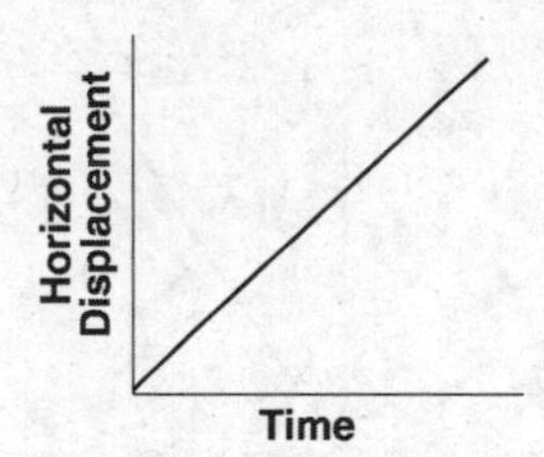

60 Velocity in the horizontal direction is constant, therefore displacement is directly proportional to time elapsed.

One credit is awarded for a straight line with a positive slope.

Note: The line may be sketched in. It must approximate a straight line, but need not pass through the origin.

61–62

Given:	Find:	Solution:
$v_{ix} = 13.0$ m/s	$t_{total} = ?$	$a = \frac{\Delta v}{t}$
$v_{iy} = 7.5$ m/s		$t = \frac{\Delta v}{a}$
$d_{yi} = d_{yf} = 0$		$t = \frac{v_f - v_i}{a}$
$a = -9.81$ m/s^2		$t = \frac{(-7.5 \text{ m/s}) - (7.5 \text{ m/s})}{(-9.81 \text{ m/s}^2)}$
$v_{fy} = -7.5$ m/s		$t = 1.5$ s

61 One credit is awarded for the correct equation and substitution of values with units.

62 One credit is awarded for the correct answer with units or for an answer, with units, that is consistent with the answer to question 61.

Note: Students will not be penalized more than 1 credit for errors in units in questions 61 and 62.

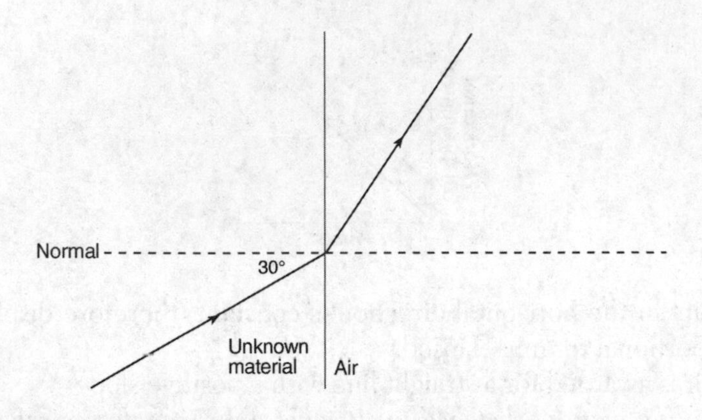

63 One credit is awarded for an answer of 56° ± 2°.

64–65

Given:

$\theta_{unknown} = 30°$

$\theta_{air} = 56°$

$n_{air} = 1.00$

Find:

$n_{unknown} = 1.00$

Solution:

$n_1 \sin \theta_1 = n_2 \sin \theta_2$

$n_{unknown} \sin \theta_{unknown} = n_{air} \sin \theta_{air}$

$$n_{unknown} = \frac{n_{air} \sin \theta_{air}}{\sin \theta_{unknown}}$$

$$n_{unknown} = \frac{(1.00)(\sin 56°)}{\sin 30°}$$

$$n_{unknown} = 1.7 \text{ or } 1.6$$

64 One credit is awarded for the correct equation and substitution of values with units.

65 One credit is awarded for the correct answer with units or for an answer, with units, that is consistent with the answer to question 64.
Note: Students will not be penalized more than 1 credit for errors in units in questions 64 and 65.

PART C

66

Given:	Find:	Solution:
$m = 0.200$ kg	$F_w = ?$	$F_w = mg$
$g = 9.81$ m/s^2		$F_w = (0.200\ \text{kg})(9.81\ \text{m/s}^2)$
		$F_w = 1.96$ N

One credit is awarded for an answer of 1.96 N or 2.0 N or 1.9 N.

67–68

Given:	Find:	Solution:
$\mu_k = 0.36$	$F_f = ?$	$F_f = \mu_k F_N$
$F_w = 1.96$ N		$F_f = (0.36)(1.96\ \text{N})$
$F_w = F_N$		$F_f = 0.71$ N

67 One credit is awarded for the correct equation and substitution of values with units.

68 One credit is awarded for the correct answer with units (0.71 N or 0.70 N) or for an answer, with units, that is consistent with the answer to question 67. *Note: Students will not be penalized more than 1 credit for errors in units in questions 67 and 68.*

69

Given:	Find:	Solution:
$F_f = 0.71$ N	$F_{net} = ?$	$F_{net} = F_{applied} - F_f$
$F_{applied} = 4.0$ N		$F_{net} = (4.0\ \text{N}) - (0.71\ \text{N})$
		$F_{net} = 3.3$ N

One credit is awarded for an answer of 3.3 N or for an answer that is consistent with the answer to question 68.

70 Given that there is a net force acting on the block, the velocity of the block would increase as the block slides across the table.

One credit is awarded for an acceptable answer as above or for an answer that is consistent with the answer to question 69.

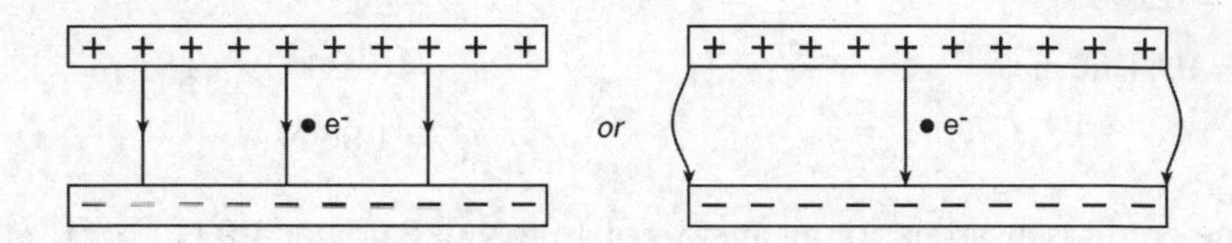

71 One credit is awarded for at least 3 arrows pointing away from the positive plate and toward the negative plate.

Note: Use of a straightedge is not necessary to draw the field lines. Field lines near the edge of the plate may be curved.

72 The direction of the electrostatic force on the electron would be toward the positive plate and away from the negative plate, i.e., opposite the direction of the field.

One credit is awarded for an answer consistent with the above explanation.

Given:	Find:	Solution:
$d = 5.0 \times 10^{-3}$ m	$E = ?$	$E = \frac{V}{d}$ or $E = \frac{F_e}{q}$
$V = 12$ V		$E = \frac{12 \text{ V}}{5.0 \times 10^{-3} \text{ m}}$
		$E = \frac{3.8 \times 10^{-16} \text{ N}}{1.6 \times 10^{-19} \text{ C}}$
$q = 1.6 \times 10^{-19}$ C		$E = 2.4 \times 10^{3}$ V/m
		$E = 2.4 \times 10^{3}$ N/C
$F = 3.8 \times 10^{-16}$ N		$E = 2.4 \times 10^{3}$ N/C

73–74 One credit is awarded for the correct equation and substitution of values with units.

74 One credit is awarded for a correct answer with units or for an answer, with units, that is consistent with the answer to question 73.

Note: Students will not be penalized more than 1 credit for errors in units in questions 73 and 74.

75 The magnitude of the net electrostatic force on the electron remains constant. One credit is awarded for an answer consistent with the explanation above.

76 Refer to *Reference Table F*. The difference in energy between level b and level f is equal to 3.06 eV.

One credit is awarded for f or energy level f.

77

Given:	Find:	Solution:
$E_{photon} = 3.06 \text{ eV}$	E_{photon} in joules	$3.06 \text{ eV}(1.6 \times 10^{-19} \text{ J/eV}) = 4.90 \times 10^{-19} \text{ J}$
$1 \text{ eV} = 1.6 \times 10^{-19} \text{ J}$		

One credit is awarded for an answer of 4.90×10^{-19} J or 4.89×10^{-19} J.

78–79

Given:	Find:	Solution:
$E_{photon} = 4.90 \times 10^{-19} \text{ J}$	$f = ?$	$E_{photon} = hf$
$h = 6.63 \times 10^{-34} \text{ J/s}$		$f = \dfrac{E_{photon}}{h}$
		$f = \dfrac{4.90 \times 10^{-19} \text{ J}}{6.63 \times 10^{-34} \text{ J/s}}$
		$f = 7.39 \times 10^{14} \text{ Hz}$

78 One credit is awarded for the correct equation and substitution of values with units.

79 One credit is awarded for a correct answer with units or for an answer, with units, that is consistent with the answer to question 78.

Note: Students will not be penalized more than 1 credit for errors in units in questions 78 and 79.

80 Refer to the *Electromagnetic Spectrum* diagram in *Reference Table D*. An $f = 7.39 \times 10^{14}$ Hz is within the visible light spectrum, specifically visible violet light.

One credit is awarded for an answer of visible light or violet light or for an answer that is consistent with the answer to question 79.

81–82

Given:	Find:	Solution:
$V = 120$ V	$R = ?$	$P = \frac{V^2}{R}$
$P = 40$ W		$R = \frac{V^2}{P}$
		$R = \frac{(120\ V)^2}{40\ W}$
		$R = 360\ \Omega$

81 One credit is awarded for the correct equation and substitution of values with units.

82 One credit is awarded for a correct answer with units or for an answer, with units, that is consistent with the answer to question 81.

Note: Students will not be penalized more than 1 credit for errors in units in questions 81 and 82.

83 The voltage across the lamp and its resistance are unchanged regardless of whether the other lamps are functional or not; therefore, there would be no change in the power dissipated by the 100 W lamp.

One credit is awarded for an answer indicating that there is no change in the power dissipated.

84 The total resistance in a parallel circuit would increase if a resistor is removed from the circuit. Therefore, if the 60 W lamp were to burn out, that would be akin to removing a resistor from a parallel circuit and the total resistance in the circuit would increase.

One credit is awarded for an answer indicating that the equivalent resistance would increase.

85 In a series circuit, the total resistance of the circuit is equal to the sum of the individual resistances. Therefore, the total resistance in a series circuit would be greater than the total resistance of the resistors connected in parallel.

One credit is awarded for an answer indicating that the equivalent resistance of the series circuit would be greater than the equivalent resistance of the parallel circuit.

Topic	Question Numbers (total)	Wrong Answers (x)	Grade
Math Skills	2, 3, 6, 7, 10, 13, 16–18, 21, 30, 31, 36, 37, 39, 40, 44–48, 53–56, 58–60, 64–68, 72–79, 81, 82: (43)		$\frac{100\ (43 - x)}{43} = \%$
Mechanics	1–7, 9–12, 22, 33, 38–41, 57–59, 61, 62, 68, 70: (24)		$\frac{100\ (24 - x)}{24} = \%$
Energy	14, 15, 42, 44, 45, 51, 52: (7)		$\frac{100\ (7 - x)}{7} = \%$
Electricity/ Magnetism	13, 19, 83, 84: (4)		$\frac{100\ (4 - x)}{4} = \%$
Waves	8, 20, 21, 23–28, 32, 34, 35, 46, 53, 54: (15)		$\frac{100\ (15 - x)}{15} = \%$
Modern Physics	29, 47, 50, 55, 56, 76–79 (9)		$\frac{100\ (9 - x)}{9} = \%$

Examination June 2016

Physics: The Physical Setting

PART A

Answer all questions in this part.

Directions (1–35): For *each* statement or question, select the *number* of the word or expression that, of those given, best completes the statement or answers the question. Some questions may require the use of the *2006 Edition Reference Tables for Physical Setting/Physics.* Record your answers in the spaces provided.

1 Which quantity is a vector?

(1) power
(2) kinetic energy
(3) speed
(4) weight

1 _____

2 A 65.0-kilogram astronaut weighs 638 newtons at the surface of Earth. What is the mass of the astronaut at the surface of the Moon, where the acceleration due to gravity is 1.62 meters per second squared?

(1) 10.7 kg
(2) 65.0 kg
(3) 105 N
(4) 638 N

2 _____

3 When the sum of all the forces acting on a block on an inclined plane is zero, the block

(1) must be at rest
(2) must be accelerating
(3) may be slowing down
(4) may be moving at constant speed

3 _____

4 The greatest increase in the inertia of an object would be produced by increasing the

(1) mass of the object from 1.0 kg to 2.0 kg
(2) net force applied to the object from 1.0 N to 2.0 N
(3) time that a net force is applied to the object from 1.0 s to 2.0 s
(4) speed of the object from 1.0 m/s to 2.0 m/s 4 ____

5 A 100-kilogram cart accelerates at 0.50 meter per second squared west as a horse exerts a force of 60. newtons west on the cart. What is the magnitude of the force that the cart exerts on the horse?

(1) 10 N (3) 60 N
(2) 50 N (4) 110 N 5 ____

6 Sound waves are described as

(1) mechanical and transverse
(2) mechanical and longitudinal
(3) electromagnetic and transverse
(4) electromagnetic and longitudinal 6 ____

7 An electrical force of 8.0×10^{-5} newton exists between two point charges, $q1$ and $q2$. If the distance between the charges is doubled, the new electrical force between the charges will be

(1) 1.6×10^{-4} N (3) 3.2×10^{-4} N
(2) 2.0×10^{-5} N (4) 4.0×10^{-5} N 7 ____

8 A blue lab cart is traveling west on a track when it collides with and sticks to a red lab cart traveling east. The magnitude of the momentum of the blue cart before the collision is 2.0 kilogram • meters per second, and the magnitude of the momentum of the red cart before the collision is 3.0 kilogram • meters per second. The magnitude of the total momentum of the two carts after the collision is

(1) 1.0 kg • m/s (3) 3.0 kg • m/s
(2) 2.0 kg • m/s (4) 5.0 kg • m/s 8 ____

9 The diagram below represents the path of a thrown ball through the air.

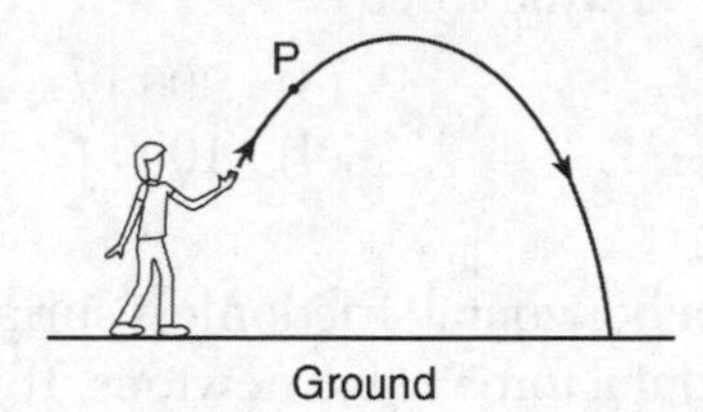

Which arrow best represents the direction in which friction acts on the ball at point *P*?

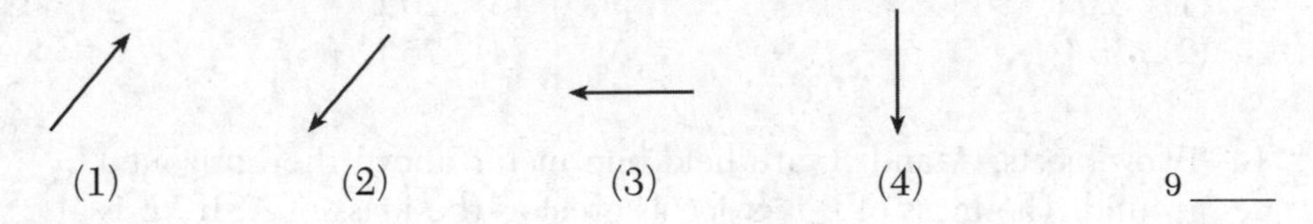

9 ____

10 A magnetic field would be produced by a beam of

(1) x rays
(2) gamma rays
(3) protons
(4) neutrons

10 ____

11 The diagram below represents the electric field in the region of two small charged spheres, *A* and *B*.

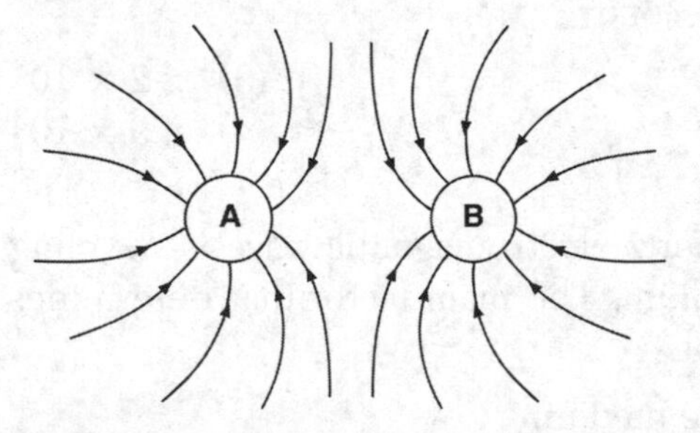

What is the sign of the net charge on *A* and *B*?

(1) *A* is positive and *B* is positive.
(2) *A* is positive and *B* is negative.
(3) *A* is negative and *B* is negative.
(4) *A* is negative and *B* is positive.

11 ____

12 A horizontal force of 20 newtons eastward causes a 10-kilogram box to have a displacement of 5 meters eastward. The total work done on the box by the 20-newton force is

(1) 40 J
(2) 100 J
(3) 200 J
(4) 1000 J

12 ____

13 A block initially at rest on a horizontal, frictionless surface is accelerated by a constant horizontal force of 5.0 newtons. If 15 joules of work is done on the block by this force while accelerating it, the kinetic energy of the block increases by

(1) 3.0 J
(2) 15 J
(3) 20 J
(4) 75 J

13 ____

14 Two objects, *A* and *B*, are held one meter above the horizontal ground. The mass of *B* is twice as great as the mass of *A*. If *PE* is the gravitational potential energy of *A* relative to the ground, then the gravitational potential energy of *B* relative to the ground is

(1) PE
(2) $2PE$
(3) $\frac{PE}{2}$
(4) $4PE$

14 ____

15 What is the kinetic energy of a 55-kilogram skier traveling at 9.0 meters per second?

(1) 2.5×10^2 J
(2) 5.0×10^2 J
(3) 2.2×10^3 J
(4) 4.9×10^3 J

15 ____

16 A 5.09×10^{14}-hertz electromagnetic wave is traveling through a transparent medium. The main factor that determines the speed of this wave is the

(1) nature of the medium
(2) amplitude of the wave
(3) phase of the wave
(4) distance traveled through the medium

16 ____

17 A motor does a total of 480 joules of work in 5.0 seconds to lift a 12-kilogram block to the top of a ramp. The average power developed by the motor is

(1) 8.0 W
(2) 40. W
(3) 96 W
(4) 2400 W

17 _____

18 A 5.8×10^4-watt elevator motor can lift a total weight of 2.1×10^4 newtons with a maximum constant speed of

(1) 0.28 m/s
(2) 0.36 m/s
(3) 2.8 m/s
(4) 3.6 m/s

18 _____

19 A stationary police officer directs radio waves emitted by a radar gun at a vehicle moving toward the officer. Compared to the emitted radio waves, the radio waves reflected from the vehicle and received by the radar gun have a

(1) longer wavelength
(2) higher speed
(3) longer period
(4) higher frequency

19 _____

20 A light wave strikes the Moon and reflects toward Earth. As the light wave travels from the Moon toward Earth, the wave carries

(1) energy, only
(2) matter, only
(3) both energy and matter
(4) neither energy nor matter

20 _____

21 The time required to produce one cycle of a wave is known as the wave's

(1) amplitude
(2) frequency
(3) period
(4) wavelength

21 _____

22 A magnetic compass is placed near an insulated copper wire. When the wire is connected to a battery and a current is created, the compass needle moves and changes its position. Which is the best explanation for the production of a force that causes the needle to move?

(1) The copper wire magnetizes the compass needle and exerts the force on the compass needle.
(2) The compass needle magnetizes the copper wire and exerts the force on the compass needle.
(3) The insulation on the wire becomes charged, which exerts the force on the compass needle.
(4) The current in the wire produces a magnetic field that exerts the force on the compass needle.

22 _____

23 A beam of monochromatic light ($f = 5.09 \times 10^{14}$ Hz) has a wavelength of 589 nanometers in air. What is the wavelength of this light in Lucite?

(1) 150 nm
(2) 393 nm
(3) 589 nm
(4) 884 nm

23 _____

24 If the amplitude of a sound wave is increased, there is an increase in the sound's

(1) loudness
(2) pitch
(3) velocity
(4) wavelength

24 _____

25 In the diagram below, point P is located in the electric field between two oppositely charged parallel plates.

+ + + + +

• P

– – – – –

Compared to the magnitude and direction of the electrostatic force on an electron placed at point P, the electrostatic force on a proton placed at point P has

(1) the same magnitude and the same direction
(2) the same magnitude, but the opposite direction
(3) a greater magnitude, but the same direction
(4) a greater magnitude and the opposite direction 25 ____

26 The effect produced when two or more sound waves pass through the same point simultaneously is called

(1) interference
(2) diffraction
(3) refraction
(4) resonance 26 ____

27 A gamma ray photon and a microwave photon are traveling in a vacuum. Compared to the wavelength and energy of the gamma ray photon, the microwave photon has a

(1) shorter wavelength and less energy
(2) shorter wavelength and more energy
(3) longer wavelength and less energy
(4) longer wavelength and more energy 27 ____

28 According to the Standard Model of Particle Physics, a neutrino is a type of

(1) lepton
(2) photon
(3) meson
(4) baryon 28 ____

29 Which combination of quarks produces a neutral baryon?

(1) cts
(2) dsb
(3) uds
(4) uct

29 _____

30 When 2.0×10^{-16} kilogram of matter is converted into energy, how much energy is released?

(1) 1.8×10^{-1} J
(2) 1.8×10^{1} J
(3) 6.0×10^{-32} J
(4) 6.0×10^{-8} J

30 _____

31 A ball is hit straight up with an initial speed of 28 meters per second. What is the speed of the ball 2.2 seconds after it is hit? [Neglect friction.]

(1) 4.3 m/s
(2) 6.4 m/s
(3) 22 m/s
(4) 28 m/s

31 _____

32 A particle with a charge of 3.00 elementary charges moves through a potential difference of 4.50 volts. What is the change in electrical potential energy of the particle?

(1) 1.07×10^{-19} eV
(2) 2.16×10^{-18} eV
(3) 1.50 eV
(4) 13.5 eV

32 _____

33 Which circuit has the largest equivalent resistance?

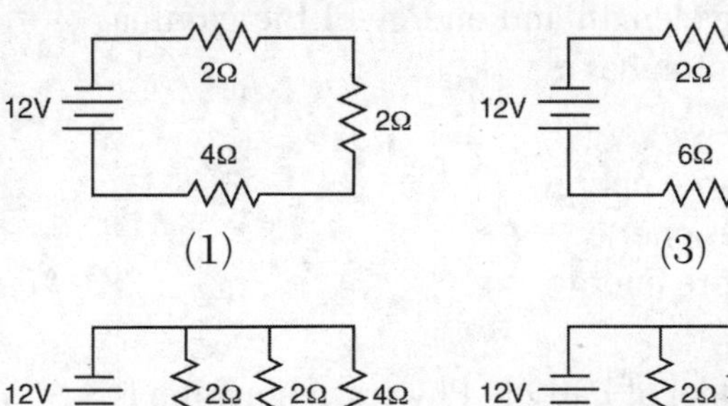

12V 2Ω 2Ω 4Ω

(2)

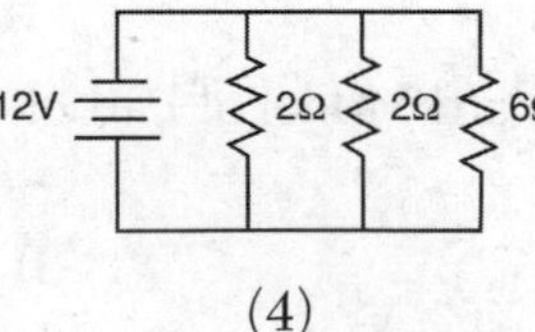

(4)

33 _____

34 A transverse wave is moving toward the right in a uniform medium. Point X represents a particle of the uniform medium. Which diagram represents the direction of the motion of particle X at the instant shown?

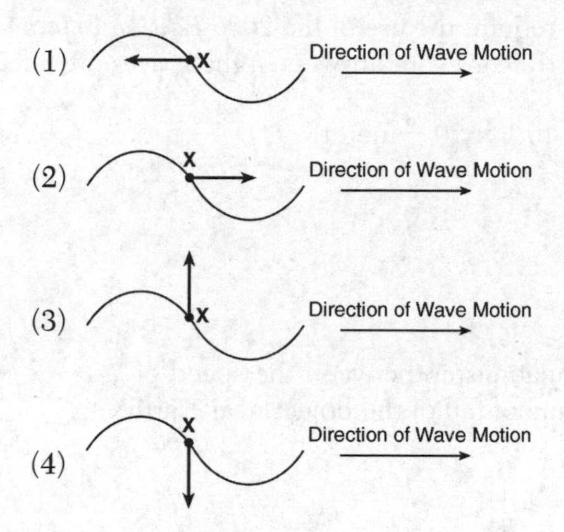

34 ____

35 Which diagram represents magnetic field lines between two north magnetic poles?

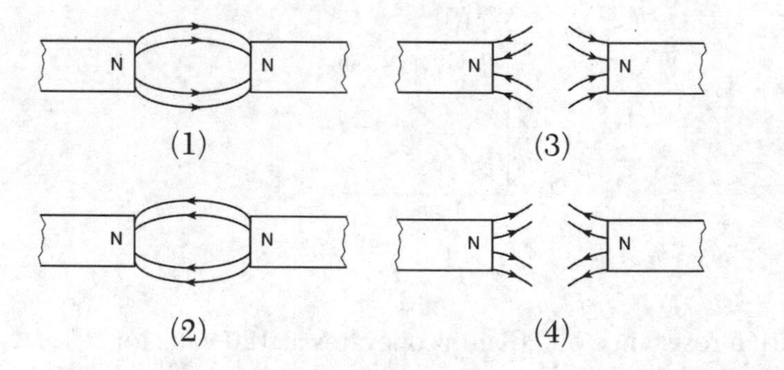

35 ____

PART B–1

Answer all questions in this part.

Directions (36–50): For *each* statement or question, select the *number* of the word or expression that, of those given, best completes the statement or answers the question. Some questions may require the use of the *2006 Edition Reference Tables for Physical Setting/Physics.* Record your answers in the spaces provided.

36 Which measurement is closest to 1×10^{-2} meter?

(1) diameter of an atom
(2) width of a student's finger
(3) length of a football field
(4) height of a schoolteacher

36 ____

37 Which graph represents the relationship between the speed of a freely falling object and the time of fall of the object near Earth's surface?

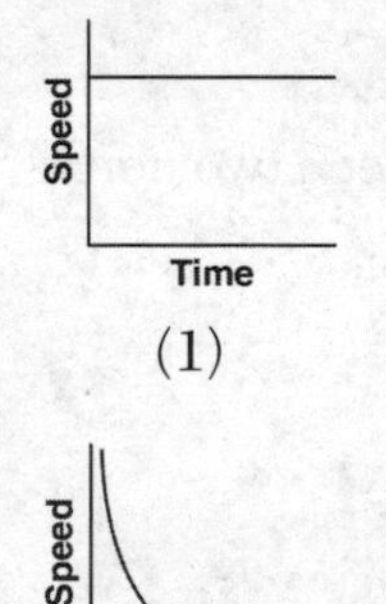

(1) (2)

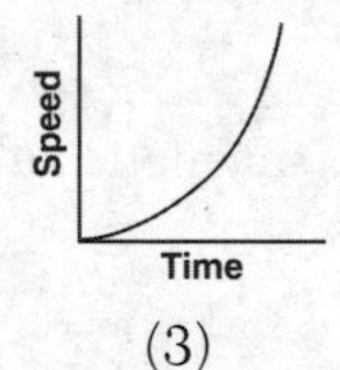

(3)

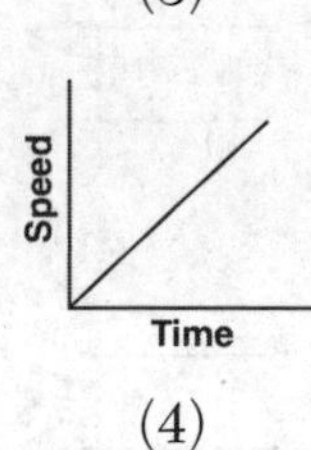

(4)

37 ____

38 A hair dryer with a resistance of 9.6 ohms operates at 120 volts for 2.5 minutes. The total electrical energy used by the dryer during this time interval is

(1) 2.9×10^{3} J
(2) 3.8×10^{3} J
(3) 1.7×10^{5} J
(4) 2.3×10^{5} J

38 ____

39 A box weighing 46 newtons rests on an incline that makes an angle of 25° with the horizontal. What is the magnitude of the component of the box's weight perpendicular to the incline?

(1) 19 N
(2) 21 N
(3) 42 N
(4) 46 N

39 ____

40 Which graph represents the motion of an object traveling with a positive velocity and a negative acceleration?

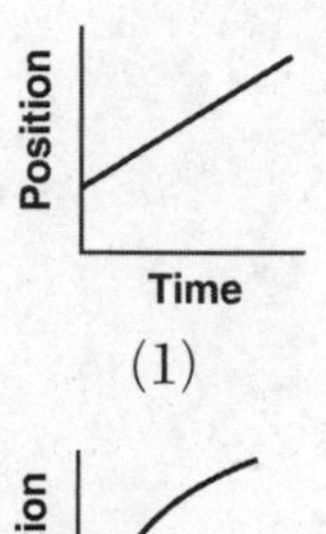

(1)

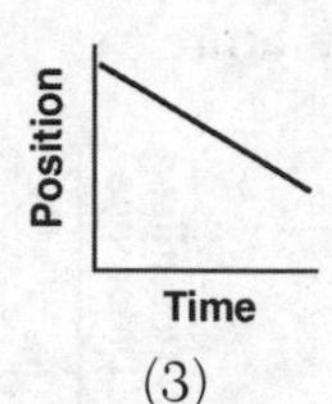

(3)

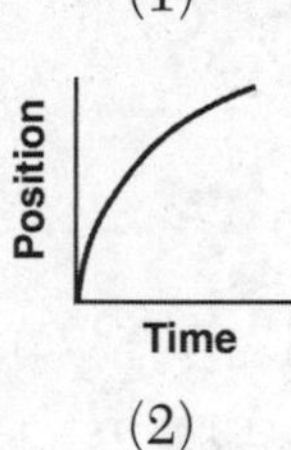

(2)

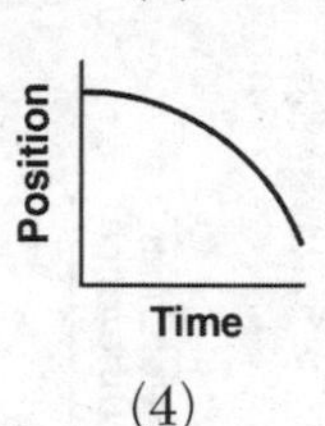

(4)

40 ____

41 Car *A*, moving in a straight line at a constant speed of 20. meters per second, is initially 200 meters behind car *B*, moving in the same straight line at a constant speed of 15 meters per second. How far must car *A* travel from this initial position before it catches up with car *B*?

(1) 200 m
(2) 400 m
(3) 800 m
(4) 1000 m

41 ____

42 A 2700-ohm resistor in an electric circuit draws a current of 2.4 milliamperes. The total charge that passes through the resistor in 15 seconds is

(1) 1.6×10^{-4} C
(2) 3.6×10^{-2} C
(3) 1.6×10^{-1} C
(4) 3.6×10^{1} C

42 ____

43 A 1000–kilogram car traveling 20.0 meters per second east experiences an impulse of 2000 newton • seconds west. What is the final velocity of the car after the impulse has been applied?

(1) 18.0 m/s east (3) 20.5 m/s west
(2) 19.5 m/s east (4) 22.0 m/s west 43 _____

44 Which graph represents the relationship between the potential difference applied to a copper wire and the resulting current in the wire at constant temperature?

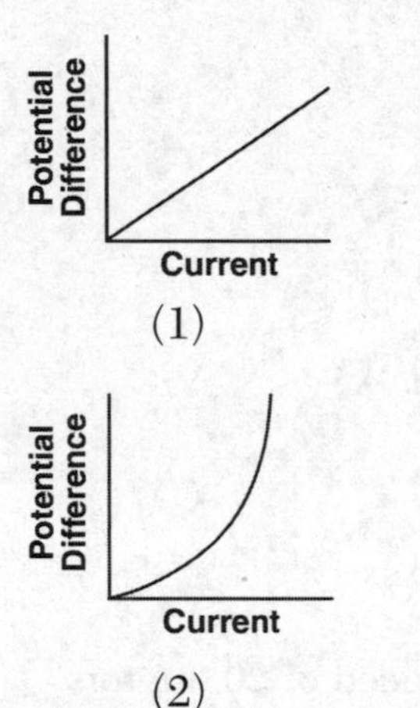

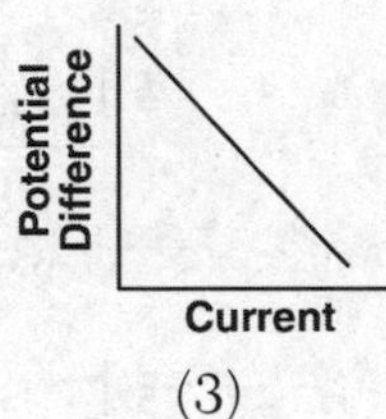

(3)

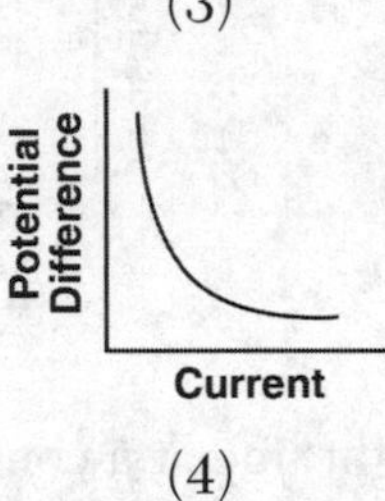

(4) 44 _____

45 A tungsten wire has resistance R at 20°C. A second tungsten wire at 20°C has twice the length and half the cross-sectional area of the first wire. In terms of R, the resistance of the second wire is

(1) $\frac{R}{2}$ (3) $2R$
(2) R (4) $4R$ 45 _____

46 After an incandescent lamp is turned on, the temperature of its filament rapidly increases from room temperature to its operating temperature. As the temperature of the filament increases, what happens to the resistance of the filament and the current through the filament?

(1) The resistance increases and the current decreases.
(2) The resistance increases and the current increases.
(3) The resistance decreases and the current decreases.
(4) The resistance decreases and the current increases. 46 _____

47 Parallel wave fronts are incident on an opening in a barrier. Which diagram shows the configuration of wave fronts and barrier opening that will result in the greatest diffraction of the waves passing through the opening? [Assume all diagrams are drawn to the same scale.]

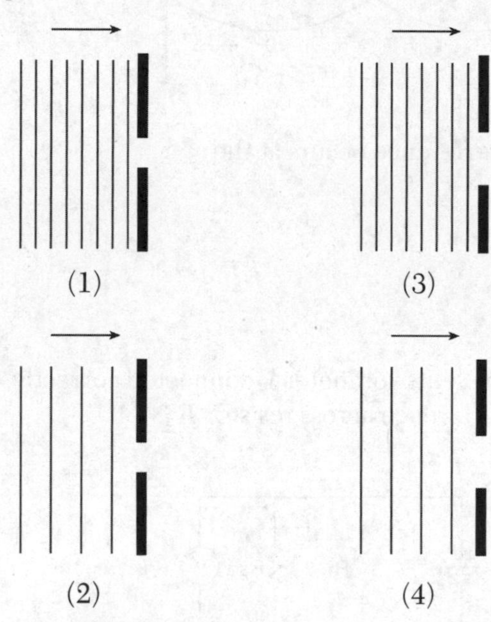

47 _____

48 A singer demonstrated that she could shatter a crystal glass by singing a note with a wavelength of 0.320 meter in air at STP. What was the natural frequency of the glass?

(1) 9.67×10^{-4} Hz
(2) 1.05×10^{2} Hz
(3) 1.03×10^{3} Hz
(4) 9.38×10^{8} Hz

48 _____

49 The diagram below represents a standing wave in a string.

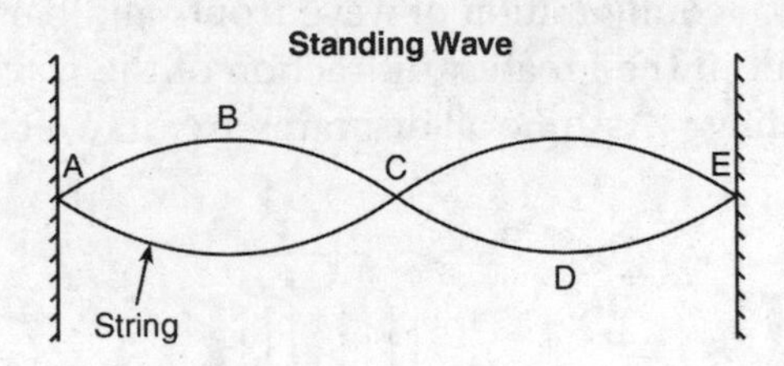

Maximum constructive interference occurs at the

(1) antinodes A, C, and E
(2) nodes A, C, and E
(3) antinodes B and D
(4) nodes B and D

49 _____

50 Which circuit diagram represents voltmeter V connected correctly to measure the potential difference across resistor R_2?

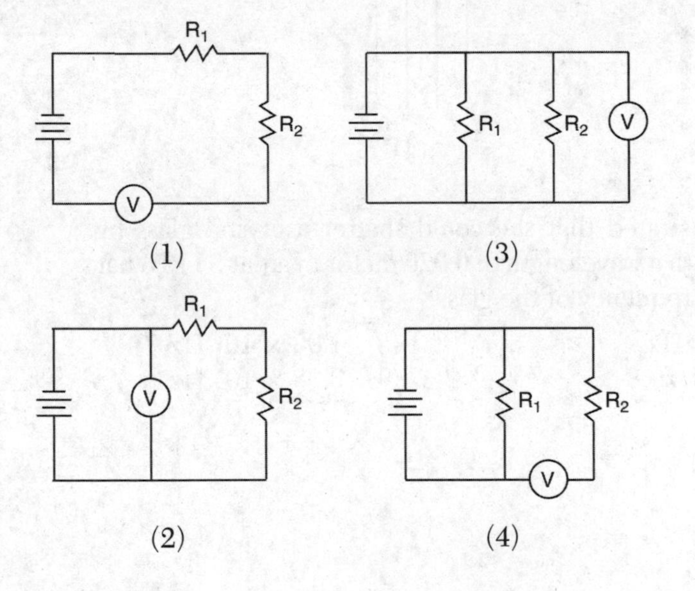

50 _____

PART B–2

Answer all questions in this part.

Directions (51–65): Record your answers on the answer sheet provided after the questions. Some questions may require the use of the *2006 Edition Reference Tables for Physical Setting/Physics.*

Base your answers to questions 51 through 53 on the information and diagram below and on your knowledge of physics.

As represented in the diagram below, a constant 15-newton force, F, is applied to a 2.5-kilogram box, accelerating the box to the right at 2.0 meters per second squared across a rough horizontal surface.

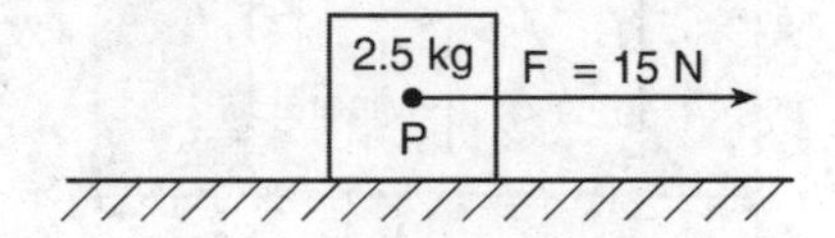

51–52 Calculate the magnitude of the net force acting on the box. [Show all work, including the equation and substitution with units.] [2]

53 Determine the magnitude of the force of friction on the box. [1]

Base your answers to questions 54 and 55 on the information and diagram below and on your knowledge of physics.

A ray of light ($f = 5.09 \times 10^{14}$ Hz) is traveling through a mineral sample that is submerged in water. The ray refracts as it enters the water, as shown in the diagram below.

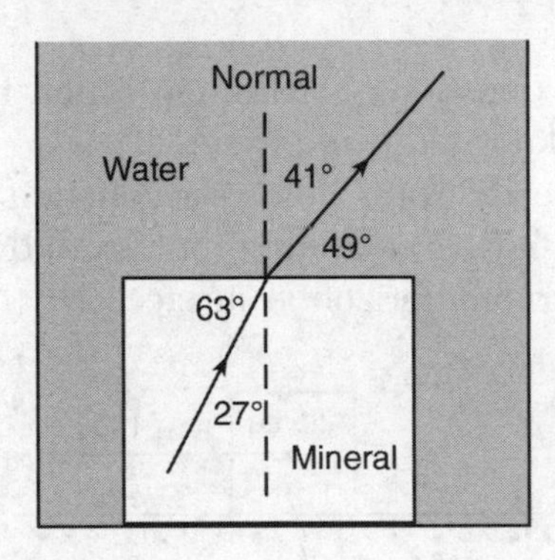

54–55 Calculate the absolute index of refraction of the mineral. [Show all work, including the equation and substitution with units.] [2]

Base your answers to questions 56 through 58 on the information below and on your knowledge of physics.

A ball is rolled twice across the same level laboratory table and allowed to roll off the table and strike the floor. In each trial, the time it takes the ball to travel from the edge of the table to the floor is accurately measured. [Neglect friction.]

56–57 In trial *A*, the ball is traveling at 2.50 meters per second when it reaches the edge of the table. The ball strikes the floor 0.391 second after rolling off the edge of the table. Calculate the height of the table. [Show all work, including the equation and substitution with units.] [2]

58 In trial *B*, the ball is traveling at 5.00 meters per second when it reaches the edge of the table. Compare the time it took the ball to reach the floor in trial *B* to the time it took the ball to reach the floor in trial *A*. [1]

Base your answers to questions 59 through 61 on the information and diagram below and on your knowledge of physics.

A toy airplane flies clockwise at a constant speed in a horizontal circle of radius 8.0 meters. The magnitude of the acceleration of the airplane is 25 meters per second squared. The diagram shows the path of the airplane as it travels around the circle.

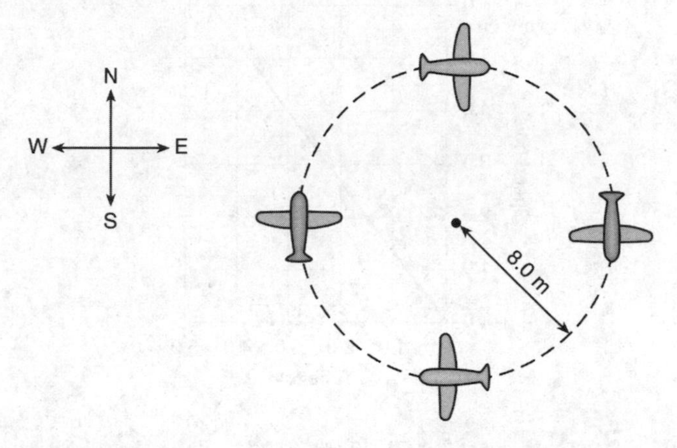

59–60 Calculate the speed of the airplane. [Show all work, including the equation and substitution with units.] [2]

61 State the direction of the velocity of the airplane at the instant the acceleration of the airplane is southward. [1]

Base your answers to questions 62 through 64 on the information and graph below and on your knowledge of physics.

The graph below represents the speed of a marble rolling down a straight incline as a function of time.

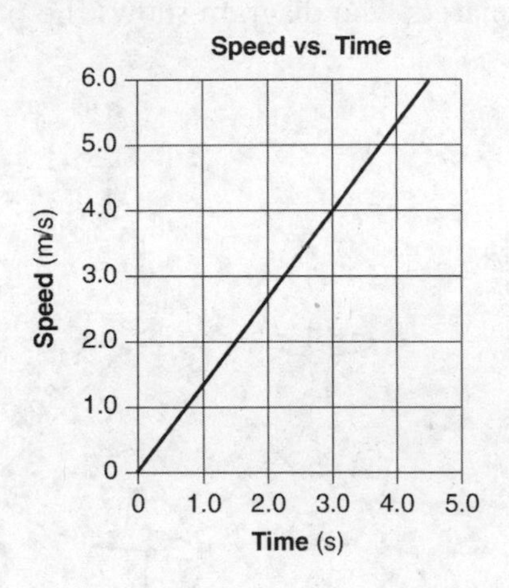

62 What quantity is represented by the slope of the graph? [1]

63–64 Calculate the distance the marble travels during the first 3.0 seconds. [Show all work, including the equation and substitution with units.] [2]

65 The graph below represents the relationship between weight and mass for objects on the surface of planet *X*.

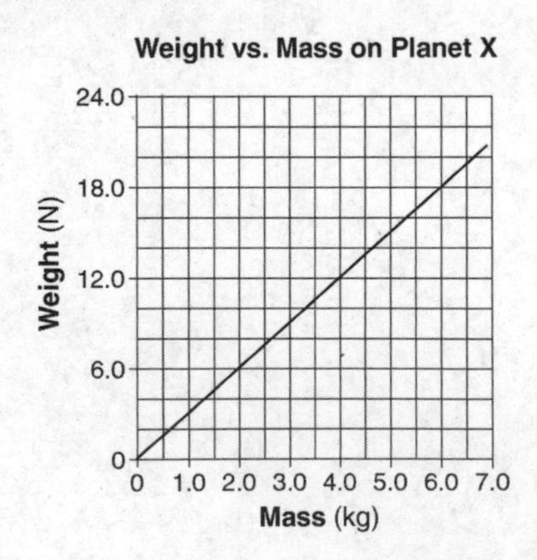

Determine the acceleration due to gravity on the surface of planet *X*. [1]

PART C

Answer all questions in this part.

Directions (66–85): Record your answers on the sheet provided in the back. Some questions may require the use of the *2006 Edition Reference Tables for Physical Setting/Physics*.

Base your answers to questions 66 through 69 on the information and vector diagram below and on your knowledge of physics.

A hiker starts at point *P* and walks 2.0 kilometers due east and then 1.4 kilometers due north. The vectors in the diagram below represent these two displacements.

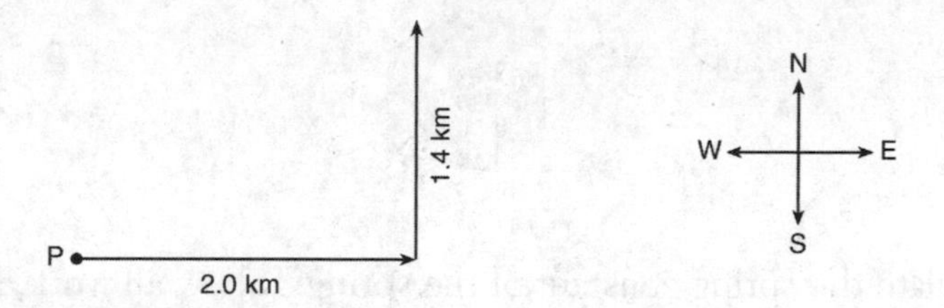

66 Using a metric ruler, determine the scale used in the vector diagram. [1]

67 On the diagram *on your answer sheet*, use a ruler to construct the vector representing the hiker's resultant displacement. [1]

68 Determine the magnitude of the hiker's resultant displacement. [1]

69 Using a protractor, determine the angle between east and the hiker's resultant displacement. [1]

Base your answers to questions 70 through 74 on the information and diagram below and on your knowledge of physics.

A jack-in-the-box is a toy in which a figure in an open box is pushed down, compressing a spring. The lid of the box is then closed. When the box is opened, the figure is pushed up by the spring. The spring in the toy is compressed 0.070 meter by using a downward force of 12.0 newtons.

70–71 Calculate the spring constant of the spring. [Show all work, including the equation and substitution with units.] [2]

72–73 Calculate the total amount of elastic potential energy stored in the spring when it is compressed. [Show all work, including the equation and substitution with units.] [2]

74 Identify *one* form of energy to which the elastic potential energy of the spring is converted when the figure is pushed up by the spring. [1]

Base your answers to questions 75 through 80 on the information below and on your knowledge of physics.

A 12-volt battery causes 0.60 ampere to flow through a circuit that contains a lamp and a resistor connected in parallel. The lamp is operating at 6.0 watts.

75 Using the circuit symbols shown on the *Reference Tables for Physical Setting/Physics*, draw a diagram of the circuit in the space provided *on your answer sheet*. [1]

76–77 Calculate the current through the lamp. [Show all work, including the equation and substitution with units.] [2]

78 Determine the current in the resistor. [1]

79–80 Calculate the resistance of the resistor. [Show all work, including the equation and substitution with units.] [2]

Base your answers to questions 81 through 85 on the information below and on your knowledge of physics.

The Great Nebula in the constellation Orion consists primarily of excited hydrogen gas. The electrons in the atoms of excited hydrogen have been raised to higher energy levels. When these atoms release energy, a frequent electron transition is from the excited $n = 3$ energy level to the $n = 2$ energy level, which gives the nebula one of its characteristic colors.

81 Determine the energy, in electronvolts, of an emitted photon when an electron transition from $n = 3$ to $n = 2$ occurs. [1]

82 Determine the energy of this emitted photon in joules. [1]

83–84 Calculate the frequency of the emitted photon. [Show all work, including the equation and substitution with units.] [2]

85 Identify the color of light associated with this photon. [1]

Answer Sheet
June 2016
Physics: The Physical Setting

PART B–2

51–52

53 ______________________ N

54–55

56–57

58 ______________________________________

____

59–60

61 ____________________

62 ____________________

63–64

65 ____________________ **m/s²**

PART C

66 1.0 cm = ______________________ **km**

67

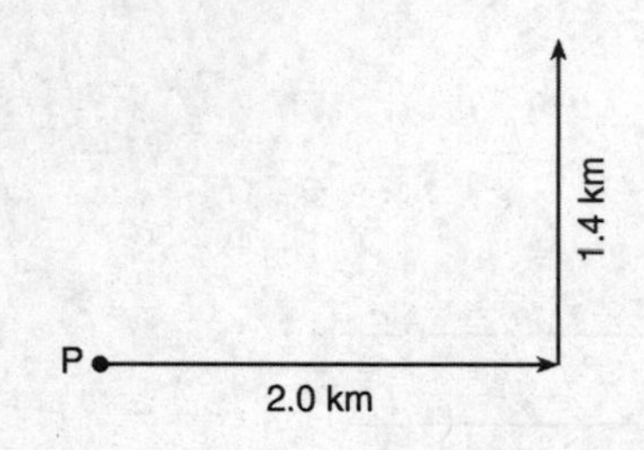

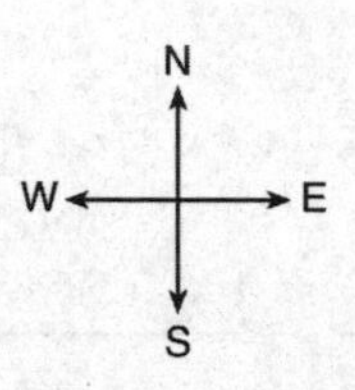

68 ______________________ **km**

69 ______________________ °

70–71

72–73

74 ______________________________

75

76–77

78 ____________________ A

79–80

81 ______________________ **eV**

82 ______________________ **J**

83–84

85 ______________________________

Answers June 2016

Physics: The Physical Setting

Answer Key

PART A

1. 4	**8.** 1	**15.** 3	**22.** 4	**29.** 3
2. 2	**9.** 2	**16.** 1	**23.** 2	**30.** 2
3. 4	**10.** 3	**17.** 3	**24.** 1	**31.** 2
4. 1	**11.** 3	**18.** 3	**25.** 2	**32.** 4
5. 3	**12.** 2	**19.** 4	**26.** 1	**33.** 3
6. 2	**13.** 2	**20.** 1	**27.** 3	**34.** 3
7. 2	**14.** 2	**21.** 3	**28.** 1	**35.** 4

PART B–1

36. 2	**39.** 3	**42.** 2	**45.** 4	**48.** 3
37. 4	**40.** 2	**43.** 1	**46.** 1	**49.** 3
38. 4	**41.** 3	**44.** 1	**47.** 2	**50.** 3

PART B–2 and **PART C**. *See* **Answers Explained**.

Answers Explained

PART A

1 Weight is a vector quantity, it has magnitude and direction dependent on the direction of acceleration due to gravity.

WRONG CHOICES EXPLAINED:
(1) power, (2) kinetic energy, and (3) speed are all scalar quantities with only magnitude and no direction.

2 The mass of the astronaut on the surface of the Moon is the same as the mass of the astronaut on the surface of the Earth, 65.0 kg. Mass is constant and does not change. Weight is dependent on the force of gravity.

3 If the sum of all forces on an object is equal to zero (i.e., there is no net force), then according to Newton's First Law of Motion, the object is either at rest or moving at constant speed, no acceleration.

WRONG CHOICES EXPLAINED:
Choice (1) is incorrect because it states that the block *must* be at rest. Had it said that the block *may* be at rest, it would have been a correct choice. Choices (2) and (3) are incorrect because, in order to be accelerating or decelerating, there must be a net force (i.e., the sum of all forces acting on the object are not zero).

4 Mass is the quantitative measure of inertia; therefore, increasing mass increases inertia.

WRONG CHOICES EXPLAINED:
(2) Increasing net force would increase acceleration
(3) Increasing time that a net force is applied results in a greater impulse
(4) Increasing the speed of the object would increase its kinetic energy

5 According to Newton's Third Law of Motion, if object A exerts a force on object B, then object B exerts an equal but opposite force on object A.

6 Sound waves are longitudinal mechanical waves.

7 Refer to the *Mechanics* section of *Reference Table K*:

$$F_e = \frac{kq_1q_2}{r^2}$$

Electrostatic force is inversely proportional to the square of the distance between the centers of the point charges. If the electrostatic force is equal to 8.0×10^{-5} and if $r = d$, then

$$8.0 \times 10^{-5} = \frac{kq_1q_2}{d^2}$$

If $r = 2d$, then

$$F_e = \frac{kq_1q_2}{(2d)^2}$$

$$F_e = \frac{kq_1q_2}{4d^2}$$

$$F_e = \left(\frac{1}{4}\right)\left(\frac{kq_1q_2}{d^2}\right)$$

$$F_e = \frac{1}{4}8.0 \times 10^{-5}$$

$$F_e = 2.0 \times 10^{-5}$$

8 When dealing with a word problem requiring a mathematical solution, it is most helpful to translate the words into what quantities/variables are explicitly given and what constants or other variables can be deduced from the situation or words. We also write down the variable that needs to be solved for or found. Then it is much easier to determine which formula is required (refer to the appropriate section of *Reference Table K*), and then use it to solve the problem. It is also easier, and there is less chance of making an error, if we first solve the equation algebraically for the unknown and then substitute the given variables with units into the equation in order to calculate the answer.

Given:	Find:	Solution:
$p_{blue\ i} = 2.0$ kg • m/s west	$p_{total\ f} = ?$	$p_{total\ i\ +} = p_{total\ f}$
$p_{red\ i} = 3.0$ kg • m/s east		$p_{blue\ i\ +}\ p_{red\ i} = p_{total\ f}$
		2.0 kg • m/s west + 3.0 kg • m/s east = $p_{total\ f}$ 1.0 kg • m/s east = $p_{total\ f}$

9 The direction of a frictional force is opposite to the direction of motion. At point *P*, the ball is moving diagonally upward to the right; therefore, friction acts in a downward direction to the left.

10 A magnetic field can be generated by a moving charged particle like a proton.

WRONG CHOICES EXPLAINED:

Choices (1), (2), and (4) are incorrect because all these particles have no charge.

11 Field lines represent the path that a very small positive charge takes while in the field. The arrows on the field lines indicate the direction. The field lines in the diagram point toward *A*; therefore, *A* must be negative. The lines point toward *B* as well; therefore, *B* must be negative as well.

12 Write out the problem. Refer to the *Mechanics* section of *Reference Table K*.

Given:	Find:	Solution:
$F = 20$ N	$W = ?$	$W = Fd$
$m_{box} = 10$ kg		$W = (20\text{ N})(5\text{ m})$
$d = 5$ m		$W = 100\text{ Nm} = 100\text{ J}$

13 Work done on an object increases its kinetic energy. $W = \Delta KE$. Since 15 J of work is done, the kinetic energy increases by 15 J.

14 Refer to the *Mechanics* section of *Reference Table K*:

$$PE = \Delta mgh$$

Potential energy is proportional to mass. If the mass of A is equal to m, then

$$\Delta PE_A = \Delta mg\Delta h$$

If mass $B = 2m$, then

$$\Delta PE_B = (2m)g\Delta h$$
$$\Delta PE_B = (2mg\Delta h)$$
$$\Delta PE_B = 2\Delta PE_A$$

15 Write out the problem. Refer to the *Mechanics* section of *Reference Table K*:

Given:	Find:	Solution:
$m = 55$ kg	$KE = ?$	$KE = \frac{1}{2}mv^2$
$v = 9.0$ m/s		$KE = \frac{1}{2}(55\ \text{kg})(9.0\ \text{m/s})^2$
		$KE = 2.2 \times 10^3 \frac{kgm^2}{s^2}$
		$KE = 2.2 \times 10^3\ \text{J}$

16 The speed of an electromagnetic wave is less in a material medium than in a vacuum and is dependent on the nature of the medium and the frequency of the light.

17 Write out the problem. Refer to the *Mechanics* section of *Reference Table K*:

Given:	Find:	Solution:
$W = 480\ \text{J}$	$P = ?$	$P = \frac{W}{t}$
$t = 5.0\ \text{s}$		$P = \frac{480\ \text{J}}{5.0\ \text{s}}$
$m = 12\ \text{kg}$		$P = 96\frac{\text{J}}{\text{s}} = 96\ \text{W}$

18 Write out the problem. Refer to the *Mechanics* section of *Reference Table K*:

Given:	Find:	Solution:
$P = 5.8 \times 10^4\ \text{W}$	$\bar{v} = ?$	$P = F\bar{v}$
$Fw = 2.1 \times 10^4\ \text{N}$		$\bar{v} = \frac{P}{F}$

$$\bar{v} = \frac{5.8 \times 10^4\ \text{W}}{2.1 \times 10^4\ \text{N}}$$

$$\bar{v} = 2.8\frac{\text{W}}{\text{N}} = 2.8\ \frac{\text{J/s}}{\text{N}} = 2.8\frac{\text{Nm/s}}{\text{N}} = 2.8\ \text{m/s}$$

19 This scenario is an example of the Doppler effect. The Doppler effect is the apparent change in frequency that results when a wave source and an observer are in relative motion with respect to each other. In this case, the vehicle is moving toward the stationary police officer. The apparent pitch or frequency is decreased as the sound source moves away from and increased as the sound source moves toward a stationary observer.

20 Waves transfer energy without the transfer of mass.

21 The period is the time required for periodic motion to repeat itself. Choice (3) is correct.

WRONG CHOICES EXPLAINED:

(1) Amplitude is the maximum displacement of any particle in a medium relative to its rest position.

(2) Frequency is the number of repetitions produced per unit time in periodic phenomena.

(4) Wavelength is the length of one complete wave cycle.

22 Moving electric charges produce magnetic fields. The current in the wire generates a magnetic field. The magnetic field exerts a force on the compass needle, which is itself a magnet.

23 Write out the problem. Refer to the *Modern Physics* section of *Reference Table K* and to *Reference Table E*:

Given:

$\lambda_{air} = 589 \text{ nm}$

$n_{air} = 1.00$

$n_{Lucite} = 1.50$

Find:

$\lambda_{Lucite} = ?$

Solution:

$$\frac{n_2}{n_1} = \frac{\lambda_1}{\lambda_2}$$

$$\frac{n_{Lucite}}{n_{air}} = \frac{\lambda_{air}}{\lambda_{Lucite}}$$

$$\lambda_{Lucite} = \frac{(\lambda_{air})(n_{air})}{n_{Lucite}}$$

$$\lambda_{Lucite} = \frac{(589 \text{ nm})(1.00)}{1.50}$$

$$\lambda_{Lucite} = 393 \text{ nm}$$

WRONG CHOICES EXPLAINED:

If you selected choice (4), you substituted into the equation incorrectly and/or multiplied by 1.50 instead of dividing by 1.50

24 Amplitude is a measure of the energy transmitted by a wave and in sound waves this corresponds to loudness. Choice (1) is correct.

WRONG CHOICES EXPLAINED:

(2) Pitch in sound waves corresponds to frequency.
(3) Velocity of a sound wave is dependent on the properties of the medium.
(4) Wavelength is inversely proportional to the frequency of a wave.

25 Refer to the *Electricity* section in *Reference Table K*:

$$E = \frac{F_e}{q}$$

Protons and electrons have equal but opposite charges; therefore, the magnitude of the force on a electron will be the same as the magnitude on the proton, but it will be in the opposite direction.

26 The effect produced when two or more sound waves pass through the same point simultaneously is called interference. Choice (1) is correct.

WRONG CHOICES EXPLAINED:

(2) Diffraction is the bending of a wave around a barrier.

(3) Refraction is the change in direction of a wave when it passes obliquely from one medium into another in which it moves at a different speed.

(4) Resonance is the spontaneous vibration of an object at a frequency equal to that of the wave that initiates that resonant vibration.

27 Refer to the *Electromagnetic Spectrum* table in *Reference Table D*: Microwave photons have longer wavelengths than gamma ray photons. Refer to the *Modern Physics* section in *Reference Table K*:

$$E_{photon} = \frac{hc}{\lambda}$$

Therefore, microwave photons have less energy than gamma ray photons.

28 Refer to the *Particles of the Standard Model* chart in *Reference Table H*. The neutrino is a type of lepton.

29 Refer to the *Particles of the Standard Model* chart in the *Reference Table H*: A combination of an up, down, and strange quarks will result in a neutral baryon. Choice (3) is correct.

WRONG CHOICES EXPLAINED:

(1) A combination of a charm, top, and strange quarks will result in a particle with a charge of +1.

(2) A combination of a down, strange, and bottom quarks will result in a particle with a charge of −1.

(4) A combination of an up, charm, and top quark will result in a particle with a charge of +2.

30 Write out the problem. Refer to the *Modern Physics* section of *Reference Table K*.

Given:	Find:	Solution:
$m = 2.0 \times 10^{-16}$ kg	$E = ?$	$E = mc^2$
$c = 3.0 \times 10^8$ m/s		$E = (2.0 \times 10^{-16}\text{ kg})(3.0 \times 10^8\text{ m/s})^2$
		$E = 1.8 \times 10^1$ J

31 Write out the problem. Refer to the *Modern Physics* section of *Reference Table K*:

Given:	Find:	Solution:
$v_i = 28$ m/s	$v_f = ?$	$v_f = v_i + at$
$a = -9.81$ m/s^2		$v_f = \left(28\frac{\text{m}}{\text{s}}\right) + \left(-9.81\frac{\text{m}}{\text{s}^2}\right)(2.2\text{ s})$
$t = 2.2$ s		$v_f = \left(28\frac{\text{m}}{\text{s}}\right) + \left(-21.582\frac{\text{m}}{\text{s}}\right)$
		$v_f = 6.4\frac{\text{m}}{\text{s}}$

32 Refer to the *List of Physical Constants* table (*Reference Table A*): One elementary charge is equivalent to 1.60×10^{-19} C. Write out the problem. Refer to the *Electricity* section of the *Reference Table K*:

Given:	Find:	Solution:
$q = 4.8 \times 10^{-19}$ C	$W = ?$	$V = \frac{W}{q}$
$V = 4.50$ V		$W = Vq$
		$W = (4.50\ \text{V})(4.8 \times 10^{-19}\ \text{C})$
		$W = 2.16 \times 10^{-18}\ \text{J}$

Refer to the *List of Physical Constants* table (*Reference Table A*): 1 electronvolt (eV) is equivalent to 1.60×10^{-19} J.

$$W = \left(2.16 \times 10^{-18}\ \text{J}\right)\left(\frac{1\ \text{eV}}{1.60 \times 10^{-19}\ \text{J}}\right) = 13.5\ \text{eV}$$

WRONG CHOICE EXPLAINED:

If you selected choice (2), you forgot to convert joules to electronvolts.

33 In a series circuit, the total equivalent resistance of the circuit is equal to the sum of all the individual resistances. By contrast, in a parallel circuit, the equivalent resistance is less than any single resistance in the circuit.

We can eliminate choices (2) and (4) as they contain the same resistances as choices (1) and (3), respectively, but in parallel versus series circuits, and, thus, their total equivalent resistances will be less.

Choice (3) is correct as this circuit has an equivalent resistance of 10 Ω, whereas choice (1) only has an equivalent resistance of 8 Ω.

34 In a transverse wave, the particles of the medium vibrate or exhibit simple harmonic motion (SHM) about a rest position perpendicular to the direction of motion of the wave. This eliminates choices (1) and (2). Because the wave shown is moving toward the right of the page, at the instant shown, point *X* is moving upward along the crest of the wave. Choice (3) is correct.

35 By convention, magnetic field lines point away from north and towards the south. Since all of the diagrams shown are of 2 north poles of magnets, only the diagram in choice 4 shows the field lines pointing away from both north poles. Choice (4) is correct.

PART B–1

36 This is an order of magnitude question. Ten to the –2 power is equivalent to one hundredth. One hundredth of a meter is equal to 1 cm, which is a reasonable approximation for the width of a very small student's finger. One centimeter is not a reasonable approximation of the diameter of an atom, the length of a football field, or the height of a schoolteacher.

37 The graph in choice (4) illustrates a direct proportional relationship between speed and time. Choice (4) is the correct answer as an object in free fall experiences a constant acceleration due to gravity and speed increases 9.81 m/s every second.

WRONG CHOICES EXPLAINED:

(1) This graph illustrates constant speed—no change in speed over time.

(2) This graph illustrates an inverse proportional relationship between speed and time.

(3) This graph illustrates an exponential relationship between speed and time (i.e., not constant acceleration, but increasing acceleration).

38 Write out the problem. Refer to the *Electricity* section of the *Reference Table K*:

Given:	Find:	Solution:
$R = 9.6\ \Omega$	$W = ?$	$W = \frac{V^2t}{R}$
$V = 120\ \text{V}$		$W = \frac{(120\ V)^2\,(150\ \text{s})}{9.6\ \Omega}$
$t = 2.5\ \text{min} = 150\ \text{s}$		$W = 2.3 \times 10^5\ \text{J}$

WRONG CHOICE EXPLAINED:

If you selected choice (2), you forgot to convert 2.5 min into 150 seconds before substituting into the equation.

39 Write out the problem. Refer to the *Mechanics* section of the *Reference Table K*:

Given:	Find:	Solution:
$F_W = 46$ N	$F_\perp = ?$	$F_\perp = F_W \cos\theta$
$\theta = 25°$ to the horizontal		$F_\perp = (46\text{ N})\cos(25°)$
		$F_\perp = 42$ N

WRONG CHOICE EXPLAINED:

If you selected choice (1), you used sin instead of cos or you found the cos of the complementary angle 65°. This answer is equivalent to the component of the weight parallel to the surface of the inclined plane.

40 This graph shows an exponential proportional relationship between position and time; there is a positive velocity with a negative acceleration. Position is increasing over time, but the rate at which it is increasing decreases as time goes by.

WRONG CHOICES EXPLAINED:

(1) This graph shows a directly proportional relationship between position and time. Constant positive velocity (i.e., no acceleration).

(3) This graph shows a directly proportional relationship between position and time. Constant negative velocity (i.e., no acceleration).

(4) This graph shows an exponential proportional relationship between position and time, but there is a negative velocity with a negative acceleration.

41 Create a Data Table to compare position of Car *A* to Car *B* over time.

Time (s)	Position of Car *A* (m)	Position of Car *B* (m)
0	0	200
10	200	350
20	400	500
30	600	650
40	800	800

42 Write out the problem. Refer to the *Electricity* section of *Reference Table K*:

Given:	Find:	Solution:
$R = 2700\ \Omega$	$q = ?$	$I = \frac{q}{t}$
$I = 2.4\ \text{mA} = 2.4 \times 10^{-3}\ \text{A}$		$q = It$
$t = 15\ \text{s}$		$q = (2.4 \times 10^{-3}\ \text{A})(15\ \text{s})$
		$q = 3.6 \times 10^{-2}\ \text{C}$

WRONG CHOICE EXPLAINED:

If you selected choice (4), you did not convert milliamperes to amperes before substituting into the equation.

43 Write out the problem. Refer to the *Mechanics* section of *Reference Table K*:

Given:	Find:	Solution:
$m = 1000\ \text{kg}$	$v_f = ?$	$J = \Delta p$
$v_i = 20.0\ \text{m/s east}$		$J = m\Delta v$
$J = 2000\ \text{N·s west}$		$J = m(v_f - v_i)$

$$J = mv_f - mv_i$$

$$mv_f = J + mv_i$$

$$v_f = \frac{J + mv_i}{m}$$

$$v_f = \frac{(-2000Ns) + (1000\ \text{kg})\left(20.0\frac{\text{m}}{\text{s}}\right)}{1000\ \text{kg}}$$

$$v_f = +18.0\frac{\text{m}}{\text{s}} = 18\ \text{m/s east}$$

WRONG CHOICE EXPLAINED:

(4) The impulse is applied in the opposite direction as the initial velocity. You did not account for this by making either the velocity or the impulse negative and as a result obtained this value.

44 Refer to the *Electricity* section of *Reference Table K*: At constant temperature

$$R = \frac{\rho L}{A} \qquad R = \frac{V}{I}$$

Resistance is constant and potential difference is directly proportional to the current. The graph in choice (1) correctly represents a direct proportional relationship.

WRONG CHOICES EXPLAINED:

(2) This graph represents an exponentially proportional relationship.

(3) This graph shows that decreased potential difference results in increased current.

(4) This graph represents an inversely proportional relationship.

45 Refer to the *Electricity* section of *Reference Table K*: If the length of the first wire equals L, and its cross-sectional area equals A at 20°C, then

$$R = \frac{\rho L}{A}$$

If the length of the second wire is $2L$, and the cross-sectional area is ½A, then

$$R_2 = \frac{\rho(2L)}{(0.5A)}$$

$$R_2 = \frac{2(\rho L)}{0.5(A)}$$

$$R_2 = \frac{4(\rho L)}{(A)} = 4R$$

46 For most metallic conductors, as the temperature increases, resistance increases.

Refer to the *Electricity* section of *Reference Table K*:

$$R = \frac{V}{I} \qquad I = \frac{V}{R}$$

Current is inversely proportional to resistance. The greater the resistance, the smaller the current.

47 Diffraction is the bending of a wave around an obstacle. The longer the wavelength of the wave, the larger the amount of diffraction. The greatest bending or diffraction occurs when the width of the gap or opening is approximately the same size as the wavelength of the wave. The diagram in choice (2) accurately represents these criteria.

48 Refer to the *List of Physical Constants* table in *Reference Table A*. Write out the problem. Refer to the *Waves and Optics* section of *Reference Table K*:

Given:	Find:	Solution:
$\lambda = 0.320$ m	$f = ?$	$v = f\lambda$
$v = 3.31 \times 10^2$ m/s		$f = \frac{v}{\lambda}$
		$f = \frac{3.31 \times 10^2 \text{ m/s}}{0.320\text{m}}$
		$f = 1.03 \times 10^3 \text{ Hz}$

49 The points on a standing wave that appear to stand still are called nodes and are the result of total destructive interference. The areas of maximum constructive interference, alternating crest/troughs are the antinodes: in this diagram, points *B* and *D*.

50 A voltmeter measures the potential difference across 2 points in a circuit. If we want to measure the potential difference across resistor R_2, then the voltmeter needs to be connected to both ends of resistor R_2. This is correctly represented in choice (3).

PART B–2

51–52 Write out the problem. Refer to the *Mechanics* section of *Reference Table K*:

Given:

$m = 2.5$ kg

$a = 2.0$ m/s^2

Find:

$F_{net} = ?$

Solution:

$$a = \frac{F_{net}}{m}$$

$$F_{net} = ma$$

$$F_{net} = (2.5 \text{ kg})(2.0 \text{ m/s}^2)$$

51 One credit is awarded for the correct equation and substitution of values with units.

52 One credit is awarded for the correct answer with units or for an answer, with units, that is consistent with the answer to question 51.

Note: Students will not be penalized more than 1 credit for errors in units in questions 51 and 52.

53 The net force is the sum of all the applied forces. The force of friction is always in the opposite direction of motion so we subtract the friction force from the applied force to get the net force.

$$F_{net} = F_{applied} - F_f$$

$$F_f = F_{applied} - F_{net}$$

$$F_f = 15 \text{ N} - 5.0 \text{ N}$$

$$F_f = 10 \text{ N}$$

One credit is awarded for the correct answer or for an answer that is consistent with the difference between 15 N and the student's answer to question 52.

54–55 Write out the problem. Refer to the *Waves and Optics* section and the *Absolute Indices of Refraction* table (*Reference Table E*):

Given:

$\theta_{mineral} = 27°$

$\theta_{water} = 41°$

$n_{water} = 1.33$

Find:

$n_{mineral} = ?$

Solution:

$$n_1 \sin\theta_1 = n_2 \sin\theta_2$$

$$n_{mineral} \sin\theta_{mineral} = n_{water} \sin\theta_{water}$$

$$n_{mineral} = \frac{n_{water} \sin\theta_{water}}{\sin\theta_{mineral}}$$

$$n_{mineral} = \frac{(1.33)(\sin\ 41°)}{(\sin\ 27°)}$$

$$n_{mineral} = 1.9$$

54 One credit is awarded for the correct equation and substitution of values with units.

55 One credit is awarded for the correct answer or for an answer that is consistent with the answer to question 54.
Note: Students will not be penalized more than 1 credit for errors in units in questions 54 and 55.

56–57 Write out the problem. Refer to the *Mechanics* section of *Reference Table K*:

Given:

$v_{ix} = 2.50$ m/s

$t = 0.391$ s

$v_{iy} = 0$ m/s

$a_y = 9.81$ m/s^2

Find:

$d_{table} = ?$

Solution:

$$d_y = v_{iy}t + \frac{1}{2}a_y t^2$$

$$d_y = \left(0\frac{\text{m}}{\text{s}}\right)(0.391\text{ s}) + \frac{1}{2}\left(9.81\frac{\text{m}}{\text{s}^2}\right)(0.391\text{ s})^2$$

$$d = 0\text{ m} + 0.749881305$$

$$d_y = 0.750\text{ m}$$

56 One credit is awarded for the correct equation and substitution of values with units.

57 One credit is awarded for the correct answer with units or for an answer, with units, that is consistent with the answer to question 56.

Note: Students will not be penalized more than 1 credit for errors in units in questions 56 and 57.

58 Because we are dealing with a level table, the initial speed of the ball given in both cases is the horizontal speed. The initial vertical speed of the ball in both cases is 0 m/s. Since the table does not change height and the ball accelerates due to gravity the same way in both trials, the time it takes the ball to hit the ground is the same in both trials. What would be different is how far the ball would travel in the horizontal direction—the distance away from the table or the range.

One credit is awarded for stating that the measured times are the same.

59–60 Write out the problem. Refer to the *Mechanics* section of *Reference Table K*:

Given:	Find:	Solution:
$r = 8.0\text{ m}$	$v = ?$	$a_c = \frac{v^2}{r}$
$a_c = 25\text{ m/s}^2$		$v^2 = a_c r$
		$v = \sqrt{a_c r}$
		$v = \sqrt{\left(25\frac{\text{m}}{\text{s}^2}\right)(8.0\text{ m})}$
		$v = \sqrt{200\frac{\text{m}^2}{\text{s}^2}}$
		$v = 14\text{ m/s}$

59 One credit is awarded for the correct equation and substitution of values with units.

60 One credit is awarded for the correct answer with units or for an answer, with units, that is consistent with the answer to question 59.

Note: Students will not be penalized more than 1 credit for errors in units in questions 58 and 59.

61 The centripetal acceleration is directed exactly southward when the plane is at the top of the circle or the most northward point of the circle. The direction of velocity for an object traveling in circular motion is to the circle. In this instance, directly to the right or east.

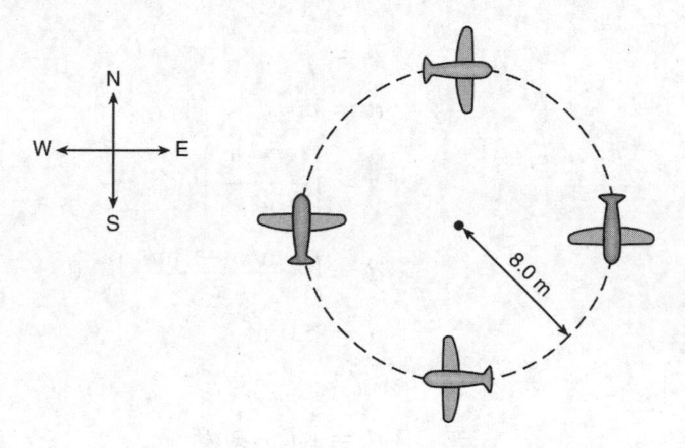

One credit is awarded for indicating/stating that the direction of the airplane's velocity is east or to the right.

62 The slope of a line is defined as the change in y over the change in x. In this graph, y is speed and x is time; thus, the slope on a speed versus time graph is equal to the change in speed over the change in time, which is equal to the acceleration.

$$slope = \frac{\Delta y}{\Delta x} = \frac{\Delta v}{\Delta t} = a$$

One credit is awarded for an answer of acceleration *or* rate of change of velocity (speed).

63–64 Write out the problem. Refer to the *Mechanics* section of *Reference Table K*:

Given:	Find:	Solution:
$t = 3.0\ \text{s}$	$d = ?$	$\bar{v} = \frac{d}{t}$
$v_i = 0\ \text{m/s}$		$d = \bar{v}t$
$v_f = 4\ \text{m/s}$		$d = \left(\frac{v_f + v_i}{2}\right)t$
		$d = \left(\frac{0\ \text{m/s} + 4\ \text{m/s}}{2}\right)(3.0\ \text{s})$
		$d = 6.0\ \text{m}$

or

$$d = v_i t + \frac{1}{2}at^2$$

$$a = \frac{v}{t}$$

$$d = v_i t + \frac{1}{2}\left(\frac{v}{t}\right)t^2$$

$$a = \left(0\frac{\text{m}}{\text{s}}\right)(3.0\ \text{s}) + \frac{1}{2}\left(4.0\ \frac{\text{m}}{\text{s}} - 0\frac{\text{m}}{\text{s}}\right)(3.0\ \text{s})$$

$$d = 6.0\ \text{m}$$

or

The area under the curve in a speed-versus-time graph is equal to the displacement.

Refer to the *Geometry and Trigonometry* section of *Reference Table K*:

$$A = ½bh$$

$$A = ½(3.0\ \text{s})(4.0\ \text{m/s})$$

$$A = 6.0\ \text{m}$$

63 One credit is awarded for a correct equation and substitution of values with units.

64 One credit is awarded for the correct answer with units or for an answer, with units, that is consistent with the answer to question 63.

Note: Students will not be penalized more than 1 credit for errors in units in questions 63 and 64.

65 Refer to the *Mechanics* section of *Reference Table K*:

$$g = \frac{F_g}{m}$$

The slope of a line is defined as the change in y over the change in x. In this graph, y is weight or the force due to gravity and x is mass; thus, the slope on a weight-versus-mass graph is equal to the change in weight over the change in mass, which is equal to the gravitational acceleration.

$$slope = \frac{\Delta y}{\Delta x} = \frac{\Delta F_g}{m} = g$$

$$g = \frac{\Delta F_g}{\Delta m} = \frac{(6.0\ \text{N} - 0\ \text{N})}{(2.0\ \text{kg} - 0\ \text{kg})} = 3.0\frac{\text{N}}{\text{kg}} = 3.0\frac{\text{m}}{\text{s}^2}$$

One credit is awarded for an answer of $3.0\ \text{m/s}^2 \pm 0.25\ \text{m/s}^2$.

PART C

66 If we measure the 2.0-km line using a metric ruler, we find the line to measure approximately 5.0 cm long. 2.0 km/5.0 cm equals 0.4 km/cm.

One credit is awarded for an answer of 0.40 km ± 0.02 km.

67

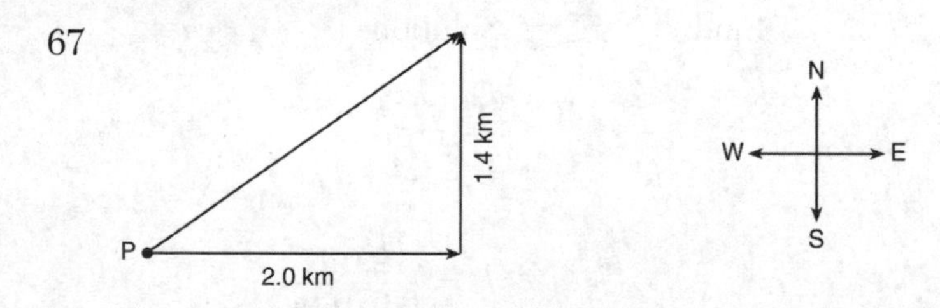

One credit is awarded for constructing an arrow starting at point P and ending with an arrowhead at the top of the 1.4-km vector or for a vector 6.1 cm ± 0.2 cm long at an angle of 35° north of the 2.0-km vector.

68 If we measure the vector drawn with a metric ruler, we measure approximately 6.0 cm, and according to our scale 1.0 cm = 0.4 km, this translates (6.0 cm × 0.4k m/cm) to 2.4 km.

Alternatively, we use the technique of vector addition. We recognize that the vectors form a right triangle and the resultant vector would be the hypotenuse of a right triangle. We can use the Pythagorean theorem to solve algebraically for the resultant vector.

Refer to the *Geometry and Trigonometry* section of *Reference Table K*:

$$a^2 + b^2 = c^2$$
$$c = \sqrt{a^2 + b^2}$$
$$c = \sqrt{(2.0 \text{ km})^2 + (1.4 \text{ km})^2}$$
$$c = 2.4 \text{ km}$$

One credit is awarded for an answer of 2.4 ± 0.2 km or for an answer that is the product of the length of the resultant (in centimeters) drawn in question 67 and the response to question 66.

69 Assuming the vector was drawn correctly in question 67, a protractor properly placed at point *P* will yield an answer of 35°.

One credit is awarded for an answer of 35° ± 2° or for an answer that is the angle between the 2.0-km vector and the response to question 67.

70–71 Write out the problem. Refer to the *Mechanics* section of *Reference Table K*:

Given:	Find:	Solution:
$x = 0.070 \text{ m}$	$k = ?$	$F_s = kx$
$F_s = 12.0 \text{ N}$		$k = \frac{F_s}{x}$
		$k = \frac{12.0 \text{ N}}{0.070 \text{ m}}$
		$k = 170 \text{ N/m}$

70 One credit is awarded for a correct equation and substitution of values with units.

71 One credit is awarded for the correct answer with units or for an answer, with units, that is consistent with the answer to question 70.

Note: Students will not be penalized more than 1 credit for errors in units in questions 70 and 71.

72–73 Write out the problem. Refer to the *Mechanics* section of *Reference Table K*:

Given:	Find:	Solution:
$k = 170$ N/m	$PE_s = ?$	$PE_s = \frac{1}{2}kx^2$
$x = 0.070$ m		$PE_s = \frac{1}{2}\left(170\frac{\text{N}}{\text{m}}\right)(0.070\text{ m})^2$
		$PE_s = 0.42\text{ Nm} = 0.42\text{ J}$

72 One credit is awarded for a correct equation and substitution of values with units.

73 One credit is awarded for the correct answer with units or for an answer, with units, that is consistent with the answer to question 72.

Note: Students will not be penalized more than 1 credit for errors in units in questions 72 and 73.

74 One credit is awarded for an acceptable response. Acceptable responses include, but are not limited to,

- kinetic energy (energy that an object possesses because of its motion)
- sound (as the lid opens and it pops out there would be sounds)
- internal energy (thermal energy—the total kinetic and potential energy associated with the atoms and molecules of an object)
- gravitational potential energy (the energy that an object acquires as a result of the work done in moving the object against a gravitational force)

75 Refer to the *Circuit Symbols* section in *Reference Table I*.

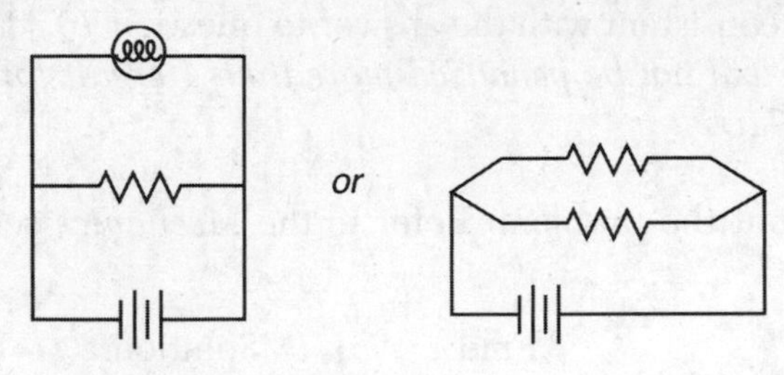

One credit is awarded for a correct circuit diagram.

Note: Credit is allowed if the diagram uses a cell symbol instead of a battery symbol.

76–77 Write out the problem. Refer to the *Electricity* section of *Reference Table K*:

Given:	Find:	Solution:
$V = 12.0\text{ V}$	$I_{lamp} = ?$	$P_{lamp} = VI_{lamp}$
$I_{total} = 0.60\text{ A}$		$I = \frac{P}{V}$
$P_{lamp} = 6.0\text{ W}$		$I = \frac{6.0\text{ W}}{12.0\text{ V}}$
		$I = 0.50\text{ A}$

76 One credit is awarded for a correct equation and substitution of values with units.

77 One credit is awarded for the correct answer with units or for an answer, with units, that is consistent with the answer to question 76.

Note: Students will not be penalized more than 1 credit for errors in units in questions 76 and 77.

78 This is a parallel circuit. The total current is equal to the sum of the currents in the individual pathways.

Given:	Find:	Solution:
$I_{total} = 0.60\text{ A}$	$I_{resistor} = ?$	$I_{total} = I_{lamp} + I_{resistor}$
$I_{lamp} = 0.50\text{ A}$		$I_{resistor} = I_{total} - I_{lamp}$
		$I_{resistor} = 0.60\text{ A} - 0.50\text{ A}$
		$I_{resistor} = 0.10\text{ A}$

One credit is awarded for an answer of 0.10 A or for an answer that is the difference between the response to question 77 and 0.60 A or for an answer that is consistent with the response to question 75.

79–80 Write out the problem. Refer to the *Electricity* section of *Reference Table K*:

Given:	Find:	Solution:
$V = 12.0\text{ V}$	$R_{resistor} = ?$	$V = IR$
$I_{resistor} = 0.10\text{ A}$		$R = \frac{V}{I}$
		$R = \frac{12.0\text{ V}}{0.10\text{ A}}$
		$R = 120\ \Omega$

79 One credit is awarded for a correct equation and substitution of values with units or for an answer that is consistent with the response given in question 78.

80 One credit is awarded for the correct answer with units or for an answer, with units, that is consistent with the answer to question 79.

Note: Students will not be penalized more than 1 credit for errors in units in questions 79 and 80.

81 Refer to the *Energy Level Diagram for Hydrogen* in *Reference Table F*. Find the difference in energy between level 3 and level 2.

$$\left(-1.51 - (-3.40) = 1.89\text{ eV}\right)$$

One credit is awarded for answer of 1.89 eV.

82 Write out the problem. Refer to *Reference Table A*:

Given:	Find:	Solution:
$E_{photon} = 1.89$ eV	E_{photon} in Joules	$1.89 \text{ eV}(1.6 \times 10^{-19} \text{ J/eV}) =$
$1 \text{ eV} = 1.6 \times 10^{-19}$ J		3.02×10^{-19} J

One credit is awarded for an answer of 3.02×10^{-19} J or for an answer that is the product of the response to question 81 and 1.6×10^{-19} J/eV

83–84 Write out the problem. Refer to the *Modern Physics* section of *Reference Table K*.
Refer to *Reference Table A*:

Given:	Find:	Solution:
$E_{photon} = 3.02 \times 10^{-19}$ J	$f = ?$	$E_{photon} = hf$
$h = 6.63 \times 10^{-34}$ Js		

$$f = \frac{E_{photon}}{h}$$

$$f = \frac{3.02 \times 10^{-19} \text{ J}}{6.63 \times 10^{-34} \text{ Js}}$$

$$f = 4.56 \times 10^{14} \text{ Hz}$$

83 One credit is awarded for the correct equation and substitution of values with units or for an answer with units that is consistent with the response to question 82.

84 One credit is awarded for a correct answer with units or for an answer, with units, that is consistent with the answer to question 83.
Note: Students will not be penalized more than 1 credit for errors in units in questions 83 and 84.

85 Refer to the *Electromagnetic Spectrum* diagram in *Reference Table D*. A $f = 4.56 \times 10^{14}$ Hz is within the visible light spectrum, specifically visible red light.
One credit is awarded for an answer of red light or for an answer that is consistent with the answer to question 84.

Topic	Question Numbers (total)	Wrong Answers (x)	Grade
Math Skills	2, 5, 7, 8, 12, 13, 14, 15, 17, 18, 23, 30, 31, 32, 33, 36, 37, 38, 39, 40, 41, 42, 43, 44, 45, 46, 48, 51, 52, 53, 54, 55, 56, 57, 59, 60, 62, 63, 64, 65, 66, 68, 70, 71, 72, 73, 76, 77, 78, 79, 80, 81, 82, 83, 84: (55)		$\frac{100(55-x)}{55} = \%$
Mechanics	1–5, 8, 9, 25, 31, 37, 39, 51–53, 56–64, 65, 67–69 70, 71: (29)		$\frac{100(29-x)}{29} = \%$
Energy	12–15, 17, 18, 72, 73, 74: (9)		$\frac{100(9-x)}{9} = \%$
Electricity/ Magnetism	7, 10, 11, 22, 32, 33, 35, 36, 38, 42, 44, 45, 46, 50, 75, 76, 77, 78, 80: (19)		$\frac{100(19-x)}{19} = \%$
Waves	6, 16, 19, 20, 21, 24, 23, 26, 34, 47, 48, 49: (12)		$\frac{100(12-x)}{12} = \%$
Modern Physics	27, 28, 29, 30, 81, 82, 83, 84, 85: (9)		$\frac{100(9-x)}{9} = \%$

Examination June 2017

Physics: The Physical Setting

PART A

Answer all questions in this part.

Directions (1–35): For *each* statement or question, select the *number* of the word or expression that, of those given, best completes the statement or answers the question. Some questions may require the use of the *2006 Edition Reference Tables for Physical Setting/Physics*. Record your answers in the spaces provided.

1 A unit used for a vector quantity is

(1) watt
(2) newton
(3) kilogram
(4) second

1 _____

2 A displacement vector with a magnitude of 20 meters could have perpendicular components with magnitudes of

(1) 10 m and 10 m
(2) 12 m and 8.0 m
(3) 12 m and 16 m
(4) 16 m and 8.0 m

2 _____

3 A hiker travels 1.0 kilometer south, turns and travels 3.0 kilometers west, and then turns and travels 3.0 kilometers north. What is the total distance traveled by the hiker?

(1) 3.2 km
(2) 3.6 km
(3) 5.0 km
(4) 7.0 km

3 _____

4 A car with an initial velocity of 16.0 meters per second east slows uniformly to 6.0 meters per second east in 4.0 seconds. What is the acceleration of the car during this 4.0-second interval?

(1) 2.5 m/s^2 west
(2) 2.5 m/s^2 east
(3) 4.0 m/s^2 west
(4) 4.0 m/s^2 east

4 _____

5 On the surface of planet *X*, a body with a mass of 10 kilograms weighs 40 newtons. The magnitude of the acceleration due to gravity on the surface of planet *X* is

(1) 4.0×10^3 m/s^2
(2) 4.0×10^2 m/s^2
(3) 9.8 m/s^2
(4) 4.0 m/s^2

5 ____

6 A car traveling in a straight line at an initial speed of 8.0 meters per second accelerates uniformly to a speed of 14 meters per second over a distance of 44 meters. What is the magnitude of the acceleration of the car?

(1) 0.41 m/s^2
(2) 1.5 m/s^2
(3) 3.0 m/s^2
(4) 2.2 m/s^2

6 ____

7 An object starts from rest and falls freely for 40 meters near the surface of planet *P*. If the time of fall is 4.0 seconds, what is the magnitude of the acceleration due to gravity on planet *P*?

(1) 0 m/s^2
(2) 1.3 m/s^2
(3) 5.0 m/s^2
(4) 10 m/s^2

7 ____

8 If a block is in equilibrium, the magnitude of the block's acceleration is

(1) zero
(2) decreasing
(3) increasing
(4) constant, but not zero

8 ____

9 The diagram below shows a light ray striking a plane mirror.

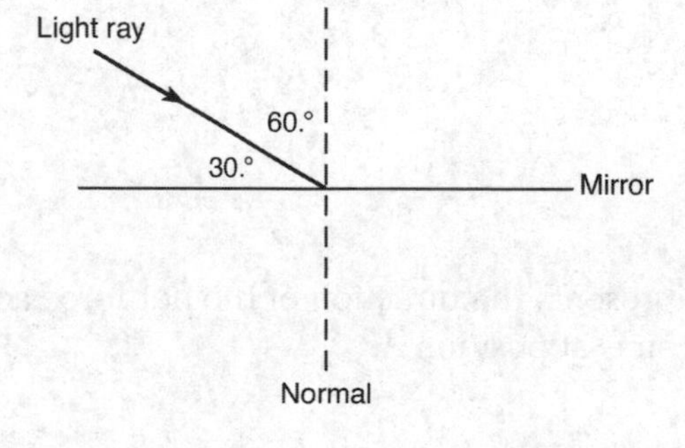

What is the angle of reflection?

(1) 30°
(2) 60°
(3) 90°
(4) 120°

9 ____

10 An electric field exerts an electrostatic force of magnitude 1.5×10^{-14} newton on an electron within the field. What is the magnitude of the electric field strength at the location of the electron?

(1) 2.4×10^{-33} N/C
(2) 1.1×10^{-5} N/C
(3) 9.4×10^{4} N/C
(4) 1.6×10^{16} N/C

10 ____

11 A 7.0-kilogram cart, *A*, and a 3.0-kilogram cart, *B*, are initially held together at rest on a horizontal, frictionless surface. When a compressed spring attached to one of the carts is released, the carts are pushed apart. After the spring is released, the speed of cart *B* is 6.0 meters per second, as represented in the diagram below.

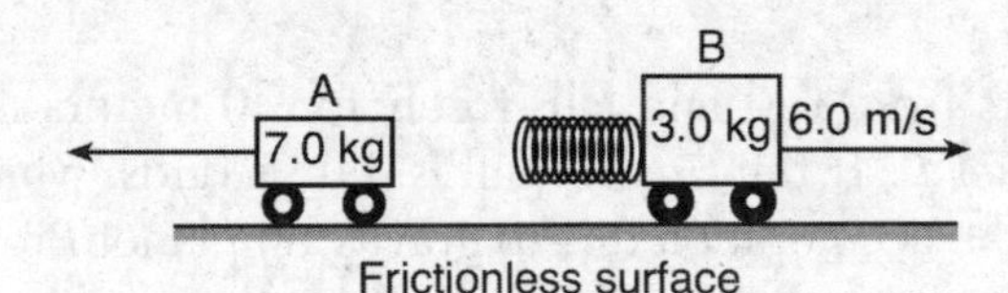

What is the speed of cart A after the spring is released?

(1) 14 m/s
(2) 6.0 m/s
(3) 3.0 m/s
(4) 2.6 m/s

11 ____

12 An electron in a magnetic field travels at constant speed in the circular path represented in the diagram below.

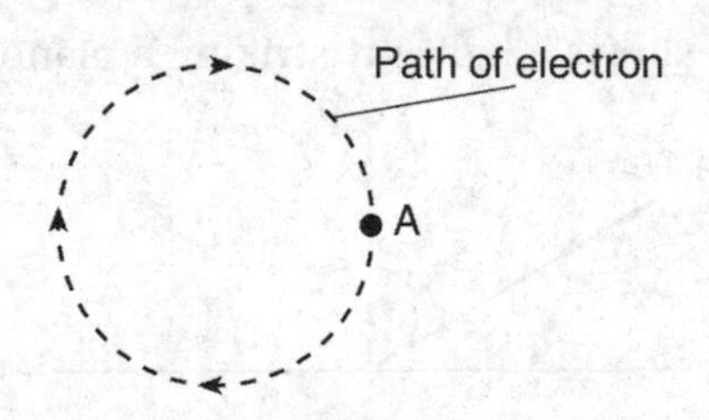

Which arrow represents the direction of the net force acting on the electron when the electron is at position *A*?

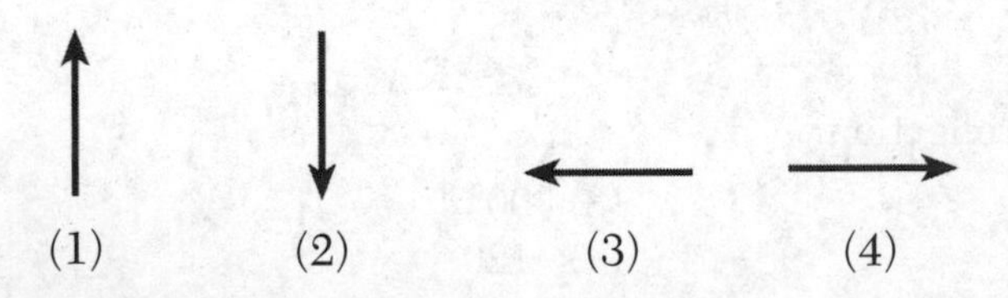

(1) (2) (3) (4)

12 ____

13 The potential difference between two points, *A* and *B*, in an electric field is 2.00 volts. The energy required to move a charge of 8.00×10^{-19} coulomb from point A to point *B* is

(1) 4.00×10^{-19} J
(2) 1.60×10^{-18} J
(3) 6.25×10^{17} J
(4) 2.50×10^{18} J

13 ____

14 Which statement describes the gravitational force and the electrostatic force between two charged particles?

(1) The gravitational force may be either attractive or repulsive, whereas the electrostatic force must be attractive.
(2) The gravitational force must be attractive, whereas the electrostatic force may be either attractive or repulsive.
(3) Both forces may be either attractive or repulsive.
(4) Both forces must be attractive.

14 ____

15 An electrostatic force exists between two $+3.20 \times 10^{-19}$-coulomb point charges separated by a distance of 0.030 meter. As the distance between the two point charges is *decreased*, the electrostatic force of

(1) attraction between the two charges decreases
(2) attraction between the two charges increases
(3) repulsion between the two charges decreases
(4) repulsion between the two charges increases

15 ____

16 What is the energy of the photon emitted when an electron in a mercury atom drops from energy level *f* to energy level *b*?

(1) 8.42 eV
(2) 5.74 eV
(3) 3.06 eV
(4) 2.68 eV

16 ____

17 An observer counts 4 complete water waves passing by the end of a dock every 10 seconds. What is the frequency of the waves?

(1) 0.40 Hz
(2) 2.5 Hz
(3) 40 Hz
(4) 4.0 Hz

17 ____

18 Copper is a metal commonly used for electrical wiring in houses. Which metal conducts electricity better than copper at 20°C?

(1) aluminum
(2) gold
(3) nichrome
(4) silver

18 ____

19 A motor does 20 joules of work on a block, accelerating the block vertically upward. Neglecting friction, if the gravitational potential energy of the block increases by 15 joules, its kinetic energy

(1) decreases by 5 J
(2) increases by 5 J
(3) decreases by 35 J
(4) increases by 35 J

19 ____

20 When only one lightbulb blows out, an entire string of decorative lights goes out. The lights in this string must be connected in

(1) parallel with one current pathway
(2) parallel with multiple current pathways
(3) series with one current pathway
(4) series with multiple current pathways

20 ____

21 An electric toaster is rated 1200 watts at 120 volts. What is the total electrical energy used to operate the toaster for 30 seconds?

(1) 1.8×10^3 J
(2) 3.6×10^3 J
(3) 1.8×10^4 J
(4) 3.6×10^4 J

21 ____

22 What is the rate at which work is done in lifting a 35-kilogram object vertically at a constant speed of 5.0 meters per second?

(1) 1700 W
(2) 340 W
(3) 180 W
(4) 7.0 W

22 ____

23 When a wave travels through a medium, the wave transfers

(1) mass, only
(2) energy, only
(3) both mass and energy
(4) neither mass nor energy

23 ____

24 Glass may shatter when exposed to sound of a particular frequency. This phenomenon is an example of

(1) refraction
(2) diffraction
(3) resonance
(4) the Doppler effect

24 ____

25 Which waves require a material medium for transmission?

(1) light waves
(2) radio waves
(3) sound waves
(4) microwaves

25 ____

26 Which type of oscillation would most likely produce an electromagnetic wave?

(1) a vibrating tuning fork
(2) a washing machine agitator at work
(3) a swinging pendulum
(4) an electron traveling back and forth in a wire

26 ____

27 If monochromatic light passes from water into air with an angle of incidence of 35°, which characteristic of the light will remain the same?

(1) frequency
(2) wavelength
(3) speed
(4) direction

27 ____

28 The absolute index of refraction of medium Y is twice as great as the absolute index of refraction of medium X. As a light ray travels from medium X into medium Y, the speed of the light ray is

(1) halved
(2) doubled
(3) quartered
(4) quadrupled

28 ____

29 The diagram below shows a transverse wave moving toward the right along a rope.

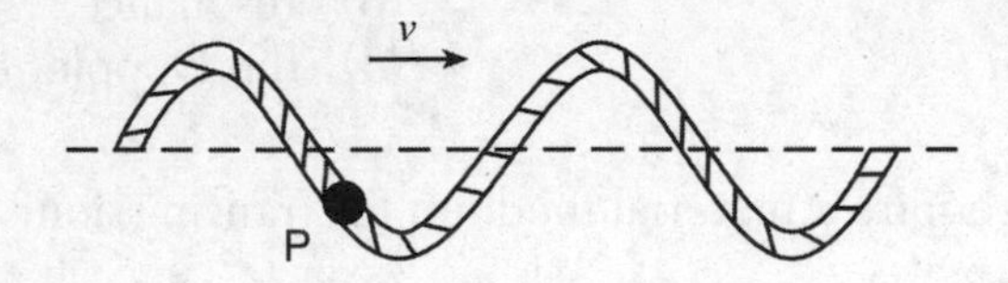

At the instant shown, point *P* on the rope is moving toward the

(1) bottom of the page
(2) top of the page
(3) left
(4) right

29 _____

30 When an isolated conductor is placed in the vicinity of a positive charge, the conductor is attracted to the charge. The charge of the conductor

(1) must be positive
(2) must be negative
(3) could be neutral or positive
(4) could be neutral or negative

30 _____

31 The quarks that compose a baryon may have charges of

(1) $+\frac{2}{3}e, +\frac{2}{3}e$, and $-\frac{1}{3}e$

(2) $+\frac{1}{3}e, +\frac{1}{3}e$, and $-\frac{2}{3}e$

(3) $-1e, -1e$, and 0

(4) $+\frac{2}{3}e, +\frac{2}{3}e$, and 0

31 _____

32 A rubber block weighing 60 newtons is resting on a horizontal surface of dry asphalt. What is the magnitude of the minimum force needed to start the rubber block moving across the dry asphalt?

(1) 32 N
(2) 40 N
(3) 51 N
(4) 60 N

32 _____

33 The data table below lists the mass and speed of four different objects.

Object	Mass (kg)	Speed (m/s)
A	2.0	6.0
B	4.0	5.0
C	6.0	4.0
D	8.0	2.0

Which object has the greatest inertia?

(1) *A*
(2) *B*
(3) *C*
(4) *D*

33 _____

34 The electroscope shown in the diagram below is made completely of metal and consists of a knob, a stem, and leaves. A positively charged rod is brought near the knob of the electroscope and then removed.

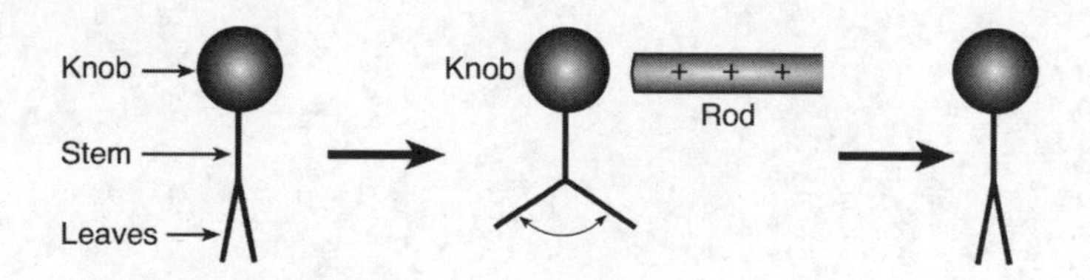

The motion of the leaves results from electrons moving from the

(1) leaves to the knob, only
(2) knob to the leaves, only
(3) leaves to the knob and then back to the leaves
(4) knob to the leaves and then back to the knob

34 _____

35 Which circuit diagram represents the correct way to measure the current in a resistor?

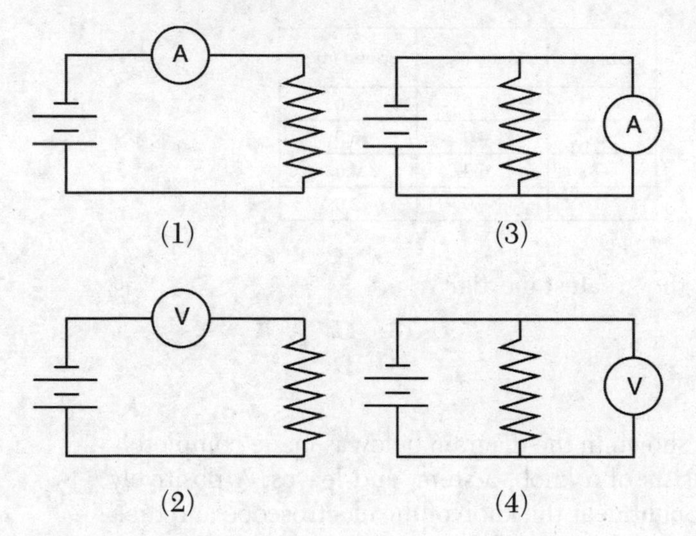

35 ____

PART B–1

Answer all questions in this part.

Directions (36–50): For *each* statement or question, select the *number* of the word or expression that, of those given, best completes the statement or answers the question. Some questions may require the use of the *2006 Edition Reference Tables for Physical Setting/Physics*. Record your answers in the spaces provided.

36 The height of a typical kitchen table is approximately

(1) 10^{-2} m
(2) 10^{0} m
(3) 10^{1} m
(4) 10^{2} m

36 _____

37 A ball is thrown with a velocity of 35 meters per second at an angle of 30° above the horizontal. Which quantity has a magnitude of zero when the ball is at the highest point in its trajectory?

(1) the acceleration of the ball
(2) the momentum of the ball
(3) the horizontal component of the ball's velocity
(4) the vertical component of the ball's velocity

37 _____

38 The graph below represents the relationship between velocity and time of travel for a toy car moving in a straight line.

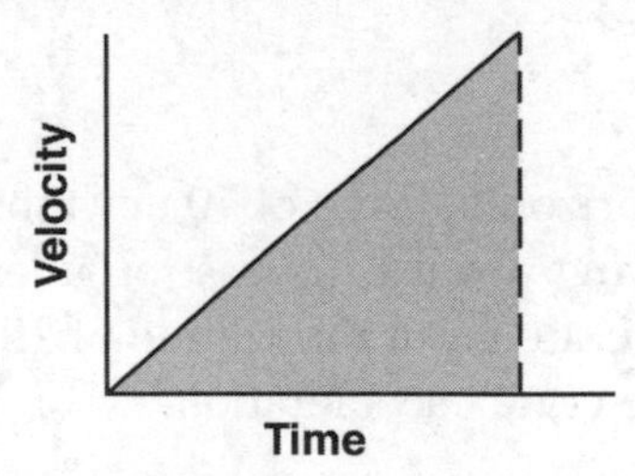

The shaded area under the line represents the toy car's

(1) displacement
(2) momentum
(3) acceleration
(4) speed

38 _____

39 A spring stores 10 joules of elastic potential energy when it is compressed 0.20 meter. What is the spring constant of the spring?

(1) 5.0×10^{1} N/m
(2) 1.0×10^{2} N/m
(3) 2.5×10^{2} N/m
(4) 5.0×10^{2} N/m

39 _____

Base your answers to questions 40 and 41 on the information below and on your knowledge of physics.

A cannonball with a mass of 1.0 kilogram is fired horizontally from a 500-kilogram cannon, initially at rest, on a horizontal, frictionless surface. The cannonball is acted on by an average force of 8.0×10^3 newtons for 1.0×10^{-1} second.

40 What is the magnitude of the change in momentum of the cannonball during firing?

(1) 0 kg • m/s
(2) 8.0×10^2 kg • m/s
(3) 8.0×10^3 kg • m/s
(4) 8.0×10^4 kg • m/s

40 ____

41 What is the magnitude of the average net force acting on the cannon?

(1) 1.6 N
(2) 16 N
(3) 8.0×10^3 N
(4) 4.0×10^6 N

41 ____

42 A metal sphere, *X*, has an initial net charge of -6×10^{-6} coulomb and an identical sphere, *Y*, has an initial net charge of $+2 \times 10^{-6}$ coulomb. The spheres touch each other and then separate. What is the net charge on sphere *X* after the spheres have separated?

(1) 0 C
(2) -2×10^{-6} C
(3) -4×10^{-6} C
(4) -6×10^{-6} C

42 ____

43 A constant eastward horizontal force of 70 newtons is applied to a 20-kilogram crate moving toward the east on a level floor. If the frictional force on the crate has a magnitude of 10 newtons, what is the magnitude of the crate's acceleration?

(1) 0.50 m/s^2
(2) 3.5 m/s^2
(3) 3.0 m/s^2
(4) 4.0 m/s^2

43 ____

44 Which graph represents the relationship between the energy of photons and the wavelengths of photons in a vacuum?

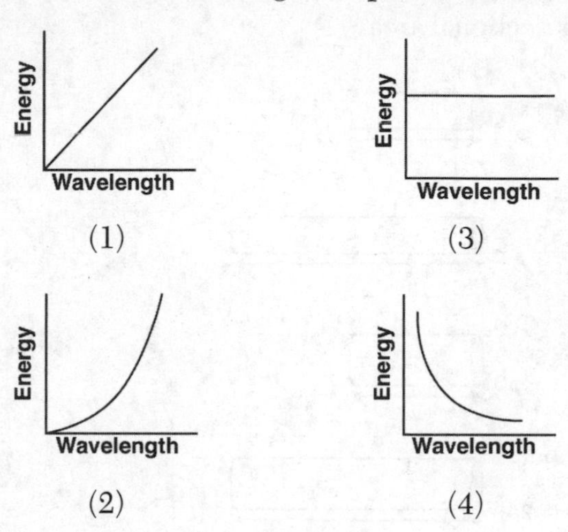

44 _____

Base your answers to questions 45 and 46 on the information and diagram below and on your knowledge of physics.

One end of a long spring is attached to a wall. A student vibrates the other end of the spring vertically, creating a wave that moves to the wall and reflects back toward the student, resulting in a standing wave in the spring, as represented in the diagram.

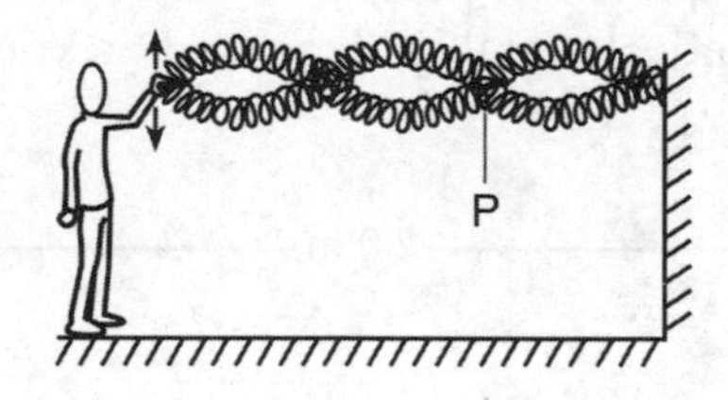

45 What is the phase difference between the incident wave and the reflected wave at point *P*?

(1) 0°
(2) 90°
(3) 180°
(4) 270°

45 _____

46 What is the total number of antinodes on the standing wave in the diagram?

(1) 6
(2) 2
(3) 3
(4) 4

46 _____

47 The diagrams below represent four pieces of copper wire at 20°C. For each piece of wire, ℓ represents a unit of length and A represents a unit of cross-sectional area.

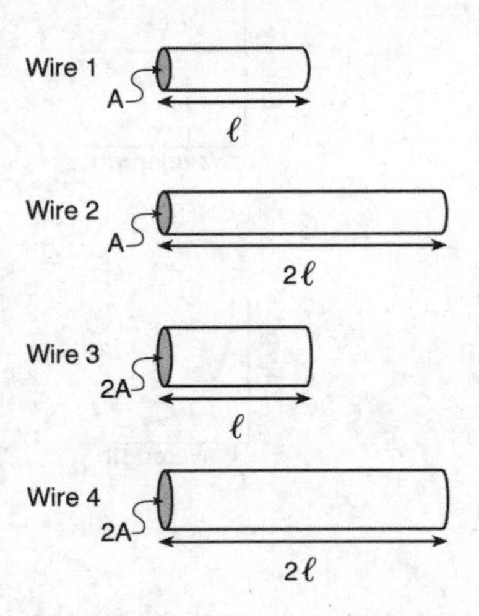

The piece of wire that has the greatest resistance is

(1) wire 1
(2) wire 2
(3) wire 3
(4) wire 4

47 ____

Base your answers to questions 48 and 49 on the diagram below, which represents two charged, identical metal spheres, and on your knowledge of physics.

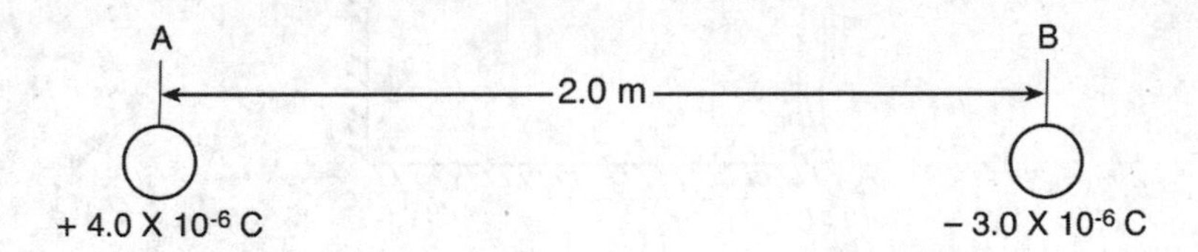

48 The number of excess elementary charges on sphere *A* is

(1) 6.4×10^{-25}
(2) 6.4×10^{-19}
(3) 2.5×10^{13}
(4) 5.0×10^{13}

48 ____

49 What is the magnitude of the electric force between the two spheres?

(1) 3.0×10^{-12} N
(2) 1.0×10^{-6} N
(3) 2.7×10^{-2} N
(4) 5.4×10^{-2} N

49 ____

50 The diagram below represents the wave fronts produced by a point source moving to the right in a uniform medium. Observers are located at points *A* and *B*.

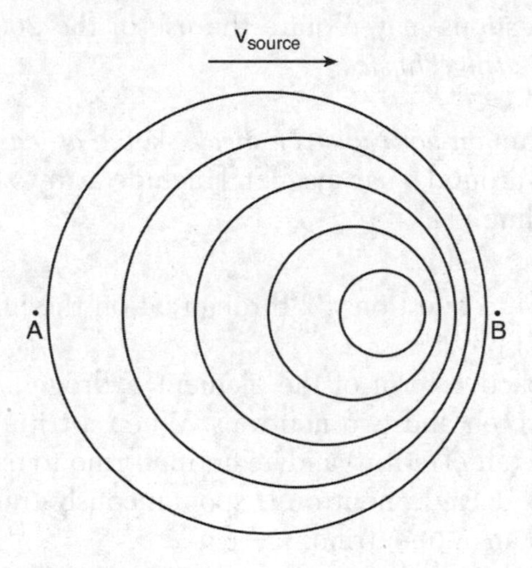

Compared to the wave frequency and wavelength observed at point *A*, the wave observed at point *B* has a

(1) higher frequency and a shorter wavelength
(2) higher frequency and a longer wavelength
(3) lower frequency and a shorter wavelength
(4) lower frequency and a longer wavelength

50 _____

PART B–2

Answer all questions in this part.

Directions (51–65): Record your answers on the answer sheet provided after the questions. Some questions may require the use of the *2006 Edition Reference Tables for Physical Setting/Physics*.

51 On the diagram *on your answer sheet*, sketch *at least four* magnetic field lines of force around a bar magnet. [Include arrows to show the direction of each field line.] [1]

Base your answers to questions 52 through 54 on the information below and on your knowledge of physics.

Tritium is a radioactive form of the element hydrogen. A tritium nucleus is composed of one proton and two neutrons. When a tritium nucleus decays, it emits a beta particle (an electron) and an antineutrino to create a stable form of helium. During beta decay, a neutron is spontaneously transformed into a proton, an electron, and an antineutrino.

52 What is the total number of quarks in a tritium nucleus? [1]

53 What is the total charge, in elementary charges, of a proton, an electron, and an antineutrino? [1]

54 What fundamental interaction is responsible for binding together the protons and neutrons in a helium nucleus? [1]

55 The diagram below represents a ball projected horizontally from a cliff at a speed of 10 meters per second. The ball travels the path shown and lands at time t and distance d from the base of the cliff. [Neglect friction.]

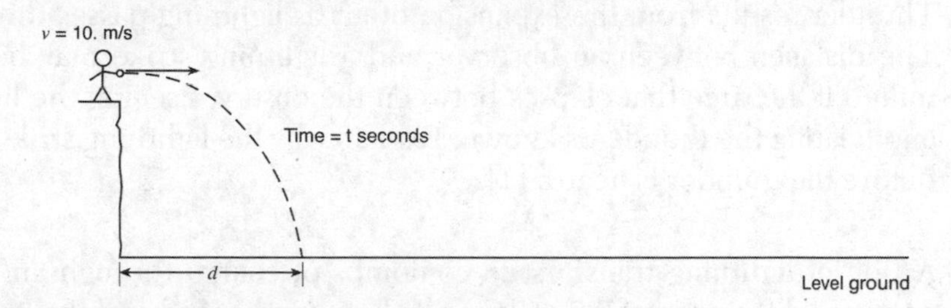

A second, identical ball is projected horizontally from the cliff at 20 meters per second. Determine the distance the second ball lands from the base of the cliff in terms of d. [1]

56–57 An operating television set draws 0.71 ampere of current when connected to a 120-volt outlet. Calculate the time it takes the television to consume 3.0×10^5 joules of electric energy. [Show all work, including the equation and substitution with units.] [2]

58–59 On the centimeter grid *on your answer sheet*, draw *at least one* cycle of a periodic transverse wave with an amplitude of 2.0 centimeters and a wavelength of 6.0 centimeters. [2]

60 The diagram below represents a 35-newton block hanging from a vertical spring, causing the spring to elongate from its original length.

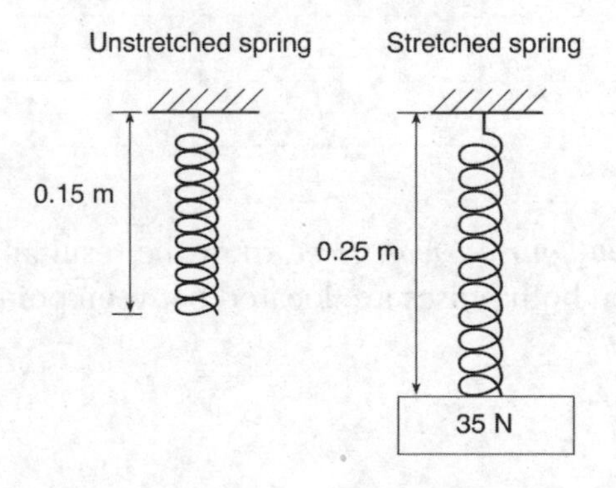

Determine the spring constant of the spring. [1]

61 Determine the amount of matter, in kilograms, that must be converted to energy to yield 1.0 gigajoule. [1]

62 Thunder results from the expansion of air as lightning passes through it. The distance between an observer and a lightning strike may be determined if the time that elapses between the observer seeing the lightning and hearing the thunder is known. Explain why the lightning strike is seen before the thunder is heard. [1]

63–64 A bolt of lightning transfers 28 coulombs of charge through an electric potential difference of 3.2×10^7 volts between a cloud and the ground in 1.5×10^{-3} second. Calculate the average electric current between the cloud and the ground during this transfer of charge. [Show all work, including the equation and substitution with units.] [2]

65 The diagram below represents two pulses traveling toward each other in a uniform medium.

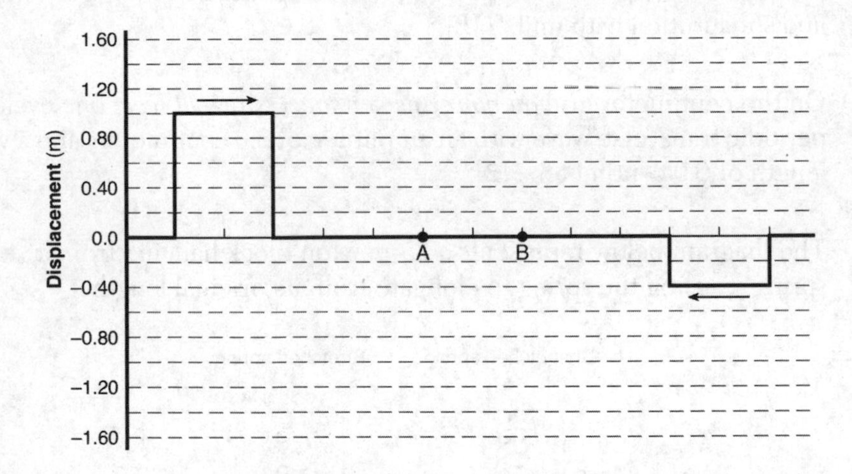

On the grid *on your answer sheet*, draw the resultant displacement of the medium when both pulses are located between points *A* and *B*. [1]

PART C

Answer all questions in this part.

Directions (66–85): Record your answers on the sheet provided in the back. Some questions may require the use of the *2006 Edition Reference Tables for Physical Setting/Physics*.

Base your answers to questions 66 through 70 on the information and diagram below and on your knowledge of physics.

As represented in the diagram, a ski area rope-tow pulls a 72.0-kilogram skier from the bottom to the top of a 40.0-meter-high hill. The rope-tow exerts a force of magnitude 158 newtons to move the skier a total distance of 230.0 meters up the side of the hill at constant speed.

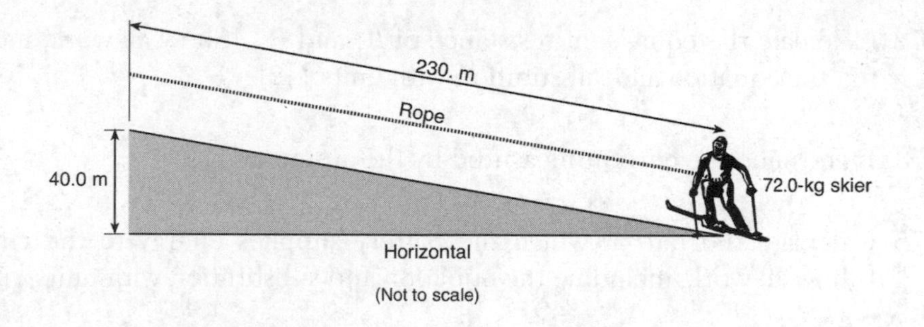

66 Determine the total amount of work done by the rope on the skier. [1]

67–68 Calculate the total amount of gravitational potential energy gained by the skier while moving up the hill. [Show all work, including the equation and substitution with units.] [2]

69 Describe what happens to the internal energy of the skier-hill system as the skier is pulled up the hill. [1]

70 Describe what happens to the total mechanical energy of the skier-hill system as the skier is pulled up the hill. [1]

Base your answers to questions 71 through 76 on the diagram and information below and on your knowledge of physics.

A 15-ohm resistor, .30-ohm resistor, and an ammeter are connected as shown with a 60-volt battery.

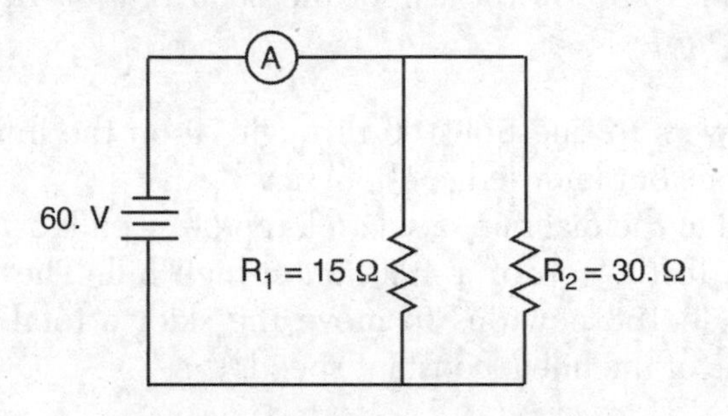

71–72 Calculate the equivalent resistance of R_1 and R_2. [Show all work, including the equation and substitution with units.] [2]

73 Determine the current measured by the ammeter. [1]

74–75 Calculate the rate at which the battery supplies energy to the circuit. [Show all work, including the equation and substitution with units.] [2]

76 If another resistor were added in parallel to the original circuit, what effect would this have on the current through resistor R_1? [1]

Base your answers to questions 77 through 80 on the information below and on your knowledge of physics.

A gas-powered model airplane has a mass of 2.50 kilograms. A student exerts a force on a cord to keep the airplane flying around her at a constant speed of 18.0 meters per second in a horizontal, circular path with a radius of 25.0 meters.

77–78 Calculate the kinetic energy of the moving airplane. [Show all work, including the equation and substitution with units.] [2]

79–80 Calculate the magnitude of the centripetal force exerted on the airplane to keep it moving in this circular path. [Show all work, including the equation and substitution with units.] [2]

Base your answers to questions 81 through 85 on the information and diagram below and on your knowledge of physics.

A ray of light with a frequency of 5.09×10^{14} hertz traveling in medium *X* is refracted at point *P*. The angle of refraction is 90°, as represented in the diagram.

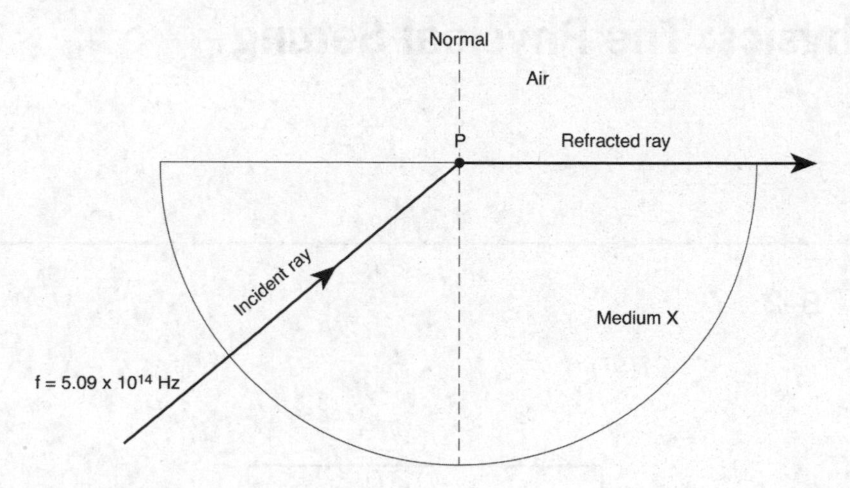

81–82 Calculate the wavelength of the light ray in air. [Show all work, including the equation and substitution with units.] [2]

83 Measure the angle of incidence for the light ray incident at point *P* and record the value *on your answer sheet.* [1]

84–85 Calculate the absolute index of refraction for medium *X*. [Show all work, including the equation and substitution with units.] [2]

Answer Sheet
June 2017
Physics: The Physical Setting

PART B–2

51

S N

52 ______________________________ **quarks**

53 ______________________________ **e**

54 ______________________________

55 ______________________________

56–57

58–59

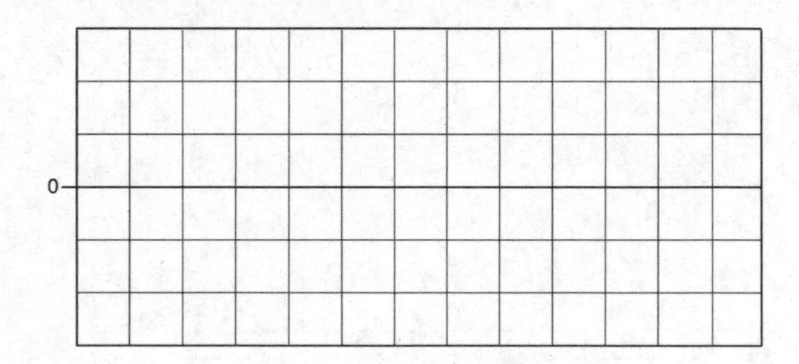

60 ____________________ **N/m**

61 ____________________ **kg**

62 ____

____

63–64

65

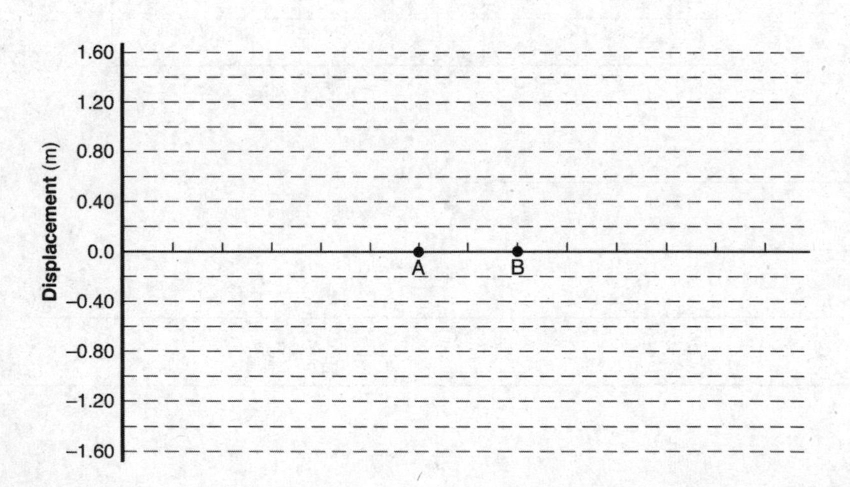

PART C

66 ______________________________ J

67–68

69 ______________________________

70 ______________________________

71–72

73 ______________________________ A

74–75

76 ____________________________________

77–78

79–80

81–82

83 ______________________________ °

84–85

Answers
June 2017
Physics: The Physical Setting

Answer Key

PART A

1. 2	**8.** 1	**15.** 4	**22.** 1	**29.** 2
2. 3	**9.** 2	**16.** 3	**23.** 2	**30.** 4
3. 4	**10.** 3	**17.** 1	**24.** 3	**31.** 1
4. 1	**11.** 4	**18.** 4	**25.** 3	**32.** 3
5. 4	**12.** 3	**19.** 2	**26.** 4	**33.** 4
6. 2	**13.** 2	**20.** 3	**27.** 1	**34.** 3
7. 3	**14.** 2	**21.** 4	**28.** 1	**35.** 1

PART B–1

36. 2	**39.** 4	**42.** 2	**45.** 3	**48.** 3
37. 4	**40.** 2	**43.** 3	**46.** 3	**49.** 3
38. 1	**41.** 3	**44.** 4	**47.** 2	**50.** 1

PART B–2 and **PART C**. *See* **Answers Explained**.

Answers Explained

PART A

1 The newton is the unit of measure for force, which is a vector quantity. Force is a vector quantity because it has both magnitude and direction.

WRONG CHOICES EXPLAINED:

(1), (3), and (4) Watt, kilogram, and second are all units representing scalar quantities. They represent power, mass, and time, respectively. These scalar quantities have magnitude but they do not have direction.

2 A displacement vector is formed by two perpendicular vectors. The three vectors form a right triangle. In this question, the resultant displacement vector measures 20 m. Any triangle whose sides are in a 3 : 4 : 5 ratio is a right triangle. The vectors 12 m, 16 m, and 20. m are in a 3 : 4 : 5 ratio, so they can be the perpendicular components and the given displacement vector.

WRONG CHOICES EXPLAINED:

(1), (2), (4) None of these answer choices can form a right triangle where the longest side measures 20 m.

3 To solve for total distance traveled, add all the individual distances traveled regardless of direction:

Solution: 1.0 km + 3.0 km + 3.0 km = 7.0 km

WRONG CHOICES EXPLAINED:

(1) This is a nonsense distracter.

(2) This is the total displacement, not the total distance traveled. Displacement is the change of position. In other words, it is the difference between the final position and the initial position.

(3) This answer adds the displacement traveled in the north-south direction and the distance traveled west. It is not the total distance traveled.

4 Refer to the *Mechanics* section of *Reference Table K.*

Given: $v_i = 16.0$ m/s

$v_f = 6.0$ m/s

$t = 4.0$ s

Find: $a = ?$

Solution:
$$a = \frac{\Delta v}{t}$$
$$= \frac{v_f - v_i}{t}$$
$$= \frac{(6.0 \text{ m/s}) - (16.0 \text{ m/s})}{4.0 \text{ s}}$$
$$= -2.5 \text{ m/s}^2$$

The negative acceleration (deceleration) indicates that the direction of acceleration is opposite to the direction of motion, which is why the car slows down. So the acceleration can be described either as -2.5 m/s^2 east or as 2.5 m/s^2 west.

5 Remember that weight is the force due to gravity. Refer to the *Mechanics* section of *Reference Table K.*

Given: $m = 10$ kg

$F_g = 40$ N

Find: $g = ?$

Solution:
$$g = \frac{F_g}{m}$$
$$= \frac{40 \text{ N}}{10 \text{ kg}}$$
$$= 4.0 \text{ m/s}^2$$

6 Refer to the *Mechanics* section of *Reference Table K.* Rearrange the equation to solve for acceleration before substituting the values with units:

Given: $v_i = 8.0 \text{ m/s}$

$v_f = 14 \text{ m/s}$

Find: $d = 44 \text{ m}$

$a = ?$

Solution:

$$v_f^2 = v_i^2 + 2ad$$

$$a = \frac{v_f^2 - v_i^2}{2d}$$

$$= \frac{(14 \text{ m/s})^2 - (8.0 \text{ m/s})^2}{(2)(44 \text{ m})}$$

$$= 1.5 \text{ m/s}^2$$

7 Refer to the *Mechanics* section of *Reference Table K.* Rearrange the equation to solve for acceleration:

Given: $v_i = 0 \text{ m/s}$

$d = 40 \text{ m}$

$t = 4.0 \text{ s}$

Find: $a = ?$

Solution:

$$d = v_i t + \frac{1}{2}at^2$$

$$a = \frac{2(d - v_i t)}{t^2}$$

$$= \frac{2(40 \text{ m} - (0 \text{ m/s})(4.0 \text{ s}))}{(4.0 \text{ s})^2}$$

$$= 5.0 \text{ m/s}^2$$

8 Acceleration is the result of an unbalanced force acting on an object. When an object is in equilibrium, the sum of the net forces acting on the object equals zero. Therefore, when a block is in equilibrium, the magnitude of the block's acceleration is zero.

WRONG CHOICES EXPLAINED:

(2), (3), and (4) When a block is not in equilibrium, there is a net force acting on the object. Therefore, the block could experience an acceleration that is decreasing, increasing, or constant but not zero.

9 Refer to the *Waves* section of *Reference Table K*. The angle of incidence is defined as the angle between the incoming light ray and the normal (the perpendicular). The angle of reflection is the angle of the outgoing light ray and the normal (the perpendicular):

Given: $\theta_i = 60°$

Find: $\theta_r = ?$

Solution: $\theta_i = \theta_r$

$\theta_r = 60°$

Remember that the angle of incidence equals the angle of reflection when light strikes a plane mirror.

WRONG CHOICES EXPLAINED:

(1) The angle of incidence is defined as the angle that the ray makes with the normal. The angle the ray makes with the surface of the mirror, 30°, is not equal to the angle of incidence and therefore is not the angle of reflection.

(3) Since the incoming light ray is not striking the surface at 90°, the angle of reflection does not equal 90°

(4) The angle of reflection can only be an acute angle (less than 90°). An angle of 120° is too large to be an angle of reflection.

10 Refer to the *Mechanics* section of *Reference Table K*. Remember that the charge on an electron is the elementary charge, e:

Given: $F_e = 1.5 \times 10^{-14}$ N

$q = e = 1.6 \times 10^{-19}$ C

Find: $E = ?$

Solution:
$$E = \frac{F_e}{q}$$
$$= \frac{1.5 \times 10^{-14}\ \text{N}}{1.6 \times 10^{-19}\ \text{C}}$$
$$= 9.375 \times 10^4\ \text{N/C}$$
$$= 9.4 \times 10^4\ \text{N/C}$$

11 The total momentum of the system before the spring is released must equal the total momentum of the system after the spring is released. Let a positive sign indicate movement to the left and a negative sign indicate movement to the right. Refer to the *Mechanics* section of *Reference Table K*.

Given: $m_A = 7.0$ kg

$m_B = 3.0$ kg

$v_{A_i} = 0.0$ m/s

$v_{B_i} = 0.0$ m/s

$v_{B_f} = -6.0$ m/s

Find: $v_{A_f} = ?$

Solution:

$$p_{before} = p_{after}$$

$$p = mv$$

$$m_A v_{A_i} + m_B v_{B_i} = m_A v_{A_f} + m_B v_{B_f}$$

$$(7.0 \text{ kg})(0 \text{ m/s}) + (3.0 \text{ kg})(0 \text{ m/s}) = (7.0 \text{ kg})\left(v_{A_f}\right) + (3.0 \text{ kg})(-6.0 \text{ m/s})$$

$$0 \frac{\text{kg} \bullet \text{m}}{\text{s}} = (7.0 \text{ kg})\left(v_{A_f}\right) + \left(-18.0 \frac{\text{kg} \bullet \text{m}}{\text{s}}\right)$$

$$(-7.0 \text{ kg})\left(v_{A_f}\right) = \left(-18.0 \frac{\text{kg} \bullet \text{m}}{\text{s}}\right)$$

$$\left(v_{A_f}\right) = 2.6 \text{ m/s}$$

12 Centripetal force is always directed toward the center of the circular path. When the electron is at point A, the centripetal force is directed toward the left.

13 Refer to the *Electricity* section of *Reference Table K*. The question asks for the energy required to move a charge. So rearrange the equation to solve for W, which is the electrical energy:

Given: $q = 8.00 \times 10^{-19}$ C

$V = 2.00$ V

Find: $W = ?$

Solution:

$$V = \frac{W}{q}$$

$$W = Vq$$

$$= (2.00 \text{ V})\left(8.00 \times 10^{-19} \text{ C}\right)$$

$$= 1.6 \times 10^{-18} \text{ J}$$

14 Gravitational force is the universal attraction between two masses. So by definition, gravitational force is always attractive. Electrostatic force may be either attractive or repulsive. Opposite charges attract, but like charges repel.

15 Refer to the *Electricity* section of *Reference Table K*:

$$F_e = \frac{kq_1q_2}{r^2}$$

This equation shows that electrostatic force is inversely proportional to the square of the distance between the centers of the point charges. The question states that the distance between the point charges decreases. Therefore, the electrostatic force between the charges increases. Since both charges are positive, the force is a repulsive force.

WRONG CHOICES EXPLAINED:

(1) Since the charges are both positive, the electrostatic force is that of repulsion, not attraction. Additionally, the distance between the charges decreases. Since electrostatic force is inversely proportional to the square of that distance, the force increases as the distance decreases.

(2) Since the charges are both positive, the electrostatic force is that of repulsion, not attraction.

(3) The distance between the charges decreases. Since electrostatic force is inversely proportional to the square of that distance, the force increases as the distance decreases.

16 Refer to the *Energy Level Diagram for Mercury* on *Reference Table F*. An electron in a mercury atom at energy level *f* has –2.68 eV of energy and at energy level *b* has –5.74 eV of energy. Calculate the difference between energy levels:

$$\text{Solution:} \quad -2.68 \text{ eV} - (-5.74 \text{ eV}) = 3.06 \text{ eV}$$

17 Frequency is defined as the number of cycles per second (s^{-1} or Hz).

$$\text{Solution:} \quad f = \frac{4}{10.\text{ seconds}} = 0.40 \text{ Hz}$$

18 Refer to the *Resistivities at 20°C* chart of the *Reference Table J*. The resistivity of copper at 20°C is $1.72 \times 10^{-8}\ \Omega \bullet \text{m}$. Of the metals listed in the answer choices, only silver has a resistivity less than that of copper. Since the resistivity of silver is $1.59 \times 10^{-8}\ \Omega \bullet \text{m}$ at 20°C, silver conducts electricity better than copper.

19 The work done on the block is equal to the total change in energy of the block. Refer to the *Mechanics* section of *Reference Table K*. Since we are neglecting the effect of friction, there is no conversion of mechanical energy to heat the block and to absorb energy, therefore, the internal energy of the block does not change. So $Q = 0$:

$$\text{Solution:} \quad W = \Delta E_T$$
$$E_T = PE + KE + Q$$
$$KE = E_T - PE - Q$$
$$= 20\text{ J} - 15\text{ J} - 0$$
$$= 5\text{ J}$$

20 A series circuit, by definition, is a circuit with a single current pathway. Therefore, when only one lightbulb blows out, the circuit is effectively cut. Electricity can no longer flow through the circuit.

WRONG CHOICES EXPLAINED:

(1) By definition, a parallel circuit must have more than one pathway for the current, not one pathway. This is a nonsense distracter.

(2) When one lightbulb goes out in a parallel circuit with multiple current pathways, only the lightbulbs in the portion of the circuit that contains the broken lightbulb go out. Electricity is still able to flow through the other current pathways. The lightbulbs stay on where the electricity is still able to make a complete circuit.

(4) By definition, a series circuit has only one pathway for the current, not multiple pathways. This is a nonsense distracter.

21 Refer to the *Electricity* section of *Reference Table K*:

$$\text{Given:} \quad P = 1200\text{ W}$$
$$V = 120\text{ V}$$
$$t = 30\text{ s}$$

$$\text{Find:} \quad W = ?$$

$$\text{Solution:} \quad W = Pt$$
$$= (1200\text{ W})(30\text{ s})$$
$$= 3.6 \times 10^4\text{ J}$$

22 Refer to *Reference Table A* and *Equations ME11* and *ME23* on *Reference Table K*:

Given: $m = 35\ kg$

$\overline{v} = 5.0\text{ m/s}$

$g = 9.81\text{ m/s}^2$

Find: $P = ?$

Solution: $g = \frac{F_g}{m}$

$$F_g = mg$$

$$P = F\overline{v}$$

$$P = mg\overline{v}$$

$$= (35\text{ kg})(9.81\text{ m/s}^2)(5.0\text{ m/s})$$

$$= 1700\text{ W}$$

23 Waves only transfer energy. Waves do not transfer mass.

24 Resonance is the spontaneous vibration of an object at a frequency equal to that of the wave that initiates that resonant vibration. If that spontaneous vibration is strong enough, it can shatter or break the object.

WRONG CHOICES EXPLAINED:

(1) Refraction is the change in direction of a wave when it passes obliquely from one medium into another in which the wave moves at a different speed.

(2) Diffraction is the bending of a wave around a barrier.

(4) The Doppler effect is the apparent change in frequency that results when a wave source and an observer are in relative motion with respect to each other.

25 Mechanical waves require a medium for transmission. Sound waves are mechanical waves, so they require a medium through which to travel.

WRONG CHOICES EXPLAINED:

(1), (2), and (4) Light waves, radio waves, and microwaves are all examples of electromagnetic waves. Electromagnetic waves do not require a medium through which to travel. They can travel in a vacuum.

26 An electromagnetic wave is a periodic wave consisting of mutually perpendicular electric and magnetic fields that is radiated away from the vicinity of an accelerating charge. Since a moving electron produces both electric and magnetic fields, choice (4) would most likely produce an electromagnetic wave.

WRONG CHOICES EXPLAINED:

(1) A vibrating tuning fork produces sound waves. Sound waves are mechanical waves, not electromagnetic waves.

(2) and (3) Neither a washing machine agitator at work nor a swinging pendulum produces mutually perpendicular electric and magnetic fields. Therefore, neither one produces an electromagnetic wave.

27 When monochromatic light passes from one medium into another medium, the speed of the light wave changes. If the light enters at an oblique angle, it changes direction as it enters the new medium. As a result of the change in speed, the light's wavelength changes. The frequency of the wave does not change.

28 Refer to *Equation W6* on *Reference Table K*:

Given: $n_y = 2n_X$

Find: $v_y = ?$

Solution:

$$\frac{n_2}{n_1} = \frac{v_1}{v_2}$$

$$\frac{n_y}{n_x} = \frac{v_x}{v_y}$$

$$v_Y = \frac{(v_x)(n_x)}{n_y}$$

$$= \frac{(v_x)(n_x)}{2n_x}$$

$$= \frac{1}{2}v_x$$

As a light wave travels from medium *X* into medium *Y*, the speed of the light wave is halved.

29 In a transverse wave, the particles of the medium vibrate or exhibit simple harmonic motion about a rest point perpendicular to the direction of the motion of the wave. In the diagram, point *P* is the disturbance and the wave is moving toward the right. So point *P* moves up and down. At the instant shown in the diagram, point *P* is moving toward the top of the page since a wave crest is coming toward point *P*.

30 Like charges repel, and opposite charges attract. Therefore, the charge of the conductor can be negative. Note that the positive charge can cause a redistribution of charge in a neutral conductor. The electrons in the neutral conductor can be attracted to the side of the conductor closer to the positive charge. That side of the conductor will have a negative charge, and the opposite side will have a positive charge. The entire charge on the conductor will remain neutral, but the negative side of the conductor will be attracted to the positive charge.

WRONG CHOICES EXPLAINED:

(1) and (3) A positive conductor will be repelled by the positive charge.

(2) A negative conductor will be attracted to the positive charge. However, as explained above, a neutral conductor can also be attracted to the positive charge.

31 Refer to the *Classification of Matter* and *Particles of the Standard Model* charts of *Reference Tables* G and *H*. Baryons are composed of combinations of 3 quarks. There are 6 different quarks, but each has a charge of either $+\frac{2}{3}$e or $-\frac{1}{3}$e. The total charge on a baryon must either be neutral or be a multiple of a whole elementary charge. The total charge cannot be fractional. Only choice (1) is comprised of 3 particles that have accepted quark charge values and add up to a whole number, +1e.

WRONG CHOICES EXPLAINED:

(2) The total charge is $+\frac{2}{3}$e, which cannot occur in a baryon.

(3) Quarks have a charge of either $+\frac{2}{3}$e or $-\frac{1}{3}$e. They never have a charge of −1e or of 0.

(4) Quarks have a charge of either $+\frac{2}{3}$e or $-\frac{1}{3}$e. They never have a charge of 0.

32 Since the rubber block is originally at rest, use the coefficient of static friction to determine the minimum force needed to start the rubber block moving across dry asphalt. Remember that the normal force, F_N, equals the weight of the rubber block. Refer to the *Mechanics* section of *Reference Table K* and the *Approximate Coefficients of Friction* table of *Reference Table C*:

Given: $\mu_s = 0.85$

$F_N = 60 \text{ N}$

Find: $F_f = ?$

Solution: $F_f = \mu_s F_N$

$= (0.85)(60 \text{ N})$

$= 51 \text{ N}$

33 Mass is the quantitative measure of inertia. As the mass of an object increases, the object's inertia increases. Since object *D* has the greatest mass, 8.0 kg, it has the greatest inertia. Speed does not affect inertia. This is extraneous information.

34 Opposite charges attract. Bringing a positively charged rod near the knob of the electroscope causes a redistribution of charge within the electroscope. The negative electrons are attracted to the positive rod. So the electrons move from the leaves and toward the knob. Like charges repel; therefore, the leaves repel each other now that they are both positive from the "missing" electrons. When the rod is removed, the electrons move from the knob back to the leaves so that each portion of the electroscope becomes neutral again. The neutral leaves no longer repel each other and thus return to their original position.

35 An ammeter measures current. It needs to be placed in the current path that is being measured. In choice (1), the ammeter is in the same current path as the resistor on the right side.

WRONG CHOICES EXPLAINED:

(2) A voltmeter measures potential difference across two points in a circuit. This diagram shows a voltmeter, not an ammeter.

(3) Although the diagram shows an ammeter, it is not in the current path that includes the resistor.

(4) An ammeter measures current. This diagram shows a voltmeter, not an ammeter. The voltmeter is correctly connected to measure the potential difference of the resistor, not the current through the resistor.

PART B–1

36 This is an order of magnitude question; $10^0 = 1$. Since 1 meter is approximately equal to 3.3 feet, 10^0 m is a reasonable approximation for the height of a kitchen table.

WRONG CHOICES EXPLAINED:

(1) 10 centimeters is not a reasonable approximation of the height of a kitchen table

(3) 10 meters (approximately 33 feet) is not a reasonable approximation of the height of a kitchen table

(4) 100 meters (approximately 330 feet) is not a reasonable approximation of the height of a kitchen table

37 As the ball rises, the vertical component of its velocity decelerates due to gravity and reaches a velocity of zero at its highest point. At this point, the vertical component of the velocity begins to accelerate due to gravity in the downward directions.

WRONG CHOICES EXPLAINED:

(1) The acceleration of the ball is constant throughout the trajectory. Its acceleration is –9.81 m/s^2 or 98.1 m/s^2 in the downward direction, which is the acceleration due to gravity.

(2) Momentum is equal to mass times velocity. Horizontal velocity is constant until the ball lands on the ground and its motion stops, so the momentum is not zero at the highest point.

(3) Horizontal velocity is constant until the ball lands on the ground and its motion stops. So the horizontal velocity is not zero at the highest point.

38 The shaded area under the line in a velocity versus time graph is equal to displacement. Mathematically, the area under the line can be expressed as $d = vt$.

WRONG CHOICES EXPLAINED:

(3) The slope of the line, $\frac{\Delta v}{t}$, would be equal to the acceleration on a velocity versus time graph.

39 Refer to the *Mechanics* section of *Reference Table K*:

Given: $PE_s = 10\ \text{J}$

$x = 0.20\ \text{m}$

Find: $k = ?$

Solution: $PE_s = \frac{1}{2}Kx^2$

$$k = \frac{2PE_s}{x^2}$$

$$= \frac{2(10\ \text{J})}{(0.20\ \text{m})^2}$$

$$= 5 \times 10^2\ \text{N/m}$$

40 Refer to the *Mechanics* section of *Reference Table K*:

Given: $m_{\text{ball}} = 1.0\ \text{kg}$

$m_{\text{cannon}} = 500\ \text{kg}$

$v_i = 0\ \text{m/s}$

$F = 8.0 \times 10^3\ \text{N}$

$t = 1.0^{-1}\ \text{s}$

Find: $\Delta p = ?$

Solution: $\Delta p = F_{net}t$

$$= \left(8.0 \times 10^3\ \text{N}\right)\left(1 \times 10^{-1}\ \text{s}\right)$$

$$= 8.0 \times 10^2\ \text{kg} \bullet \text{m/s}$$

41 Newton's third law states that if object *A* exerts a force on object *B*, then object *B* exerts a force equal in magnitude but opposite in direction to the force on object *A*.

42 When two charged metal spheres of identical size come into contact with one another, the charges redistribute evenly between the spheres. To find the final charge on each, add the 2 original charges and divide by 2:

Solution: $$\left(-6\times10^{-6}\ \text{C}\right)+\left(+2\times10^{-6}\ \text{C}\right)=-4\times10^{-6}\ \text{C}$$

$$\frac{-4\times10^{-6}\ \text{C}}{2}=-2\times10^{-6}\ \text{C}$$

WRONG CHOICES EXPLAINED:

(3) You added up the 2 charges but forgot to divide by 2.

43 The net force is the sum of all the applied forces. The force of friction is always in the opposite direction of motion, so we actually subtract the force of friction from the applied force to get the net force. Refer to the *Mechanics* section of *Reference Table K*.

Given: $$F_{applied}=70\ \text{N}$$
$$m=20\ \text{kg}$$
$$F_f=10\ \text{N}$$

Find: $$a=\ ?$$

Solution: $$F_{net}=F_{applied}-F_f$$
$$=70\ \text{N}-10\ \text{N}$$
$$=60\ \text{N}$$
$$a=\frac{F_{net}}{m}$$
$$=\frac{60\ \text{N}}{20\ \text{kg}}$$
$$=3.0\ \text{m/s}^2$$

44 Refer to the *Modern Physics* section of *Reference Table K*:

$$E_{photon}=\frac{hc}{\lambda}$$

This equation shows that wavelength, *l*, is inversely proportional to the energy of a photon. The graph depicted in choice (4) correctly indicates this type of relationship.

WRONG CHOICES EXPLAINED:

(1) This graph represents a direct proportional relationship.

(2) This graph represents an exponentially proportional relationship.

(3) This graph shows that energy is always the same no matter the wavelength.

45 Point P is a node, the place in a standing wave where there is complete destructive interference. For this to occur, the waves have to be 180º out of phase.

WRONG CHOICES EXPLAINED:

(1) This would result in maximum constructive interference, which is an antinode.

46 Antinodes occur when there is maximum constructive interference between the waves. In this diagram, antinodes occur in 3 places.

47 Refer to the *Electricity* section of *Reference Table K*:

$$R = \frac{\rho L}{A}$$

This equation shows that at constant temperature, resistance is directly proportional to length and inversely proportional to cross-sectional area. Assume that wire 1 has a resistance of R. Then wire 2, which is 2 times the length of wire 1, has 2 times the resistance of wire 1.

WRONG CHOICES EXPLAINED:

(3) Assume that wire 1 has a resistance of R. Then wire 3, which has 2 times the cross-sectional area of wire 1, has half the resistance of wire 1.

(4) Assume that wire 1 has a resistance of R. Then wire 4, which has 2 times the length and 2 times the cross-sectional area of wire 1, has the same resistance as wire 1.

48 Refer to the *List of Physical Constants* on *Reference Table A* to find the value of an elementary charge:

$$e = 1.60 \times 10^{-19}\ \text{C}$$

Then convert the charge on sphere A into elementary charges:

Given: $q_A = +4.0 \times 10^{-6}$ C

Solution: $$\text{excess charges} = \frac{(+4.0 \times 10^{-6}\ \text{C})}{(1.6 \times 10^{-19}\ \text{C})}$$

$$= 2.5 \times 10^{13}$$

49 Refer to the *List of Physical Constants* (*Reference Table* A) and the *Electricity* section of *Reference Table K*:

Given: $q_A = +4.0 \times 10^{-6}$ C

$q_B = -3.0 \times 10^{-6}$ C

$r = 2.0$ m

Find: $k = 8.99 \times 10^9\ \text{N} \bullet \text{m}^2/\text{C}^2$

$F_e = ?$

Solution: $$F_e = \frac{kq_1q_2}{r^2}$$

$$= \frac{(8.99 \times 10^9\ \text{N} \bullet \text{m}^2/\text{C}^2)(4.0 \times 10^{-6}\ \text{C})(-3.0 \times 10^{-6}\ \text{C})}{(2.0\ \text{m})^2}$$

$$= -2.7 \times 10^{-2}\ \text{N}$$

The negative sign in the answers is ignored because the question asked about the magnitude of the force only.

50 This diagram illustrates the Doppler effect, which is the apparent change in frequency that results when a wave source and an observer are in relative motion to each other. The lines in the diagram represent wave crests. The spaces between the lines represent troughs. At point B, both the wave crests and the troughs are much closer together than at point A. This indicates a shorter wavelength at point B. Refer to the *Waves* section of *Reference Table K*.

$$v = f\lambda$$

This equation shows that when velocity is constant, frequency and wavelength are inversely proportional. If wavelength decreases, frequency increases and vice versa. Therefore, a shorter wavelength results in a higher frequency.

PART B–2

51 One credit is awarded for drawing at least 4 field lines. Magnetic field lines must be drawn from North to South. Each magnetic field line must have an arrowhead to indicate proper direction. Field lines drawn inside the magnet are ignored.

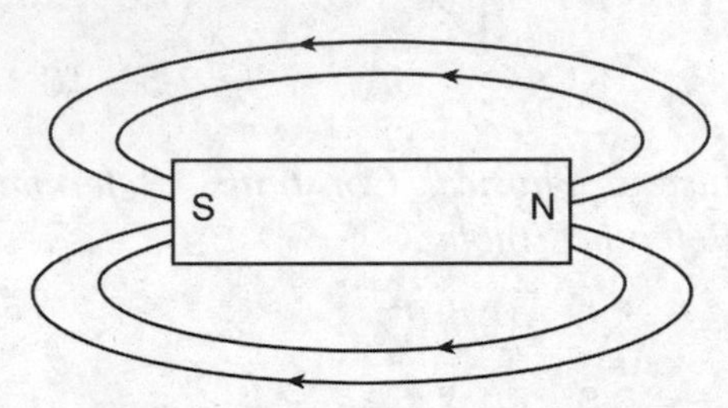

52 Refer to the *Classification of Matter* chart and *Particles of the Standard Model* chart (*Reference Tables G* and *H*). Each proton and each neutron is composed of 3 quarks. Therefore the nucleus of a tritium atom containing 1 proton and 2 neutrons has a total of 9 quarks.

One credit is awarded for an answer of 9.

53 A proton carries an elementary charge of +1. An electron carries an elementary charge of –1. Since a neutrino is neutral, it carries a charge of 0. Add those charges:

$$+1 + (-1) + 0 = 0$$

One credit is awarded for an answer of 0.

54 One credit is awarded for:

- Strong force
- Strong
- Strong nuclear force
- Strong interaction

No credit is awarded for nuclear force since it would be unclear whether you meant strong nuclear force or weak nuclear force.

55 Since both balls are launched from the same height at the same angle, they both fall at the same rate, which is the acceleration due to gravity. Therefore, the second ball travels through the air for the same amount of time as does the first ball. Refer to the *Mechanics* section of *Reference Table K*:

$$v = \frac{d}{t}$$

Since time is constant, the equation indicates that velocity is directly proportional to distance. Since the second ball has twice the velocity of the first ball, the second ball travels twice the distance in the same amount of time.

One credit is awarded for an answer of $2d$.

56–57 Refer to the *Electricity* section of *Reference Table K*. Solve for time:

Given: $I = 0.71\text{ A}$

$V = 120\text{ V}$

$W = 3.0 \times 10^5\text{ J}$

Find: $t = ?$

Solution:

$$W = VIt$$

$$t = \frac{W}{VI}$$

$$= \frac{3.0 \times 10^5\text{ J}}{(120\text{ V})(0.71\text{ A})}$$

$$= 3.5 \times 10^3\text{ s}$$

56 One credit is awarded for showing the correct equation *and* for substituting the correct values with their units.

57 One credit is awarded for the correct answer, with units, or for an answer that is consistent with the work shown in question 56.

Note: Students will not be penalized more than 1 credit for errors in units in questions 56 and 57.

58–59 Examples of 2 credit responses:

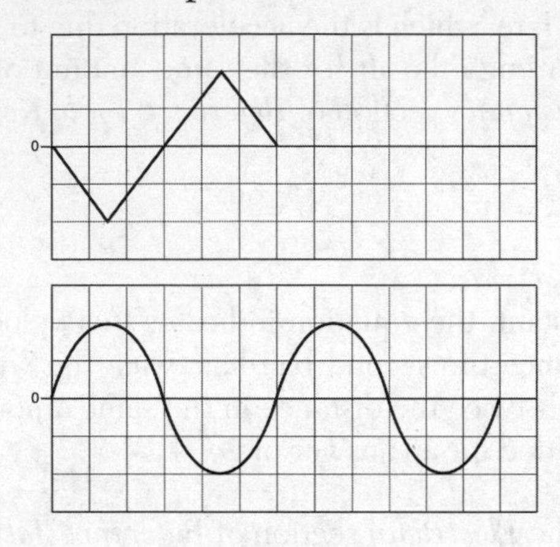

58 One credit is awarded for drawing a transverse wave with an amplitude of 2.0 cm ± 0.2 cm.

59 One credit is awarded for drawing a transverse wave with a wavelength of 6.0 cm ± 0.2 cm.

60 Refer to the *Mechanics* section of *Reference Table K*. Solve for k, which is the spring constant:

Given: $F_s = 35\text{ N}$

$x = 0.10\text{ m}$

Find: $k = ?$

Solution: $F_s = kx$

$$k = \frac{F_s}{x}$$

$$= \frac{35\text{ N}}{0.10\text{ m}}$$

$$= 350\text{ N/m}$$

One credit is awarded for an answer of 350 N/m.

61 Refer to the *Prefixes for Powers* chart (*Reference Table B*) and the *Modern Physics* section of *Reference Table K*:

Given: $\text{giga} = 10^9$

$E = 1.0 \times 10^9 \text{ J}$

$c = 3.00 \times 10^8 \text{ m/s}$

$m = ?$

Solution: $E = mc^2$

$$m = \frac{E}{c^2}$$

$$= \frac{1.0 \times 10^9 \text{ J}}{\left(3.00 \times 10^8 \text{ m/s}\right)^2}$$

$$= 1.1 \times 10^{-8} \text{ kg}$$

One credit is awarded for an answer of 1.1×10^{-8} kg.

62 The speed of light is greater than the speed of sound. Based on the *List of Physical Constants* table (*Reference Table A*), the speed of light is 3.00×10^8 m/s and the speed of sound is 3.31×10^2 m/s. In other words, light travels approximately 1 million times faster than sound!

One credit is awarded for indicating that the speed of light is greater than the speed of sound.

63–64 Refer to the *Electricity* section of *Reference Table K*:

Given: $q = 28 \text{ C}$

$V = 3.2 \times 10^7 \text{ V}$

$t = 1.5 \times 10^{-3} \text{ s}$

Find: $I = ?$

Solution: $I = \frac{\Delta q}{t}$

$$= \frac{28 \text{ C}}{1.5 \times 10^{-3} \text{ s}}$$

$$= 1.9 \times 10^4 \text{ A}$$

63 One credit is awarded for showing the correct equation *and* for substituting the correct values with their units.

64 One credit is awarded for the correct answer, with units, or for an answer that is consistent with the work shown in question 63.

Note: Students will not be penalized more than 1 credit for errors in units in questions 63 and 64.

65 The resultant displacement is the net sum of the individual displacements of each pulse: $+1.00 \text{ m} + (-0.40 \text{ m}) = +0.60 \text{ m}$.

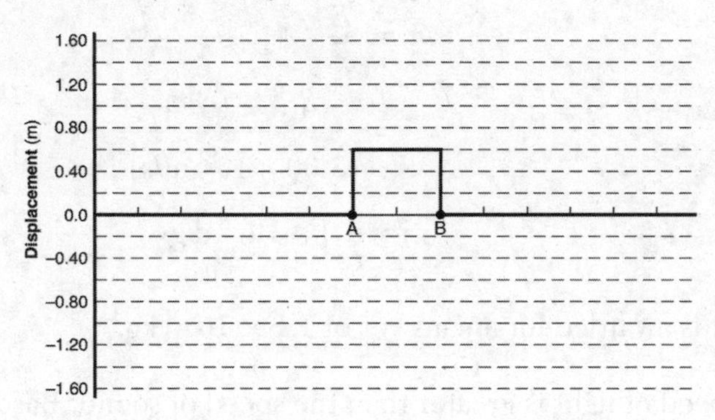

One credit is awarded for drawing a square wave between *A* and *B* that has a displacement of +0.6 m.

PART C

66 Refer to the *Mechanics* section of *Reference Table K*:

Given: $F_{rope} = 158 \text{ N}$

$d = 230 \text{ m}$

Find: $W = ?$

Solution:

$$W = Fd$$

$$= (158 \text{ N}) \times (230 \text{ m})$$

$$= 3.63 \times 10^4 \text{ J}$$

One credit is awarded for an answer of 3.63×10^4 J or 36,300 J.

67–68 Refer to *Reference Table A* and the *Mechanics* section of *Reference Table K*:

Given: $m = 72.0$ kg

$h = 40.0$ m

$g = 9.81$ m/s^2

Find: $\Delta PE = ?$

Solution:

$$\Delta PE = mg\Delta h$$

$$= (72.0 \text{ kg})(9.81 \text{ m/s}^2)(40.0 \text{ m})$$

$$= 2.83 \times 10^4 \text{ J}$$

67 One credit is awarded for showing the correct equation *and* for substituting the correct values with their units.

68 One credit is awarded for the correct answer, with units, or for an answer that is consistent with the work shown in question 67.

Note: Students will not be penalized more than 1 credit for errors in units in questions 67 and 68.

69 Refer to the *Mechanics* section of *Reference Table K*:

$$W = \Delta E_T$$

$$E_T = PE + KE + Q$$

The work, W, done by the rope is greater than the potential energy, PE, gained by the skier. There is no change in kinetic energy for the skier, $KE = 0$. Therefore, the rest of the energy acted against friction. So the internal thermal energy, Q, of the skier-hill system increased.

One credit is awarded for indicating that the internal energy increases.

70 Refer to the *Mechanics* section of *Reference Table K*:

$$W = \Delta E_T$$

The work done by the rope is increasing the mechanical energy of the system.
One credit is awarded for indicating that the total mechanical energy increases.

71–72 The resistors are connected in parallel. Refer to the *Electricity* section of *Reference Table K*:

Given: $R_1 = 15\ \Omega$

$R_2 = 30\ \Omega$

Find: $R_{eq} = ?$

Solution=:

$$\frac{1}{R_{eq}} = \frac{1}{R_1} + \frac{1}{R_2}$$

$$= \frac{1}{15\ \Omega} + \frac{1}{30\ \Omega}$$

$$= \frac{2}{30\ \Omega} + \frac{1}{30\ \Omega}$$

$$= \frac{3}{30\ \Omega}$$

$$= \frac{1}{10\ \Omega}$$

$$R_{eq} = 10\ \Omega$$

71 One credit is awarded for showing the correct equation *and* for substituting the correct values with their units.

72 One credit is awarded for the correct answer, with units, or for an answer that is consistent with the work shown in question 71.
Note: Students will not be penalized more than 1 credit for errors in units in questions 71 and 72.

73 Refer to the *Electricity* section of *Reference Table K*. Solve for I:

Given: $R_{eq} = 10\ \Omega$

$V = 60\ \text{V}$

Find: $I = ?$

Solution:
$$R = \frac{V}{I}$$
$$I = \frac{V}{R}$$
$$= \frac{60\ \text{V}}{10\ \Omega}$$
$$= 6.0\ \text{A}$$

One credit is awarded for an answer of 6.0 A or for an answer that is the result of 60 V divided by the answer obtained in question 72.

74–75 Refer to the *Electricity* section of *Reference Table K*. Based on the information originally provided with the diagram as well as the answers found in questions 71 through 73, you can determine the rate (power) in three different ways:

Given: $R_{eq} = 10\ \Omega$

$V = 60\ \text{V}$

$I = 6.0\ \text{A}$

Find: $P = ?$

Solution:
$$P = \frac{V^2}{R}$$
$$= \frac{(60\ \text{V})^2}{10\ \Omega}$$
$$= 360\ \text{W}$$

or
$$P = I^2R$$
$$= (6.0\ \text{A})^2(10\ \Omega)$$
$$= 360\ \text{W}$$

or
$$P = VI$$
$$= (60\ \text{V})(6.0\ \text{A})$$
$$= 360\ \text{W}$$

74 One credit is awarded for showing the correct equation *and* for substituting the correct values with their units.

75 One credit is awarded for the correct answer, with units, or for an answer that is consistent with the work shown in question 74.

Note: Students will not be penalized more than 1 credit for errors in units in questions 74 and 75.

76 Refer to the *Electricity* section of *Reference Table K*:

$$V = V_1 = V_2 = V_3 = \cdots$$

$$R = \frac{V}{I}$$

As shown in the first equation above, all resistors connected in parallel to each other experience the same potential difference provided by the power source. Since the potential difference doesn't change when adding an additional resistor in parallel, neither does the current through the individual resistor. [However, by adding an additional resistor in parallel, the total resistance of the circuit decreases; therefore, the total current on the circuit increases.]

One credit is awarded for:

- The current will remain the same.
- No change.

77–78 Refer to the *Mechanics* section of *Reference Table K*:

Given: $m = 2.50 \text{ kg}$

$v = 18.0 \text{ m/s}$

$r = 25.0 \text{ m}$

Find: $KE = ?$

Solution:
$$KE = \frac{1}{2}mv^2$$
$$= \frac{1}{2}(2.50 \text{ kg})(18.0 \text{ m/s})^2$$
$$= 405 \text{ J}$$

77 One credit is awarded for showing the correct equation *and* for substituting the correct values with their units.

78 One credit is awarded for the correct answer, with units, or for an answer that is consistent with the work shown in question 77.

Note: Students will not be penalized more than 1 credit for errors in units in questions 77 and 78.

79–80 Refer to the *Mechanics* section of *Reference Table K*:

Given: $m = 2.50\text{ kg}$

$v = 18.0\text{ m/s}$

$r = 25.0\text{ m}$

Find: $F_c = ?$

Solution: $F_c = ma_c$

$$a_c = \frac{v^2}{r}$$

$$F_c = \frac{mv^2}{r}$$

$$= \frac{(2.50\text{ kg})(18\text{ m/s})^2}{25.0\text{ m}}$$

$$= 32.4\text{ N}$$

79 One credit is awarded for showing the correct equation *and* for substituting the the correct values with their units.

80 One credit is awarded for the correct answer, with units, or for an answer that is consistent with the work shown in question 79.

Note: Students will not be penalized more than 1 credit for errors in units in questions 79 and 80.

81–82 Refer to *Reference Table* A and the *Waves* section of *Reference Table* K:

Given: $f = 5.09 \times 10^{14}$ Hz

$vc = 3.00 \times 10^{8}$ m/s

Find: $\lambda = ?$

Solution:

$$v = f\lambda$$

$$\lambda = \frac{v}{f}$$

$$= \frac{3.00 \times 10^{8}\ \text{m/s}}{5.09 \times 10^{14}\ \text{Hz}}$$

$$= 5.89 \times 10^{-7}\ \text{m}$$

81 One credit is awarded for showing the correct equation *and* for substituting the correct values with their units.

82 One credit is awarded for the correct answer, with units, or for an answer that is consistent with the work shown in question 81.

Note: Students will not be penalized more than 1 credit for errors in units in questions 81 and 82.

83 The angle of incidence is the angle between the incident ray and the normal. Use a protractor to measure the angle of incidence.

One credit is awarded for an answer of $50° \pm 2°$.

84–85 Refer to *Reference Table E* and the *Waves* section of *Reference Table K*. Make sure your calculator is set to degree mode:

Given: $\theta_{air} = 90°$

$\theta_x = 50°$

$n_{air} = 1.00$

Find: $n_x = ?$

Solution:

$$n_1 \sin\theta_1 = n_2 \sin\theta_2$$

$$n_x \sin\theta_x = n_{air} \sin\theta_{air}$$

$$n_x = \frac{n_{air} \sin\theta_{air}}{\sin\theta_x}$$

$$= \frac{(1.00)(\sin 90°)}{\sin 50°}$$

$$= 1.3$$

84 One credit is awarded for showing the correct equation *and* for substituting the correct values with their units.

85 One credit is awarded for the correct answer, with units, or for an answer that is consistent with the work shown in question 84.

Note: Students will not be penalized more than 1 credit for errors in units in questions 84 and 85.

Topic	Question Numbers (total)	Wrong Answers (x)	Grade
Math Skills	2–7, 10–11, 13, 15–17, 19, 21–22, 28, 31–32, 36, 38–40, 42–43, 47–49, 52–53, 55–61, 63–68, 71–75, 77–85: (56)		$\frac{100(56-x)}{56} = \%$
Mechanics	1–8, 11, 14, 32–33, 37–38, 40–41, 43, 55, 60, 79–80: (21)		$\frac{100(21-x)}{21} = \%$
Energy	19, 22, 39, 66–70, 77–78: (10)		$\frac{100(10-x)}{10} = \%$
Electricity/ Magnetism	10, 12–15, 18, 20–21, 26, 30, 34–35, 42, 47–49, 51, 56–57, 63–64, 71–76: (27)		$\frac{100(27-x)}{27} = \%$
Waves	9, 17, 23–25, 27–29, 45–46, 50, 58–59, 62, 65, 81–85: (20)		$\frac{100(20-x)}{20} = \%$
Modern Physics	16, 31, 44, 52–54, 61: (7)		$\frac{100(7-x)}{7} = \%$

Examination June 2018

Physics: The Physical Setting

PART A

Answer all questions in this part.

Directions (1–35): For *each* statement or question, select the number of the word or expression that, of those given, best completes the statement or answers the question. Some questions may require the use of the *2006 Edition Reference Tables for Physical Setting/Physics.* Record your answers in the spaces provided.

1 Which combination correctly pairs a vector quantity with its corresponding unit?

(1) weight and kg
(2) velocity and m/s
(3) speed and m/s
(4) acceleration and m^2/s

1 ____

2 A 12.0-kilogram cart is moving at a speed of 0.25 meter per second. After the speed of the cart is tripled, the inertia of the cart will be

(1) unchanged
(2) one-third as great
(3) three times greater
(4) nine times greater

2 ____

3 While taking off from an aircraft carrier, a jet starting from rest accelerates uniformly to a final speed of 40 meters per second on a runway that is 70 meters long. What is the magnitude of the acceleration of the jet?

(1) 0.29 m/s^2
(2) 0.57 m/s^2
(3) 1.8 m/s^2
(4) 11 m/s^2

3 ____

4 A 6.0-kilogram cart initially traveling at 4.0 meters per second east accelerates uniformly at 0.50 meter per second squared east for 3.0 seconds. What is the speed of the cart at the end of this 3.0 second interval?

(1) 1.5 m/s
(2) 5.5 m/s
(3) 3.0 m/s
(4) 7.0 m/s

4 ____

5 A soccer ball is kicked into the air from level ground with an initial speed of 20 meters per second and returns to ground level. At which angle above the horizontal should the ball be kicked in order for the ball to travel the greatest total horizontal distance? [Neglect friction.]

(1) 15°
(2) 30°
(3) 45°
(4) 75°

5 ____

6 Starting from rest, a car travels 18 meters as it accelerates uniformly for 3.0 seconds. What is the magnitude of the car's acceleration?

(1) 6.0 m/s^2
(2) 2.0 m/s^2
(3) 3.0 m/s^2
(4) 4.0 m/s^2

6 ____

7 A ball is rolling horizontally at 3.00 meters per second as it leaves the edge of a tabletop 0.750 meter above the floor. The ball lands on the floor 0.391 second after leaving the tabletop. What is the magnitude of the ball's acceleration 0.200 second after it leaves the tabletop? [Neglect friction.]

(1) 1.96 m/s^2
(2) 7.65 m/s^2
(3) 9.81 m/s^2
(4) 15.3 m/s^2

7 ____

8 A projectile with mass m is fired with initial horizontal velocity v_x from height h above level ground. Which change would have resulted in a greater time of flight for the projectile? [Neglect friction.]

(1) decreasing the mass to $m/2$
(2) decreasing the height to $h/2$
(3) increasing the initial horizontal velocity to $2v_x$
(4) increasing the height to $2h$

8 ____

9 A golf club hits a stationary 0.050-kilogram golf ball with an average force of 5.0×10^3 newtons, accelerating the ball to a speed of 44 meters per second. What is the magnitude of the impulse imparted to the ball by the golf club?

(1) 2.2 N • s
(2) 880 N • s
(3) 1.1×10^4 N • s
(4) 2.2×10^5 N • s

9 ____

10 A tennis player's racket applies an average force of 200. newtons to a tennis ball for 0.025 second. The average force exerted on the racket by the tennis ball is

(1) 0.025 N
(2) 5.0 N
(3) 200. N
(4) 80.0 N

10 ____

11 The diagram below represents a box sliding down an incline at constant velocity.

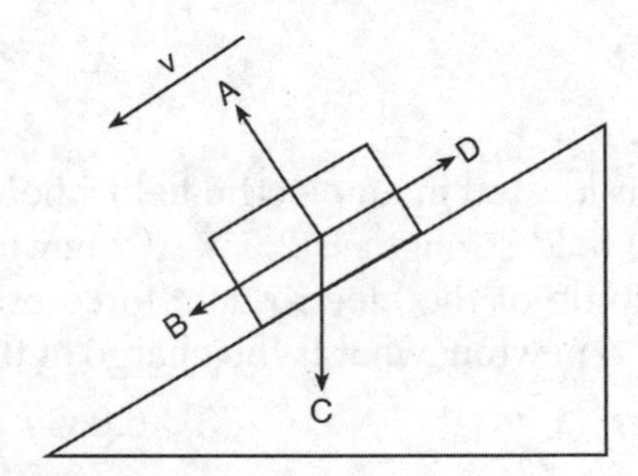

Which arrow represents the direction of the frictional force acting on the box?

(1) *A*
(2) *B*
(3) *C*
(4) *D*

11 ____

12 Which diagram represents the directions of the velocity, v, and acceleration, a, of a toy car as it moves in a clockwise, horizontal, circular path at a constant speed?

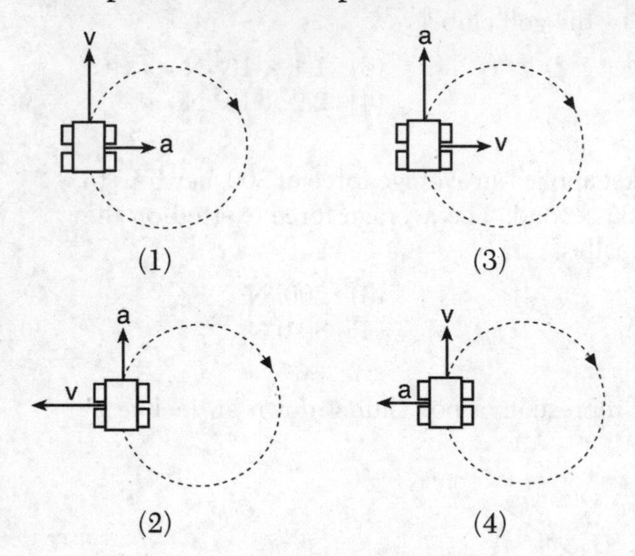

12 ____

13 A charged particle is located in an electric field where the magnitude of the electric field strength is 2.0×10^3 newtons per coulomb. If the magnitude of the electrostatic force exerted on the particle is 3.0×10^{-3} newton, what is the charge of the particle?

(1) 1.6×10^{-19} C
(2) 1.5×10^{-6} C
(3) 6.0 C
(4) 6.7×10^5 C

13 ____

14 The magnitude of the gravitational field strength near Earth's surface is represented by

(1) $\frac{F_g}{m}$
(2) G
(3) mg
(4) $\frac{Gm_1m_2}{r^2}$

14 ____

15 A car engine supplies 2.0×10^3 joules of energy during the 10 seconds it takes to accelerate the car along a horizontal surface. What is the average power developed by the car engine while it is accelerating?

(1) 2.0×10^1 W
(2) 2.0×10^2 W
(3) 2.0×10^3 W
(4) 2.0×10^4 W

15 ____

16 Which forces can be either attractive or repulsive?

(1) gravitational and magnetic
(2) electrostatic and gravitational
(3) magnetic and electrostatic
(4) gravitational, magnetic, and electrostatic

16 ____

17 Compared to the resistivity of a 0.4-meter length of 1-millimeter-diameter copper wire at 0°C, the resistivity of a 0.8-meter length of 1-millimeter-diameter copper wire at 0°C is

(1) one-fourth as great
(2) one-half as great
(3) the same
(4) four times greater

17 ____

18 The work per unit charge required to move a charge between two points in an electric circuit defines electric

(1) force
(2) power
(3) field strength
(4) potential difference

18 ____

19 A 2.0-meter length of copper wire is connected across a potential difference of 24 millivolts. The current through the wire is 0.40 ampere. The same copper wire at the same temperature is then connected across a potential difference of 48 millivolts. The current through the wire is

(1) 0.20 A
(2) 0.40 A
(3) 0.80 A
(4) 1.6 A

19 ____

20 What is the magnitude of the gravitational force of attraction between two 0.425-kilogram soccer balls when the distance between their centers is 0.500 meter?

(1) 2.41×10^{-11} N
(2) 4.82×10^{-11} N
(3) 5.67×10^{-11} N
(4) 1.13×10^{-10} N

20 ____

21 A sound wave produced by a loudspeaker can travel through water, but not through a vacuum. In comparison, a red light wave produced by a laser can travel through

(1) water, but not through a vacuum
(2) a vacuum, but not through water
(3) both water and a vacuum
(4) neither water nor a vacuum

21 ____

22 As a group of soldiers marches along a road, each soldier steps simultaneously. However, when crossing a bridge, the group does not step simultaneously in order to prevent the bridge from vibrating intensely. The phenomenon responsible for the intense vibrations is

(1) action and reaction
(2) conservation of momentum
(3) inertia
(4) resonance

22 ____

23 Which characteristics of a light wave remain constant when the light wave travels from air into corn oil?

(1) speed and frequency
(2) wavelength and frequency
(3) period and frequency
(4) wavelength and period

23 ____

24 The speed of a light ray ($f = 5.09 \times 10^{14}$ Hz) in corn oil is

(1) 1.47×10^{8} m/s
(2) 2.04×10^{8} m/s
(3) 3.00×10^{8} m/s
(4) 4.41×10^{8} m/s

24 ____

25 The spreading out of a wave after passing through an opening in a barrier is an example of

(1) diffraction
(2) Doppler effect
(3) reflection
(4) refraction

25 ____

26 A microwave with a frequency of 5.0×10^{10} hertz has a period of

(1) 2.0×10^{-11} s
(2) 6.0×10^{-3} s
(3) 1.7×10^{2} s
(4) 1.5×10^{19} s

26 ____

27 After two light waves have interfered in a vacuum, the two waves will be

(1) changed in frequency
(2) changed in velocity
(3) changed in amplitude
(4) unchanged

27 ____

28 A glass rod is rubbed with silk. During this process, a positive charge is given to the glass rod by

(1) adding electrons to the rod
(2) adding protons to the rod
(3) removing electrons from the rod
(4) removing protons from the rod

28 ____

29 A photon with an energy of 1.33×10^{-21} joule has a frequency of

(1) 5.02×10^{13} Hz
(2) 2.01×10^{12} Hz
(3) 8.82×10^{14} Hz
(4) 5.30×10^{34} Hz

29 ____

30 The speed of a car is increased uniformly from 11 meters per second to 19 meters per second. The average speed of the car during this interval is

(1) 0.0 m/s
(2) 15 m/s
(3) 30 m/s
(4) 4.0 m/s

30 ____

31 The energy equivalent of the rest mass of an electron is

(1) 2.73×10^{-22} J
(2) 8.20×10^{-14} J
(3) 1.50×10^{-10} J
(4) 1.44×10^{-2} J

31 ____

32 A spring has an unstretched length of 0.40 meter. The spring is stretched to a length of 0.60 meter when a 10-newton weight is hung motionless from one end. The spring constant of this spring is

(1) 10 N/m
(2) 17 N/m
(3) 25 N/m
(4) 50 N/m

32 _____

33 An electric circuit contains a battery, three lamps, and an open switch, as represented in the diagram below.

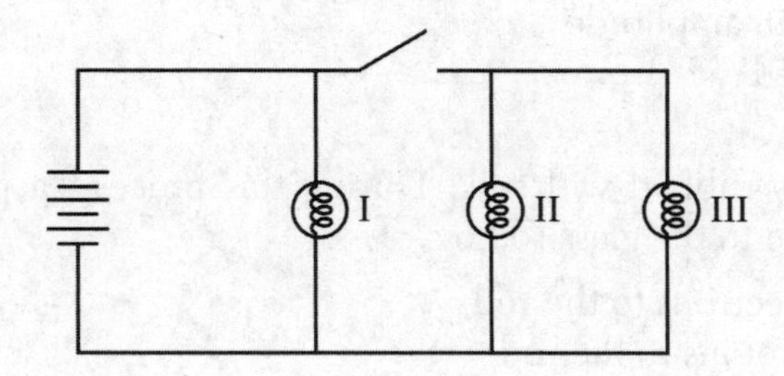

When the switch is open, there is an electric current in

(1) lamp I, only
(2) lamps II and III, only
(3) lamps I, II, and III
(4) none of the lamps

33 _____

34 Which diagram correctly represents an electric field?

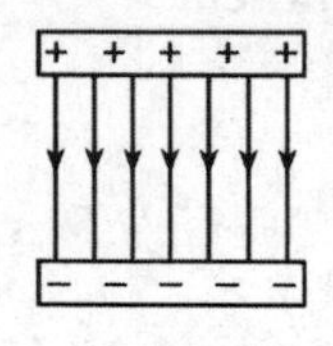

(1)

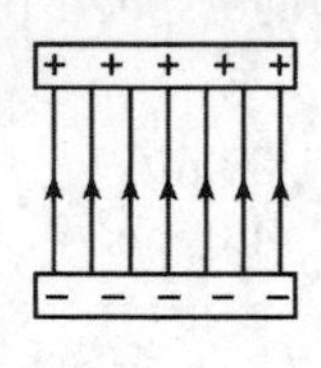

(2)

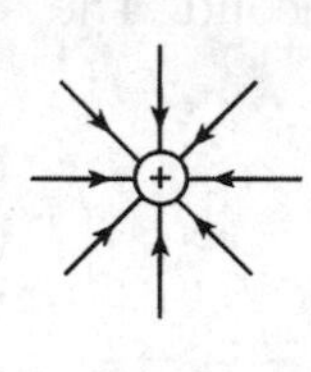

(3)

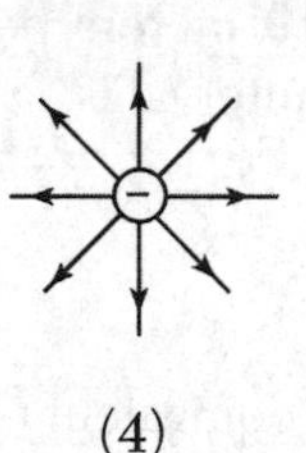

(4)

34 _____

35 Which points on the wave diagram below are 90° out of phase with each other?

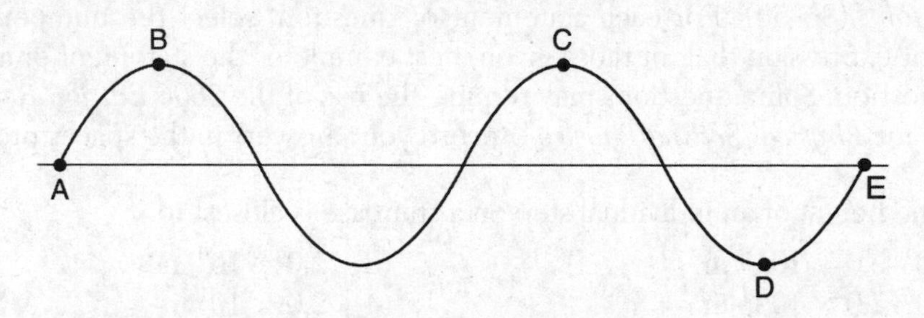

(1) *A* and *E*
(2) *B* and *C*
(3) *C* and *D*
(4) *D* and *E*

35 ____

PART B–1

Answer all questions in this part.

Directions (36–50): For each statement or question, select the number of the word or expression that, of those given, best completes the statement or answers the question. Some questions may require the use of the *2006 Edition Reference Tables for Physical Setting/Physics.* Record your answers in the spaces provided.

36 The height of an individual step on a staircase is closest to

(1) 2.0×10^{-2} m
(2) 2.0×10^{-1} m
(3) 2.0×10^{0} m
(4) 2.0×10^{1} m

36 ____

37 What is the magnitude of the electrostatic force exerted on an electron by another electron when they are 0.10 meter apart?

(1) 2.6×10^{-36} N
(2) 2.3×10^{-27} N
(3) 2.3×10^{-26} N
(4) 1.4×10^{-8} N

37 ____

38 After a 65-newton weight has fallen freely from rest a vertical distance of 5.3 meters, the kinetic energy of the weight is

(1) 12 J
(2) 340 J
(3) 910 J
(4) 1800 J

38 ____

39 A 0.500-kilogram cart traveling to the right on a horizontal, frictionless surface at 2.20 meters per second collides head on with a 0.800-kilogram cart moving to the left at 1.10 meters per second. What is the magnitude of the total momentum of the two-cart system after the collision?

(1) 0.22 kg • m/s
(2) 0.39 kg • m/s
(3) 1.98 kg • m/s
(4) 4.29 kg • m/s

39 ____

40 An object weighing 2.0 newtons is pushed across a horizontal, frictionless surface by a horizontal force of 4.0 newtons. The magnitude of the net force acting on the object is

(1) 0.0 N
(2) 2.0 N
(3) 8.0 N
(4) 4.0 N

40 ____

41 The ratio of the wavelength of AM radio waves traveling in a vacuum to the wavelength of FM radio waves traveling in a vacuum is approximately

(1) 1 to 1
(2) 2 to 1
(3) 10^2 to 1
(4) 10^8 to 1

41 ____

42 A charm quark has a charge of approximately

(1) 5.33×10^{-20} C
(2) 1.07×10^{-19} C
(3) 1.60×10^{-19} C
(4) 2.40×10^{-19} C

42 ____

43 The diagram below represents a 3.0-ohm resistor connected to a 12-volt battery. Meters *X* and *Y* are correctly connected in the circuit.

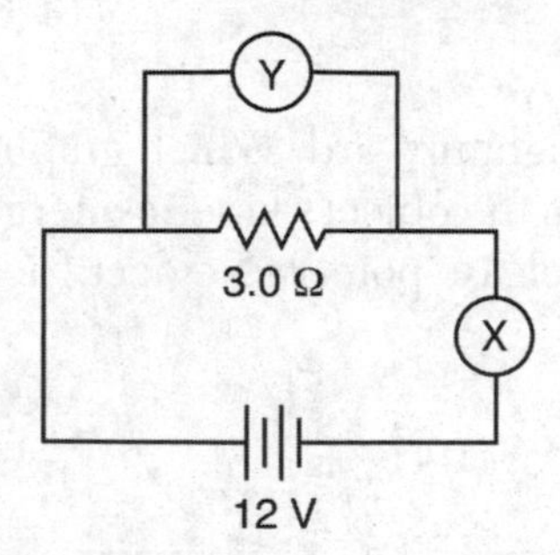

What are the readings on the meters?

(1) $X = 12$ V and $Y = 0.25$ A
(2) $X = 12$ V and $Y = 4.0$ A
(3) $X = 0.25$ A and $Y = 12$ V
(4) $X = 4.0$ A and $Y = 12$ V

43 ____

44 A toy airplane, flying in a horizontal, circular path, completes 10 complete circles in 30 seconds. If the radius of the plane's circular path is 4.0 meters, the average speed of the airplane is

(1) 0.13 m/s
(2) 0.84 m/s
(3) 1.3 m/s
(4) 8.4 m/s

44 ____

45 Which pair of graphs represents the vertical motion of an object falling freely from rest?

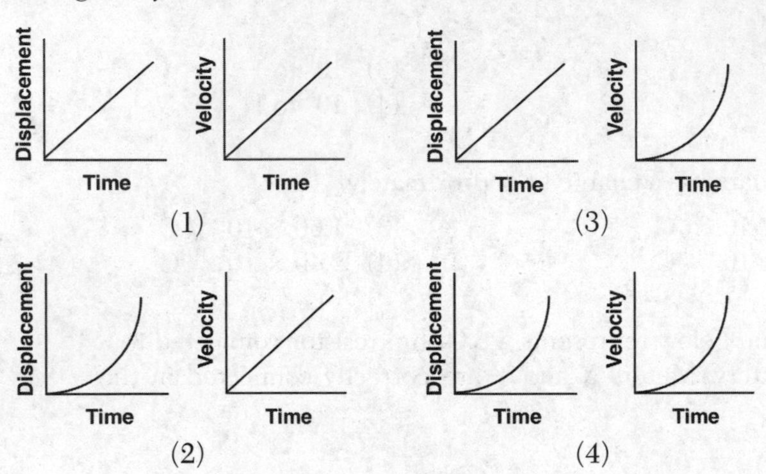

45 ____

46 An object is thrown straight upward. Which graph best represents the relationship between the object's kinetic energy and the height of the object above its release point? [Neglect friction.]

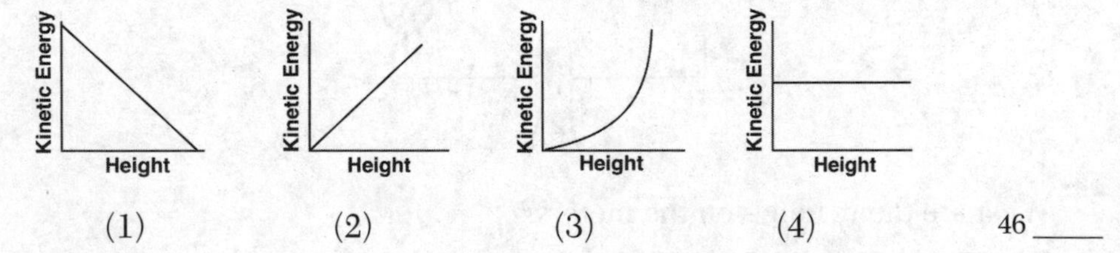

46 ____

47 In the diagram below, *X* represents a particle in a spring.

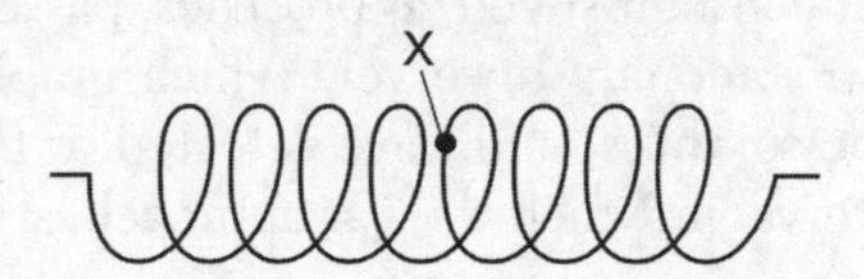

Which diagram represents the motion of particle *X* as a longitudinal wave passes through the spring toward the right?

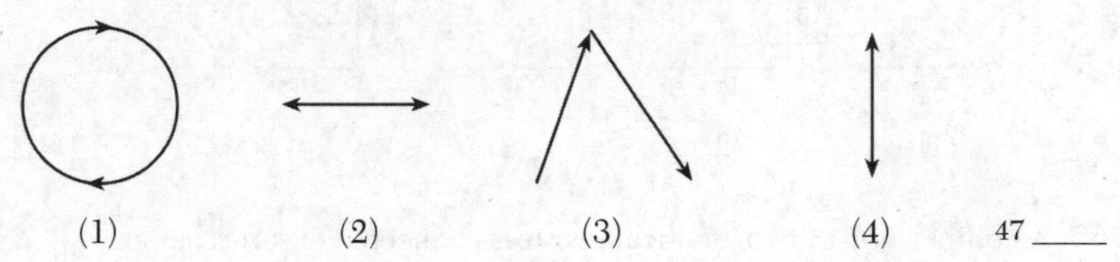

47 ____

48 As represented in the diagram below, two wave pulses, *X* and *Y*, are traveling toward each other in a rope. Both wave pulses have an amplitude of 0.30 m.

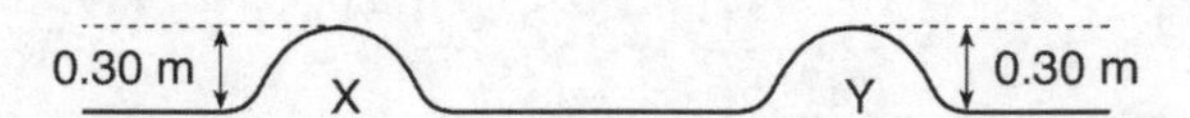

Which diagram shows the pulse produced due to the superposition of pulse *X* and pulse *Y*?

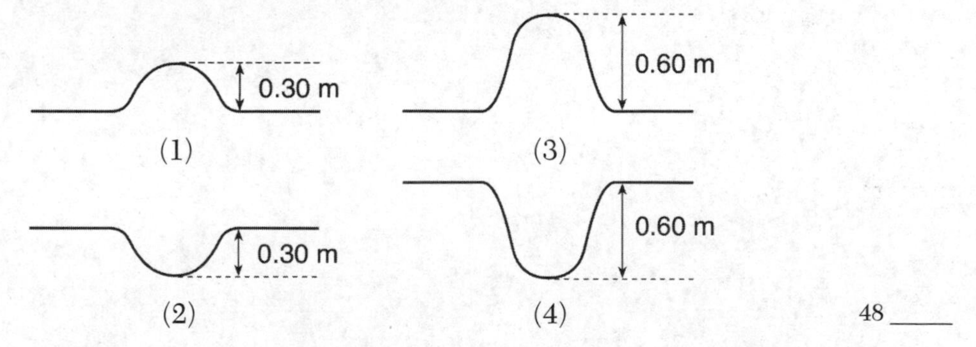

48 ____

49 The horn of a car produces a sound wave of constant frequency. The car, traveling at constant speed, approaches, passes, and then moves away from a stationary observer. Which graph best represents the frequency of this sound wave detected by the observer during the time interval in which the car approaches, passes, and moves away?

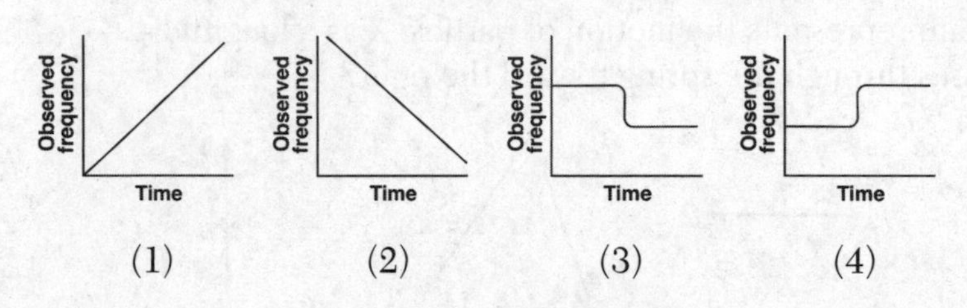

(1) (2) (3) (4) 49 ____

50 A combination of two identical resistors connected in series has an equivalent resistance of 10 ohms. What is the equivalent resistance of the combination of these same two resistors when connected in parallel?

(1) 2.5 Ω
(2) 5.0 Ω
(3) 10. Ω
(4) 20. Ω

50 ____

PART B–2

Answer all questions in this part.

Directions (51–65): Record your answers on the answer sheet provided after the questions. Some questions may require the use of the *2006 Edition Reference Tables for Physical Setting/Physics*.

Base your answers to questions 51 through 53 on the information below and on your knowledge of physics.

The scaled diagram below represents two forces acting concurrently at point *P*. The magnitude of force *A* is 32 newtons and the magnitude of force *B* is 20 newtons. The angle between the directions of force *A* and force *B* is 120°.

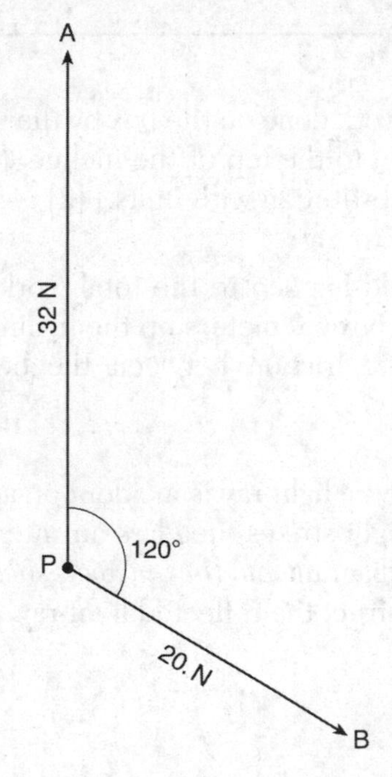

51 Determine the linear scale used in the diagram. [1]

52 On the diagram *on the answer sheet*, use a protractor and a ruler to construct a scaled vector to represent the resultant of forces *A* and *B*. Label the vector *R*. [1]

53 Determine the magnitude of the resultant force. [1]

Base your answers to questions 54 through 56 on the information and diagram below and on your knowledge of physics.

A student pushes a box, weighing 50 newtons, 6.0 meters up an incline at a constant speed by applying a force of 25 newtons parallel to the incline. The top of the incline is 2.0 meters higher than the bottom.

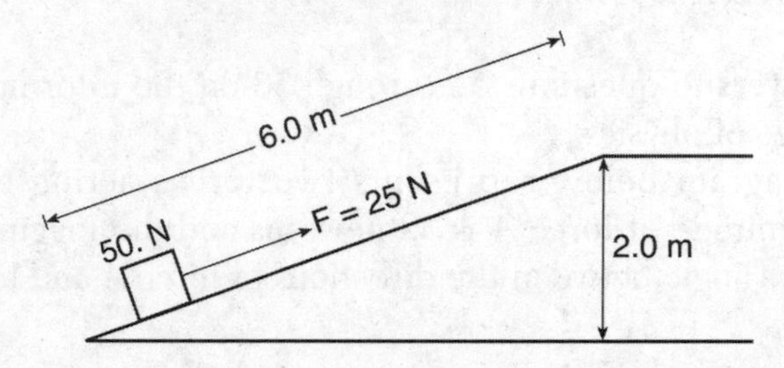

54–55 Calculate the total work done on the box by the student while pushing the box from the bottom to the top of the incline. [Show all work, including the equation and substitution with units.] [2]

56 Describe what would happen to the total work done on the box by the student to push the box 6.0 meters up the incline at constant speed if the coefficient of kinetic friction between the box and the incline were increased. [1]

57 In the diagram below, a light ray is incident on an interface between glass and air. When the light strikes the glass-air interface, some of the light is reflected. On the diagram *on the answer sheet*, use a protractor and straightedge to construct the reflected light ray. [1]

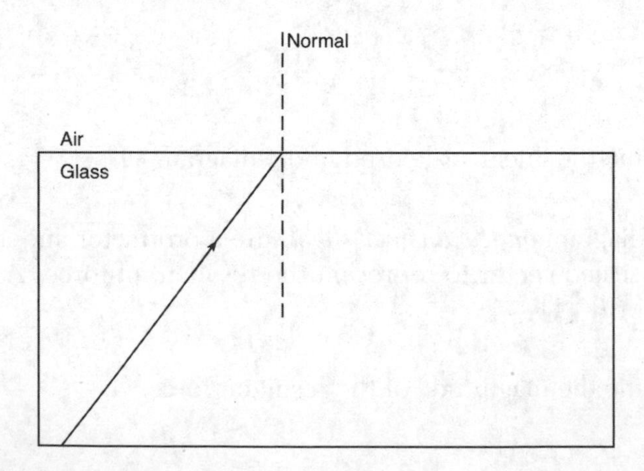

58–59 The current in a wire is 5.0 amperes. Calculate the total amount of charge, in coulombs, that travels through the wire in 36 seconds. [Show all work, including the equation and substitution with units.] [2]

60–61 A spring, with a spring constant of 100 newtons per meter, possesses 2.0 joules of elastic potential energy when compressed. Calculate the spring's change in length from its uncompressed length. [Show all work, including the equation and substitution with units.] [2]

62–63 A monochromatic ray of light ($f = 5.09 \times 10^{14}$ Hz) travels from air into medium *X*. The angle of incidence of the ray in air is 45.0° and the ray's angle of refraction in medium *X* is 29.0°. Calculate the absolute index of refraction of medium *X*. [Show all work, including the equation and substitution with units.] [2]

64–65 An argon-ion laser emits blue-green light having a wavelength of 488 nanometers in a vacuum. Calculate the energy of a photon of this light. [Show all work, including the equation and substitution with units.] [2]

PART C

Answer all questions in this part.

Directions (66–85): Record your answers on the sheet provided in the back. Some questions may require the use of the *2006 Edition Reference Tables for Physical Setting/Physics*.

Base your answers to questions 66 through 70 on the information and diagram below and on your knowledge of physics.

An incandescent lightbulb uses a length of thin tungsten wire as the filament (the part of the operating bulb that produces light).

Incandescent Lightbulb

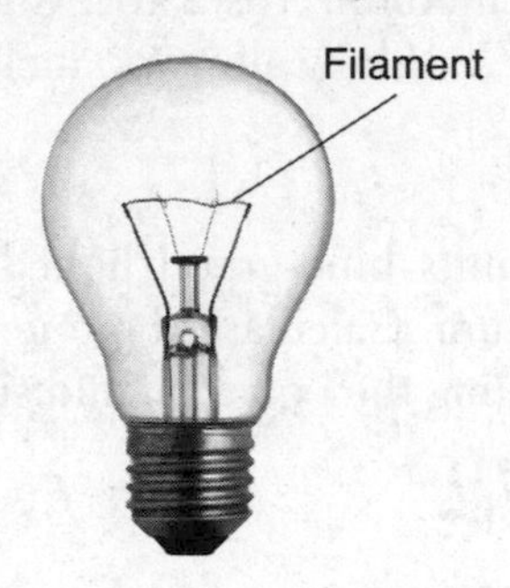

One particular lightbulb has a 0.22-meter length of the tungsten wire used as its filament. This tungsten wire filament has a resistance of 19 ohms at a temperature of 20°C. The tungsten wire filament has a resistance of 240 ohms when this bulb is operated at a potential difference of 120 volts.

66–67 Calculate the cross-sectional area of this tungsten wire filament. [Show all work, including the equation and substitution with units.] [2]

68 Explain why the resistance of the tungsten wire filament increases when the bulb is being operated compared to the resistance of the filament at 20°C. [1]

69–70 Calculate the power of this lightbulb when it is being operated at a potential difference of 120 volts. [Show all work, including the equation and substitution with units.] [2]

Base your answers to questions 71 through 75 on the information and diagram below and on your knowledge of physics.

A 150-newton force, applied to a wooden crate at an angle of 30° above the horizontal, causes the crate to travel at constant velocity across a horizontal wooden floor, as represented below.

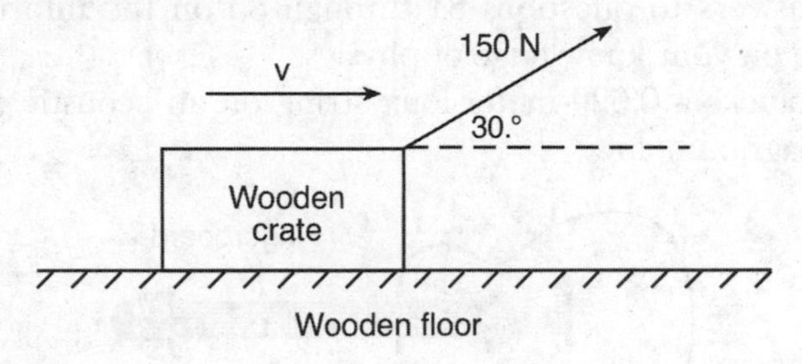

71–72 Calculate the magnitude of the horizontal component of the 150-newton force. [Show all work, including the equation and substitution with units.] [2]

73 Determine the magnitude of the frictional force acting on the crate. [1]

74–75 Calculate the magnitude of the normal force exerted by the floor on the crate. [Show all work, including the equation and substitution with units.] [2]

Base your answers to question 76 through 80 on the information below and on your knowledge of physics.

On a flat, level road, a 1500-kilogram car travels around a curve having a constant radius of 45 meters. The centripetal acceleration of the car has a constant magnitude of 3.2 meters per second squared.

76–77 Calculate the car's speed as it travels around the curve. [Show all work, including the equation and substitution with units.] [2]

78 Determine the magnitude of the centripetal force acting on the car as it travels around the curve. [1]

79 What force provides the centripetal force needed for the car to travel around the curve? [1]

80 Describe what happens to the magnitude of the centripetal force on the car as it travels around the curve if the speed of the car *decreases*. [1]

____

Base your answers to questions 81 through 85 on the information and diagram below and on your knowledge of physics.

A musician plucks a 0.620-meter-long string on an acoustic guitar, as represented in the diagram below.

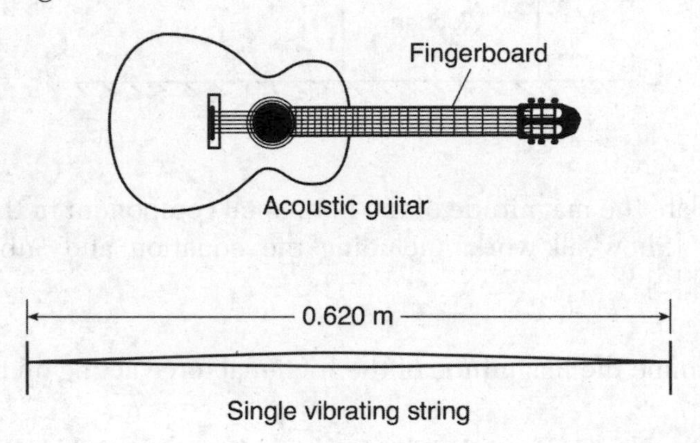

The plucked string vibrates, producing a musical note called "G." The waves traveling along the vibrating string produce a standing wave with a frequency of 196 hertz.

81 On the diagram of the standing wave *on the answer sheet*, label *one* node with the letter **N** and one antinode with the letter **A**. [1]

82 Determine the wavelength of the standing wave on the 0.620-meter-long vibrating string. [1]

83–84 Calculate the speed of the wave traveling on the vibrating string. [Show all work, including the equation and substitution with units.] [2]

85 Describe what happens to the frequency when the musician shortens the vibrating portion of the string by pinching the string against the fingerboard while the string continues to vibrate. [1]

Answer Sheet June 2018

Physics: The Physical Setting

PART B–2

51 1.0 cm = ______________________ **N**

52

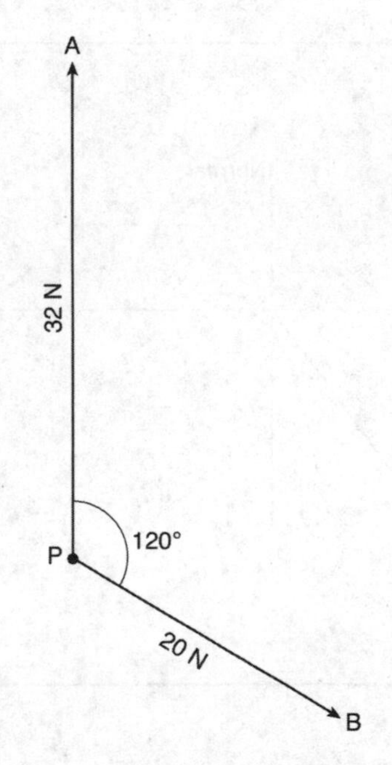

53 ____________________ N

54–55

56 ____

57

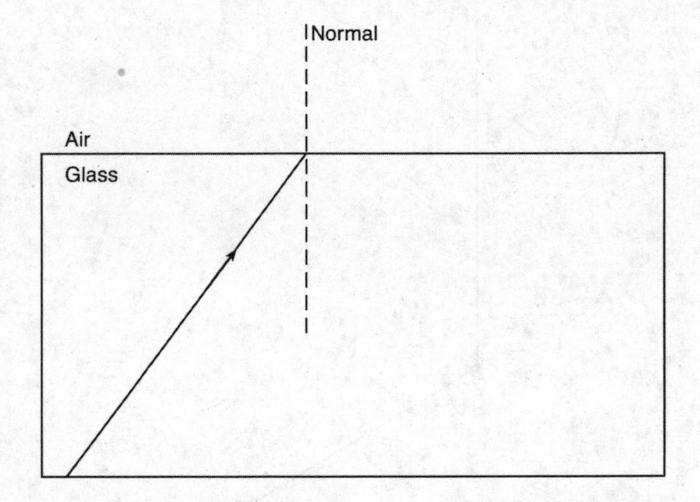

58–59

60–61

62–63

64–65

PART C

66–67

68 ____

69–70

71–72

73 ____________________ N

74–75

76–77

78 ________________ **N**

79 ______________________________________

80 ____

81

0.620 m

82 ____________________________ **m**

83–84

85 ____

Answers June 2018

Physics: The Physical Setting

Answer Key

PART A

1. 2	**8.** 4	**15.** 2	**22.** 4	**29.** 2
2. 1	**9.** 1	**16.** 3	**23.** 3	**30.** 2
3. 4	**10.** 3	**17.** 3	**24.** 2	**31.** 2
4. 2	**11.** 4	**18.** 4	**25.** 1	**32.** 4
5. 3	**12.** 1	**19.** 3	**26.** 1	**33.** 1
6. 4	**13.** 2	**20.** 2	**27.** 4	**34.** 1
7. 3	**14.** 1	**21.** 3	**28.** 3	**35.** 4

PART B–1

36. 2	**39.** 1	**42.** 2	**45.** 2	**48.** 3
37. 3	**40.** 4	**43.** 4	**46.** 1	**49.** 3
38. 2	**41.** 3	**44.** 4	**47.** 2	**50.** 1

PART B–2 and **PART C**. *See* **Answers Explained**.

Answers Explained

PART A

1 The meter/second (m/s) is the unit of measure for both speed and velocity. Velocity is the correct answer because it is a vector quantity, having both magnitude and direction.

WRONG CHOICES EXPLAINED:

(1) Weight is a vector quantity. However, its unit of measure is the newton (N), not the kilogram (kg).

(3) Speed is measured in meters/second (m/s). However, speed is a scalar quantity. It has magnitude but not direction.

(4) Acceleration is a vector quantity. However, its unit of measure is m/s^2 not m^2/s.

2 Mass is the quantitative measure of inertia. In this example, the mass of the cart has not changed. Therefore, the inertia of the cart will be unchanged.

3 Refer to the Mechanics Section of the Physics Reference Tables.

Given: $v_i = 0\text{ m/s}$

$v_f = 40\text{ m/s}$

$d = 70\text{ m}$

Find: $a = ?$

Solution: $v_f^{\,2} = v_i^{\,2} + 2ad$

$$a = \frac{v_f^{\,2} - v_i^{\,2}}{2d}$$

$$a = \frac{(40\text{ m/s})^2 - (0\text{ m/s})^2}{2(70\text{ m})}$$

$$a = 11\text{ m/s}^2$$

4 Refer to the *Mechanics* section of *Reference Table K.*

Given: $m = 6.0$ kg

$v_i = 4.0$ m/s E

$a = 0.50$ m/s^2 E

$t = 3.0$ s

Find: $v_f = ?$

Solution: $v_f = v_i + at$

$$v_f = 4.0 \text{ m/s} + \left(\left(0.50 \text{ m/s}^2\right)(3.0 \text{ s})\right)$$

$$v_f = 5.5 \text{ m/s}$$

5 The range of a projectile is determined by two parameters—the value of the horizontal component of the initial velocity and the hang time (time in the air).

Refer to the *Mechanics* section of *Reference Table K.* The horizontal component of the velocity is proportional to the velocity and the cosine of the launch angle. The larger the cosine of the angle the greater the horizontal speed.

The hang time (time of flight) is determined by the vertical component of the initial velocity and the acceleration due to gravity. The vertical component of the velocity is proportional to the velocity and the sine of the launch angle. The greater the sine of the angle, the greater the vertical speed, and the longer the time of flight or hang time.

The horizontal distance is proportional to the horizontal speed and time, the greater the horizontal speed and/or the greater the time the greater the distance traveled.

Smaller launch angles have a greater horizontal component of speed but a smaller vertical one. Larger launch angles have a greater vertical component but a smaller horizontal component. An angle of 45° results in neither category larger than the other compared to the other angles but it results it a strong enough showing in both categories to achieve the largest range.

WRONG CHOICES EXPLAINED:

(1) and (2) A smaller angle will result in a larger horizontal velocity but the range will be limited due to the shorter hang time.

(4) A larger angle will result in greater hang time but a smaller horizontal velocity and thus a shorter range.

6 Refer to the *Mechanics* section of *Reference Table K.*

Given: $v_i = 0\text{ m/s}$
$d = 18\text{ m}$
$t = 3.0\text{ s}$

Find: $a = ?$

Solution: $d = v_i t + \frac{1}{2}at^2$

$$a = \frac{2(d - v_i t)}{t^2}$$

$$a = \frac{2(18\text{ m} - ((0\text{ m/s})(3.0\text{ s})))}{(3.0\text{ s})^2}$$

$$a = 4.0\text{ m/s}^2$$

7 The only acceleration present is the acceleration in the downward direction. This is the acceleration due to gravity, which is approximately equal to 9.81 m/s^2 and is constant.

8 The time of flight, or hang time, is dependent on the motion in the vertical directly (y-direction) only. Increasing the height would increase the amount of time needed for the projectile to hit the ground.

WRONG CHOICES EXPLAINED:

(1) Changing the mass would not affect the vertical velocity in a frictionless environment.

(2) Decreasing the height would decrease the hang time.

(3) Increasing the horizontal velocity would increase the range or horizontal distance traveled but would not affect motion in the vertical direction. Therefore, the hang time would remain the same.

9 Refer to the *Mechanics* section of *Reference Table K*.

Given:

$$v_i = 0 \text{ m/s}$$
$$m = 0.050 \text{ kg}$$
$$F = 5.0 \times 10^3 \text{ N}$$
$$v_f = 44 \text{ m/s}$$

Find:

$$J = ?$$

Solution:

$$p = mv$$
$$J = \Delta p$$
$$J = m\Delta v$$
$$J = m\left(v_f - v_i\right)$$
$$J = (0.050 \text{ kg})(44 \text{ m/s} - 0 \text{ m/s})$$
$$J = 2.2 \text{ N} \bullet \text{s}$$

10 According to Newton's third law of motion, the force that the racket exerts on the tennis ball is equal but opposite in direction to the force that the tennis ball exerts on the racket. Since the racket exerts an average force of 200 N on the tennis ball, the average force exerted on the racket by the tennis ball is also 200 N.

11 The force of friction always acts in the direction opposite to the direction of motion. In the diagram, the box is moving down the incline in direction *B*. Therefore, friction acts in direction *D*.

WRONG CHOICES EXPLAINED:

(1) *A* is the direction of the normal force.
(2) *B* is the direction of motion.
(3) *C* is the direction of the weight of the box.

12 As an object moves in a circular path, its acceleration, *a*, always points toward the center of the circle. The direction of the velocity, *v*, at any point on the circular path is tangent to the circle. Only choice (1) accurately represents both of these conditions.

WRONG CHOICES EXPLAINED:

(2) The arrows are reversed for acceleration, a, and velocity, v.

(3) The arrows for both acceleration, a, and velocity, v, are pointing in incorrect directions.

(4) The arrow for velocity, v, is correct. However, the arrow for acceleration, a, is pointing outward instead of pointing toward the center of the circular path.

13 Refer to the *Electricity* section of *Reference Table K*.

Given: $E = 2.0 \times 10^3$ N/C

$F_e = 3.0 \times 10^{-3}$ N

Find: $q = ?$

Solution:

$$E = \frac{F_e}{q}$$

$$q = \frac{F_e}{E}$$

$$q = \frac{3.0 \times 10^{-3}\ \text{N}}{2.0 \times 10^3\ \text{N/C}}$$

$$q = 1.5 \times 10^{-6}\ \text{C}$$

WRONG CHOICES EXPLAINED:

(1) This is equivalent to 1 elementary charge.

(3) The magnitudes of the electric field and the electrostatic force were multiplied together instead of being divided appropriately.

14 Refer to the *Mechanics* section of *Reference Table K*.

The gravitational field strength, which is also acceleration due to gravity or g, is equal to $\frac{F_g}{m}$.

WRONG CHOICES EXPLAINED:

(2) G is the universal gravitational constant.

(3) mg is equal to weight, which is also the force due to gravity.

(4) $\frac{Gm_1m_2}{r^2}$ is equal to weight, which is also the force due to gravity.

15 Refer to the *Mechanics* section of *Reference Table K.*

Given: $W = 2.0 \times 10^3$ J
$t = 10$ s

Find: $P = ?$

Solution: $$P = \frac{W}{t}$$

$$P = \frac{2.0 \times 10^3 \text{ J}}{10 \text{ s}}$$

$$P = 2.0 \times 10^2 \text{ W}$$

16 Both magnetic and electrostatic forces can be either attractive or repulsive.

WRONG CHOICES EXPLAINED:

(1), (2), and (4) Gravitational forces are always attractive.

17 Resistivity (ρ) is dependent on a material's internal structure and temperature. In this problem, both wires are made of copper and are at 0°C. Since the internal structure and the temperature of the two wires are the same, the resistivity of the two wires is the same.

The resistance of a wire is proportional to the wire's resistivity and length and is also inversely proportional to the wire's cross-sectional area. Had the question asked about resistance instead of resistivity, doubling the length of the wire would have resulted in double the resistance.

18 Refer to the *Electricity* section of *Reference Table K.*
Work per unit charge, $\frac{W}{q}$, is equal to V, which is potential difference.

WRONG CHOICES EXPLAINED:

(1) Electrostatic force can be shown by $F_e = \frac{kq_1q_2}{r^2}$, which does not equal $\frac{W}{q}$.

(2) Power can be shown by $P = VI = I^2R = \frac{V^2}{R}$, none of which equal $\frac{W}{q}$.

(3) Field strength can be shown by $E = \frac{F_e}{q}$, which does not equal $\frac{W}{q}$.

19 Refer to the *Electricity* section of *Reference Table K.* If you have the resistance and voltage, you can find the current. However, the question does not directly give a value for the resistance in either situation.

$$R = \frac{V}{I}$$

$$R = \frac{\rho L}{A}$$

The question states that each situation uses the "same copper wire at the same temperature." This means the resistivity (ρ), length (L), and area (A) of the wire in each situation is the same. Therefore, the resistance in each situation, R_1 and R_2, must be the same, or $R_1 = R_2$.

Given: $V_1 = 24$ mV

$I_1 = 0.40$ A

$V_2 = 48$ mV

$R_1 = R_2$

Find: $I_2 = ?$

Solution:

$$R = \frac{V}{I}$$

$$\frac{V_1}{I_1} = \frac{V_2}{I_2}$$

$$I_2 = \frac{V_2 I_2}{V_1}$$

$$I_2 = \frac{(48 \text{ mV})(0.40 \text{ A})}{(24 \text{ mv})}$$

$$I_2 = 0.80 \text{ A}$$

20 Refer to *Reference Table* A and the *Mechanics* section of *Reference Table K.*

Given: $m_1 = m_2 = 0.425 \text{ kg}$

$r = 0.500 \text{ m}$

$G = 6.67 \times 10^{-11} \text{ N} \bullet \text{m}^2/\text{kg}^2$

Find: $F_g = ?$

Solution: $F_g = \dfrac{Gm_1m_2}{r^2}$

$$F_g = \frac{\left(6.67 \times 10^{-11} \text{ N} \bullet \text{m}^2/\text{kg}^2\right)(0.425 \text{ kg})(0.425 \text{ kg})}{(0.500 \text{ m})^2}$$

$F_g = 4.82 \times 10^{-11} \text{ N}$

21 Light waves are electromagnetic waves and do not need a medium through which to travel. They can travel in a vacuum as well as in a medium, such as water. Sound waves, though, are mechanical waves and require a medium through which to travel. Therefore, sound waves can travel through water but not through a vacuum.

22 Resonance is the spontaneous vibration of an object at a frequency equal to that of the wave that initiates that resonant vibration. If that vibration is strong enough, it can shatter or break the object.

WRONG CHOICES EXPLAINED:

(1) Action and reaction are terms used to describe Newton's third law of motion.

(2) The law of conservation of momentum states that if two objects that are not subjected to external forces interact, the total momentum of the objects before the interaction is equal to the total momentum of the objects after the interaction.

(3) Inertia is the property of matter that resists changes in motion.

23 The frequency of a light wave does not change as the wave travels from one medium to another. Period is the inverse of frequency and, as such, also does not change as a light wave travels from one medium to another.

WRONG CHOICES EXPLAINED:

(1) The speed of a light wave does change as the light travels from one medium to another.

(2) and (4) When the speed of a light wave changes as the wave enters a new medium, the wavelength of that light wave changes proportionally.

24 Refer to the *Reference Table A*, the *Absolute Indices of Refraction* (*Reference Table E*), and the *Waves* section of *Reference Table K*. Note that the frequency of the light ray is not needed to solve this problem.

Given: $f = 5.09 \times 10^{14}$ Hz

$n = 1.47$

$c = 3.00 \times 10^8$ m/s

Find: $v = ?$

Solution:
$$n = \frac{c}{v}$$
$$v = \frac{c}{n}$$
$$v = \frac{3.00 \times 10^8 \text{ m/s}}{1.47}$$
$$v = 2.04 \times 10^8 \text{ m/s}$$

WRONG CHOICES EXPLAINED:

(3) This is the value for the speed of light in a vacuum, not the speed of light in corn oil.

(4) You multiplied the speed of light by the index of refraction instead of dividing the two quantities.

25 Diffraction is the bending of a wave around a barrier.

WRONG CHOICES EXPLAINED:

(2) The Doppler effect is the apparent change in frequency that results when a wave source and an observer are in relative motion with respect to each other.

(3) When light is reflected from a surface, the angle that the incident light ray makes with the normal to the surface is equal to the angle that the reflected ray makes with the normal to the surface.

(4) Refraction is the change in direction of a wave when it passes obliquely from one medium into another, which causes the wave to move at a different speed.

26 Refer to the *Waves* section of *Reference Table K.*

Given: $f = 5.0 \times 10^{10}$ Hz

Find: $T = ?$

Solution: $T = \frac{1}{f}$

$$T = \frac{1}{5.0 \times 10^{10}\ \text{Hz}}$$

$$T = 2.0 \times 10^{-11}\ \text{s}$$

27 Two or more waves passing simultaneously through the same area of a medium affect the medium independently but do not affect each other. Therefore, each wave will remain unchanged.

WRONG CHOICES EXPLAINED:

(1) Apparent changes in frequency, called the Doppler effect, result when a wave source and an observer are in relative motion with respect to each other. This is not the case in this situation.

(2) Changes in velocity occur as a wave moves from one medium into another, which is not the case in this situation.

(3) Changes in amplitude occur at the point of interference. However, each wave returns to its original amplitude when the interference ends and/or after the waves have passed by each other. The question specifically states "after two light waves have interfered." Therefore, the two waves are no longer interfering with each other.

28 When a glass rod is rubbed with silk, electrons leave the rod and transfer to the silk. This leaves the rod with a net positive charge and the silk with a net negative charge.

WRONG CHOICES EXPLAINED:

(1) Adding electrons to the rod would have resulted in a net negative charge on the rod.

(2) and (4) Protons are contained in the nucleus of atoms and therefore are not free to move to or from any object.

29 Refer to *Reference Table A* and the *Modern Physics* section of *Reference Table K*.

Given: $E_{photon} = 1.33 \times 10^{-21}$ J

$h = 6.63 \times 10^{-34}$ J • s

Find: $f = ?$

Solution:

$$E_{photon} = hf$$

$$f = E_{photon}$$

$$f = \frac{1.33 \times 10^{-21} \text{ J}}{6.63 \times 10^{-34} \text{ J} \bullet \text{s}}$$

$$f = 2.01 \times 10^{12} \text{ Hz}$$

30 This question asks you to find the average of two values. To do so, add the given values and then divide by 2.

Given: $v_i = 11$ m/s

$v_f = 19$ m/s

Find: $\overline{v} = ?$

Solution:

$$\overline{v} = \frac{v_i + v_f}{2}$$

$$\overline{v} = \frac{11 \text{ m/s} + 19 \text{ m/s}}{2}$$

$$\overline{v} = 15 \text{ m/s}$$

31 Refer to *Reference Table A* and the *Modern Physics* section of *Reference Table K.*

Given: $m_e = 9.11 \times 10^{-31}$ kg

$c = 3.00 \times 10^8$ m/s

Find: $E = ?$

Solution: $E = mc^2$

$$E = \left(9.11 \times 10^{-31}\ \text{kg}\right)\left(3.00 \times 10^8\ \text{m/s}\right)^2$$

$$E = 8.20 \times 10^{-14}\ \text{J}$$

32 Refer to the *Mechanics* section of *Reference Table K.*

Given: $x_i = 0.40$ m

$x_f = 0.60$ m

$F = 10$ N

Find: $k = ?$

Solution: $F_s = kx$

$$F_s = k\Delta x$$

$$k = \frac{F_s}{\Delta x}$$

$$k = \frac{10\ \text{N}}{(0.60\ \text{m} - 0.40\ \text{m})}$$

$$k = 50.\ \text{N/m}$$

33 A closed circuit loop exists between the battery and lamp I, allowing an electric current to flow through only lamp I.

WRONG CHOICES EXPLAINED:

(2) and (3) The open switch does not allow a closed circuit loop that contains lamps II and III. Therefore, no electric current is flowing through those lamps.

(4) Since lamp I is in a closed loop circuit, electric current is flowing through that lamp.

34 A field line represents the path that a small positive charge (known as a test charge) takes while in an electric field. Arrows therefore point toward negative charges and away from positive charges in an electric field. The diagram in choice (1) correctly represents this convention.

WRONG CHOICES EXPLAINED:

(2), (3), and (4) The field lines are pointing in the wrong direction in each of these choices.

35 Points on a periodic wave are in phase if they are at equal displacements from the rest position and are experiencing identical movements. For the points to be experiencing identical movements, they must be moving in the same direction toward or away from the rest position. Points on a periodic wave are 180° degrees out of phase, or completely out of phase, if they are at equal displacements from their rest position but experiencing motion in opposite directions from each other. Points *D* and *E* are about halfway to the 180° out-of-phase positions. Therefore, points *D* and *E* are 90° out of phase with each other.

WRONG CHOICES EXPLAINED:

(1) Points *A* and E are in phase with each other.
(2) Points *B* and *C* are in phase with each other.
(3) Points *C* and *D* are 180° out of phase with each other.

PART B–1

36 This is an order of magnitude question: 2.0×10^{-1} m $\approx$ 0.2 m. Since 1 meter is approximately equal to 3.3 feet or 40 inches, 0.2 meters is approximately 8 inches. This is a reasonable approximation for the height of an individual step on a staircase.

WRONG CHOICES EXPLAINED:

(1) 2.0×10^{-1} m = 0.02 m = 2 cm, which is not a reasonable approximation for the height of an individual step.

(3) 2.0×10^{0} m = 2 m $\approx$ 6.6 ft, which is not a reasonable approximation for the height of an individual step.

(4) 2.0×10^{1} m = 20 m $\approx$ 66 ft, which is not a reasonable approximation for the height of an individual step.

37 Refer to *Reference Table A* and the *Electricity* section of *Reference Table K.*

Given: $q_1 = q_2 = e = 1.60 \times 10^{-19}$ C

$r = 0.10$ m

$k = 8.99 \times 10^9$ N • m^2/C^2

Find: $F_e = ?$

Solution: $F_e = \dfrac{kq_1q_2}{r^2}$

$$F_e = \frac{\left(8.99 \times 10^9 \text{ N} \bullet \text{m}^2/\text{C}^2\right)\left(1.6 \times 10^{-19} \text{ C}\right)\left(1.6 \times 10^{-19} \text{ C}\right)}{(0.10 \text{ m})^2}$$

$F_e = 2.3 \times 10^{-26}$ N

38 Refer to the *Mechanics* section of *Reference Table K.*

Given: $Fg = 65$ N

$h = 5.3$ m

Find: $KE = ?$

Solution: $g = \dfrac{F_g}{m}$ (Law of Conservation of Energy)

$F_g = mg$

$\Delta PE = mg\Delta h$

$KE = \Delta PE$

$KE = F_g\Delta h$

$KE = (65 \text{ N})(5.3 \text{ m})$

$KE = 340$ J

39 Refer to the *Mechanics* section of *Reference Table K*. The carts are moving in opposite directions. Let movement to the right be in the positive direction and movement to the left be in the negative direction.

Given: $m_1 = 0.500\text{ kg}$
$v_1 = 2.20\text{ m/s}$
$m_2 = 0.800\text{ kg}$
$v_2 = -1.10\text{ m/s}$

Find: $p_{after} = ?$

Solution: $p_{before} = p_{after}$

$$p = mv$$

$$p_{before} = m_1v_1 + m_2v_2$$

$$p_{before} = (0.500\text{ kg})(2.20\text{ m/s}) + (0.800\text{ kg})(-1.10\text{ m/s})$$

$$p_{before} = 0.22\text{ kg} \bullet \text{m/s}$$

40 The downward force acting on the object due to the object's weight, or the force due to gravity, is balanced by the normal force acting in the upward direction (Newton's third law of motion). A horizontal force is also acting on the object. Since the object is in a frictionless environment, no force due to friction opposes the horizontal motion. Therefore, the net force acting on the object is the 4.0 N force pushing the object.

41 Refer to the *Electromagnetic Spectrum* in *Reference Table D*. According to the chart, AM radio waves have a wavelength of approximately 10^2 m in a vacuum and FM radio waves have a wavelength of approximately 10^0 m, or 1 m, in a vacuum. Therefore in a vacuum, the ratio of the wavelength of AM radio waves to the wavelength of FM radio waves is 10^2 to 1.

42 Refer to *Reference Table A* and *Reference Table H*. A charm quark carries a charge of $+\frac{2}{3}e$. An elementary charge, e, has a value of 1.6×10^{-19} C. Therefore, a charm quark carries a charge of $\left(1.6 \times 10^{-19}\text{ C}\right)\left(\frac{2}{3}\right) = 1.07 \times 10^{-19}$ C.

WRONG CHOICES EXPLAINED:

(3) This is the value of one elementary charge, but a charm quark carries a charge of $+\frac{2}{3}e$.

(4) You multiplied the elementary charge by $\frac{3}{2}$ instead of multiplying by $\frac{2}{3}$.

43 Refer to the *Electricity* section of *Reference Table K*. Meter *X* is properly connected in series in the circuit to measure current.

Given: $V = 12\ \text{V}$

$R = 3.0\ \Omega$

Find: $I = ?$

Solution: $R = \frac{V}{I}$

$$I = \frac{V}{R}$$

$$I = \frac{12\ \text{V}}{3.0\ \Omega}$$

$$I = 4.0\ \text{A}$$

Meter *Y* is correctly connected in parallel with the resistor to measure potential difference across the resistor. The resistor is connected in series with the battery and is the only resistor in the circuit. Therefore, the potential difference across the resistor is equal to the potential difference of the battery, which is 12 V.

WRONG CHOICES EXPLAINED:

(1) *X* is an ammeter, not a voltmeter. You divided 3 by 12 instead of dividing 12 by 3 to obtain the value of the current in the circuit. Additionally, *Y* is a voltmeter, not an ammeter.

(2) *X* is an ammeter, not a voltmeter. *Y* is a voltmeter, not an ammeter.

(3) You divided 3 by 12 instead of dividing 12 by 3 to obtain the value of the current in the circuit.

44 Refer to the *Geometry and Trigonometry* section and the *Mechanics* section of *Reference Table K*.

Given: $d = 10$ complete circles

$t = 30$ s

$r = 4.0$ m

Find: $\bar{v} = ?$

Solution: $\bar{v} = \dfrac{d}{t}$

$C = 2\pi r$

$d = 10C$

$\bar{v} = \dfrac{10(2\pi r)}{t}$

$\bar{v} = \dfrac{10(2\pi \bullet 4.0\text{ m})}{30\text{ s}}$

$\bar{v} = 8.4$ m/s

45 An object falling freely experiences a constant acceleration due to gravity. The slope of a displacement versus time graph represents the velocity of the object. If an object experiences acceleration, its instantaneous velocity increases as time passes. Therefore, the curve for displacement versus time becomes steeper as time passes. The slope of a velocity versus time graph is equal to the object's acceleration. Since the acceleration due to gravity is constant, this graph should be a straight line increasing from zero as time goes by. Only the graphs depicted in choice (2) meet these criteria.

WRONG CHOICES EXPLAINED:

(1) The displacement versus time graph indicates constant velocity, while the velocity versus time graph indicates increasing velocity.

(3) The displacement versus time graph indicates constant velocity, while the velocity versus time graph indicates increasing acceleration.

(4) The displacement versus time graph indicates acceleration, while the velocity versus time graph indicates increasing acceleration.

46 As the object rises, it decelerates until it reaches its maximum height. At that point, the object's speed reaches 0 m/s (at which point it begins accelerating downward). As its speed decreases as it rises, the kinetic energy of the object decreases until the kinetic energy reaches zero at the object's maximum height.

WRONG CHOICES EXPLAINED:

(2) This graph indicates that as height increases, kinetic energy increases.

(3) This graph indicates that as height increases, kinetic energy increases exponentially.

(4) This graph indicates that as height increases, kinetic energy remains the same.

47 In a longitudinal wave, a disturbance causes the particles of the material to vibrate in simple harmonic motion in a direction parallel to the direction of motion of the wave. Since the wave is moving to the right, parallel motion would be toward the left and right, which is indicated in the diagram in choice (2).

WRONG CHOICES EXPLAINED:

(1) This would be the type of motion seen for a particle experiencing motion in a surface wave.

(3) These arrows do not indicate any type of simple harmonic motion.

(4) This would be the type of motion observed in a particle experiencing motion in a tranverse wave.

48 The principle of superposition states that the resultant displacement of any point in the medium is the algebraic sum of the displacements of all the individual waves. In this case 0.30 m + 0.30 m = 0.60 m.

WRONG CHOICES EXPLAINED:

(1) This is the displacement due to *X* or due to *Y* but not due to both.

(2) This is a displacement equal in magnitude to *X* or to *Y* but in the opposite direction.

(4) This is a displacement equal to the sum of the *X* and *Y* displacements but in the opposite direction.

49 The Doppler effect is the apparent change in frequency that results when a wave source and an observer are in relative motion with respect to each other. As the distance between the source and the observer decreases, the frequency of the source, as perceived by the observer, is increased. As the distance increases, the observed frequency is decreased.

WRONG CHOICES EXPLAINED:

(1) This graph indicates that as time goes by, the observed frequency increases.

(2) This graph indicates that as time goes by, the observed frequency decreases.

(4) This graph indicates that as time goes by, the observed frequency is initially lower and then the observed frequency is higher.

50 Refer to the *Electricity* section of *Reference Table K.*

Given: $R_{T_{series}} = 10\ \Omega$

$R_1 = R_2$

Find: $R_{T_{parallel}} = ?$

Solution:

$$R_1 = R_2 = \frac{R_{T_{series}}}{2}$$

$$R_1 = R_2 = \frac{10}{2}$$

$$R_1 = R_2 = 5\ \Omega$$

$$\frac{1}{R_{T_{parallel}}} = \frac{1}{R_1} + \frac{1}{R_2}$$

$$\frac{1}{R_{T_{parallel}}} = \frac{1}{5\ \Omega} + \frac{1}{5\ \Omega}$$

$$\frac{1}{R_{T_{parallel}}} = \frac{2}{5\ \Omega}$$

$$R_{T_{parallel}} = \frac{5\ \Omega}{2}$$

$$R_{T_{parallel}} = 2.5\ \Omega$$

WRONG CHOICES EXPLAINED:

(2) You forgot to divide by 2 and assumed that both resistors were 10 ohms.

(3) You didn't do any calculations and assumed that both series and parallel circuits with 2 identical resistors have the same resistance.

PART B–2

51 The answer booklet asks for the equivalent to 1.0 cm. Using a centimeter ruler, measure the length of either vector *PB* or vector *PA*. Vector *PB* is approximately 5.0 cm in length and 20 N ÷ 5.0 cm = 4.0 N/cm. Vector *PA* is approximately 8.0 cm in length and 32 N ÷ 8.0 cm = 4.0 N/cm.

One credit is awarded for an answer of 4.0 N ± 0.2 N.

52

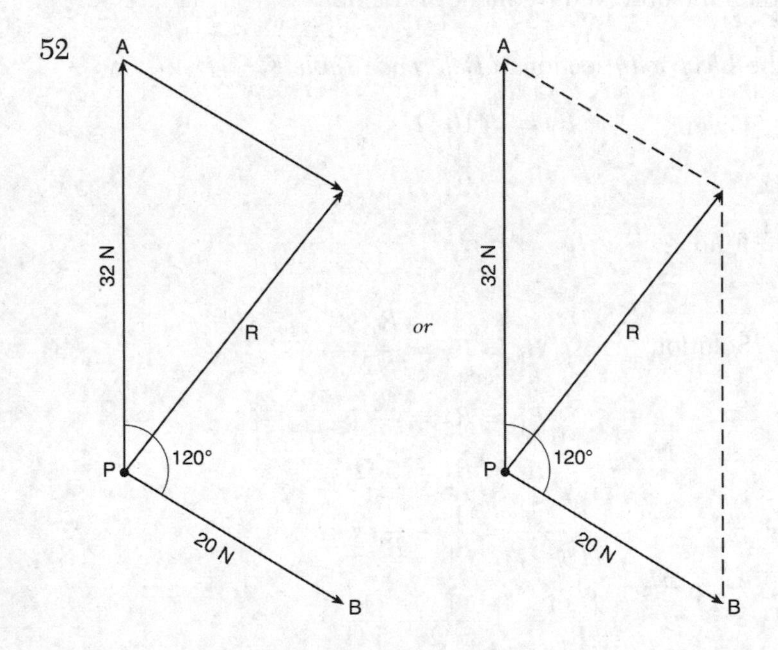

One credit is awarded for constructing the resultant 7.0 cm ± 0.2 cm long at an angle of 38° ± 2° clockwise from force *A*. The resultant vector need not be labeled to receive this credit.

53 Use a centimeter ruler to find the length of resultant vector *R*. Its length is approximately 7.0 cm. Using the scale factor determined in question 51, which is 1.0 cm for every 4.0 N, we can determine that resultant *R* has a value of 7.0 cm × 4.0 N/cm = 28 N.

One credit is awarded for an answer of 28 N ± 2 N or for an answer that is consistent with your response to questions 51 and 52.

54–55 Refer to the *Mechanics* section of *Reference Table K.*

Given: $F_g = 50$ N
$d = 6.0$ m
$F_{applied} = 25$ N
$h = 2.0$ m

Find: $W = ?$

Solution: $W = Fd$
$W = (25 \text{ N})(6.0 \text{ m})$
$W = 150 \text{ N} \bullet \text{m}$ or 150 J

54 One credit is awarded for a correct equation and for substitution of values with units.

55 One credit is awarded for the correct answer with units or for an answer, with units, that is consistent with your answer to question 54.

Note: Students will not be penalized more than 1 credit for errors in units in questions 54 and 55.

56 If the coefficient of kinetic friction increased, the force of friction would increase. If the force of friction increased, a greater force would have to be applied to overcome the force of friction to move the box at constant speed. Hence, the total work done would increase.

One credit is granted for stating that the total work would increase.

57 When you measure the angle that the incident ray makes with the normal, you obtain a value of approximately 37°. Next, draw a reflected ray at an angle of 37° from the normal.

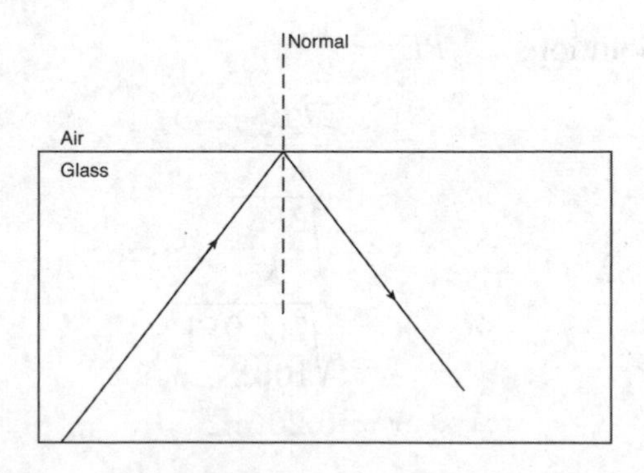

One credit is awarded for a light ray drawn at an angle of 37° ± 2°.

Note: No arrowhead is required on the reflected ray. The direction of the ray is implied by the arrowhead on the incident ray. (Rays are not vectors.)

58–59 Refer to the *Electricity* section of *Reference Table K*.

Given: $I = 5.0\text{ A}$

$t = 36\text{ s}$

Find: $q = ?$

Solution:

$$I = \frac{\Delta q}{t}$$

$$\Delta q = It$$

$$\Delta q = (5.0\text{ A})(36\text{ s})$$

$$\Delta q = 180\text{ A} \bullet \text{s} \quad \text{or} \quad 180\text{ C}$$

58 One credit is awarded for a correct equation and substitution of values with units.

59 One credit is awarded for the correct answer with units or for an answer, with units, that is consistent with your answer to question 58.

Note: Students will not be penalized more than 1 credit for errors in units in questions 58 and 59.

60–61 Refer to the *Mechanics* section of *Reference Table K*.

Given: $k = 100\text{ N/m}$

$PE_s = 2.0\text{ J}$

Find: $\Delta x = ?$

Solution:

$$PE_s = \frac{1}{2}kx^2$$

$$x^2 = \frac{2PE_s}{k}$$

$$x = \sqrt{\frac{2PE_s}{k}}$$

$$x = \sqrt{\frac{2(2.0)\text{ J}}{100\text{ N/m}}}$$

$$x = 0.20\text{ m}$$

60 One credit is awarded for a correct equation and substitution of values with units.

61 One credit is awarded for the correct answer with units or for an answer, with units, that is consistent with your answer to question 60.

Note: Students will not be penalized more than 1 credit for errors in units in questions 60 and 61.

62–63 Refer to *Reference Table E* and the *Waves* section of *Reference Table K.*

Given:

$$\theta_{air} = 45.0°$$
$$\theta_x = 29.0°$$
$$n_{air} = 1.00$$

Find:

$$n_x = ?$$

Solution:

$$n_{air} \sin \theta_{air} = n_X \sin \theta_X$$
$$n_x = \frac{n_{air} \sin \theta_{air}}{\sin \theta_X}$$
$$n_x = \frac{(1.00) \sin 45.0°}{\sin 29.0°}$$
$$n_x = 1.46$$

62 One credit is awarded for a correct equation and substitution of values with units.

63 One credit is awarded for the correct answer or for an answer, without units, that is consistent with your answer to question 62.

Note: Students will not be penalized more than 1 credit for errors in units in questions 62 and 63.

64–65 Refer to *Reference Table A* and the *Modern Physics* section of *Reference Table K.*

Given: $\lambda = 488 \text{ nm} = 4.08 \times 10^{-7} \text{ m}$

$h = 6.63 \times 10^{-34} \text{ J} \bullet \text{s}$

$c = 3.00 \times 10^{8} \text{ m/s}$

Find: $E_{photon} = ?$

Solution:

$$E_{photon} = \frac{hc}{\lambda}$$

$$E_{photon} = \frac{(6.63 \times 10^{-34} \text{ J} \bullet \text{s})(3.00 \times 10^{8} \text{ m/s})}{(4.88 \times 10^{-7} \text{ m})}$$

$$E_{photon} = 4.08 \times 10^{-19} \text{ J}$$

64 One credit is awarded for a correct equation and substitution of values with units.

65 One credit is awarded for the correct answer with units or for an answer, with units, that is consistent with your answer to question 64.

Note: Students will not be penalized more than 1 credit for errors in units in questions 64 and 65.

PART C

66–67 Refer to *Reference Table J* and the *Electricity* section of *Reference Table K.*

Given: $L = 0.22 \text{ m}$

$R_{20°} = 19\ \Omega$

$\rho_{20°} = 5.60 \times 10^{-8}\ \Omega \bullet \text{m}$

Find: $A = ?$

Solution: $R = \frac{\rho L}{A}$

$$A = \frac{\rho L}{R}$$

$$A = \frac{(5.60 \times 10 - 8\ \Omega \bullet \text{m})(0.22 \text{ m})}{19}$$

$$A = 6.5 \times 10^{-10}\ \text{m}^2$$

66 One credit is awarded for a correct equation and substitution of values with units.

67 One credit is awarded for the correct answer with units or for an answer, with units, that is consistent with your answer to question 66.

Note: Students will not be penalized more than 1 credit for errors in units in questions 66 and 67.

68 Generally, the resistance of a metallic conductor increases with increasing temperature. As the bulb operates, it produces heat and the temperature rises, thus increasing the resistance in the tungsten filament.

One credit is allowed for an explanation that the filament of the operating bulb is at a higher temperature. Acceptable responses include but are not limited to:

- The operating bulb is hotter.
- The filament gets hots when the bulb is operating.
- The resistivity of the tungsten increases.
- The temperature of the filament increases.

69–70 Refer to the *Electricity* section of *Reference Table K.*

Given: $R = 240\ \Omega$

$V = 120\ \text{V}$

Find: $P = ?$

Solution: $P = \dfrac{V^2}{R}$

$$P = \frac{(120\ \text{V})^2}{240\ \Omega}$$

$$P = 60\ \text{W}$$

69 One credit is awarded for a correct equation and substitution of values with units.

70 One credit is awarded for the correct answer with units or for an answer, with units, that is consistent with your answer to question 69.

Note: Students will not be penalized more than 1 credit for errors in units in questions 69 and 70.

71–72 Refer to the *Geometry and Trigonometry* section of *Reference Table K.*

Given: $F = 150\ \text{N}$

$\theta = 30°$

Find: $F_x = ?$

Solution: $A_x = A\cos\theta$ or $\cos\theta = \dfrac{b}{c}$

$F_x = F\cos\theta$

$F_x = (150\ \text{N})(\cos 30.0°)$

$F_x = 130\ \text{N}$

or:

$b = c\cos\theta$

$F_x = F\cos\theta$

$F_x = (150\ \text{N})(\cos 30.0°)$

$F_x = 130\ \text{N}$

71 One credit is awarded for a correct equation and substitution of values with units.

72 One credit is awarded for the correct answer with units or for an answer, with units, that is consistent with your answer to question 72.

Note: Students will not be penalized more than 1 credit for errors in units in questions 71 and 72.

73 The wooden crate is moving at constant velocity in the horizontal direction, and it is not accelerating. Therefore, no net force is acting on the box and $F_f = F_x = 130$ N.

One credit is awarded for 130 N or for an answer that is consistent with your response to question 72.

74–75 Refer to *Reference Table C* and the *Mechanics* section of *Reference Table K.*

Given: $F_f = 130 \text{ N}$

$\mu = 0.30$

Find: $F_N = ?$

Solution: $F_f = \mu F_N$

$$F_N = \frac{F_f}{\mu}$$

$$F_N = \frac{130 \text{ N}}{0.30}$$

$$F_N = 430 \text{ N}$$

74 One credit is awarded for a correct equation and substitution of values with units.

75 One credit is awarded for the correct answer with units or for an answer, with units, that is consistent with your answer to question 74.

Note: Students will not be penalized more than 1 credit for errors in units in questions 74 and 75.

76–77 Refer to the *Mechanics* section of *Reference Table K.*

Given: $r = 45 \text{ m}$

$a_c = 3.2 \text{ m/s}^2$

Find: $v = ?$

Solution:
$$a_c = \frac{v^2}{r}$$
$$v^2 = a_c r$$
$$v = \sqrt{a_c r}$$
$$v = \sqrt{(3.2 \text{ m/s}^2)(45 \text{ m})}$$
$$v = 12 \text{ m/s}$$

76 One credit is awarded for a correct equation and substitution of values with units.

77 One credit is awarded for the correct answer with units or for an answer, with units, that is consistent with your answer to question 76.

Note: Students will not be penalized more than 1 credit for errors in units in questions 76 and 77.

78 Refer to the *Mechanics* section of *Reference Table K.*

Given: $m = 1500 \text{ kg}$

$a_c = 3.2 \text{ m/s}^2$

Find: $F_c = ?$

Solution:
$$F_c = ma_c$$
$$F_c = (1500 \text{ kg})(3.2 \text{ m/s}^2)$$
$$F_c = 4800 \text{ N}$$

One credit is awarded for an answer of 4800 N.

79 The force of friction between the car and the road provides the centripetal force necessary to keep the car in a circular path.

One credit is awarded for an answer of friction, static friction, or electromagnetic. Credit is not granted for an answer of centripetal force, net force, gravitational force, weight, or normal force.

80 Refer to the *Mechanics* section of *Reference Table K*.

$$F_c = ma_c$$

$$a_c = \frac{v^2}{r}$$

$$F_c = \frac{mv^2}{r}$$

Centripetal force is proportional to the square of the speed. If the speed decreases, the centripetal force decreases.

One credit is awarded for an acceptable response. Acceptable responses include but are not limited to:

- The magnitude of the centripetal force decreases.
- It becomes less.
- It decreases.

81 In a standing wave, adjacent crests and troughs move vertically in opposite directions about points that have no motion. The points that do not move are called nodes, and the crest-trough combinations are called antinodes.

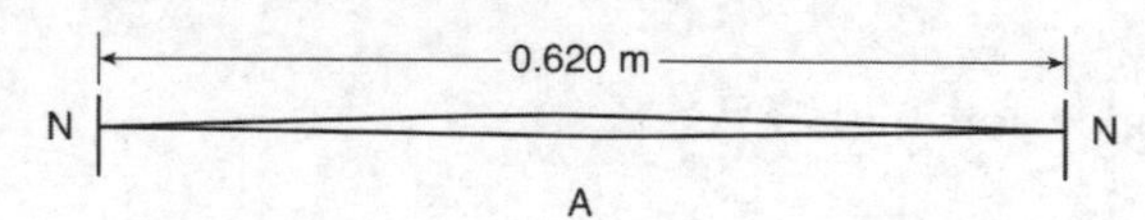

One credit is awarded for labeling one node and one antinode correctly.

82 The diagram indicates the length of a single crest or one-half wavelength. Therefore, the wavelength is $2 \times 0.620 \text{ m} = 1.24 \text{ m}$.

One credit is awarded for an answer of 1.24 m or 1.240 m.

83–84 Refer to the *Waves* section of *Reference Table K*.

Given: $f = 196 \text{ Hz}$
$\lambda = 1.24 \text{ m}$

Find: $v = ?$

Solution: $v = f\lambda$
$v = (196 \text{ Hz})(1.24 \text{ m})$
$v = 243 \text{ m/s}$

83 One credit is awarded for a correct equation and substitution of values with units.

84 One credit is awarded for the correct answer with units or for an answer, with units, that is consistent with your answer to question 83.

Note: Students will not be penalized more than 1 credit for errors in units in questions 83 and 84.

85 Refer to the *Waves* section of *Reference Table K*: $v = f\lambda$. When speed remains constant, frequency and wavelength are inversely proportional to one another. Shortening the string or shortening the wavelength will result in a higher frequency.

One credit is awarded for a statement indicating that the frequency will increase. Acceptable responses include but are not limited to:

- It increases.
- The frequency gets higher.

Topic	Question Numbers (total)	Wrong Answers (x)	Grade
Math Skills	3, 4, 6, 9, 13, 15, 18, 19, 24, 26, 29–32, 36–38, 41–46, 48–55, 57–67, 69–72, 74–78, 80, 82–84: (54)		$\frac{100(54-x)}{54} = \%$
Mechanics	1–12, 14, 20, 30, 32, 40, 44, 45, 52, 53, 71–80: (31)		$\frac{100(31-x)}{31} = \%$
Energy	15, 38, 46, 54, 55, 60, 61: (7)		$\frac{100(7-x)}{7} = \%$
Electricity/ Magnetism	13, 16–19, 28, 33, 34, 37, 43, 50, 56, 58, 59, 66–70: (19)		$\frac{100(19-x)}{19} = \%$
Waves	21–27, 35, 41, 47–49, 57, 62–65, 81–85: (22)		$\frac{100(22-x)}{22} = \%$
Modern Physics	29, 31, 42: (3)		$\frac{100(3-x)}{3} = \%$

Examination June 2019

Physics: The Physical Setting

PART A

Answer all questions in this part.

Directions (1–35): For *each* statement or question, select the *number* of the word or expression that, of those given, best completes the statement or answers the question. Some questions may require the use of the *2006 Edition Reference Tables for Physical Setting/Physics*. Record your answers in the spaces provided.

1 Which pair of quantities represent scalar quantities?

(1) displacement and velocity
(2) displacement and time
(3) energy and velocity
(4) energy and time

1 ____

2 A sailboat on a lake sails 40. meters north and then sails 40. meters due east. Compared to its starting position, the new position of the sailboat is

(1) 40. m due east
(2) 40. m due north
(3) 57 m northeast
(4) 80. m northeast

2 ____

3 A ball is thrown straight upward from the surface of Earth. Which statement best describes the ball's velocity and acceleration at the top of its flight?

(1) Both velocity and acceleration are zero.
(2) Velocity is zero and acceleration is nonzero.
(3) Velocity is nonzero and acceleration is zero.
(4) Both velocity and acceleration are not zero.

3 ____

4 As a student runs a plastic comb through her hair, the comb acquires a negative electric charge. This charge results from the transfer of

(1) protons from the comb to her hair
(2) protons from her hair to the comb
(3) electrons from the comb to her hair
(4) electrons from her hair to the comb

4 ____

5 How would the mass and weight of an object on the Moon compare to the mass and weight of the same object on Earth?

(1) Mass and weight would both be less on the Moon.
(2) Mass would be the same but its weight would be less on the Moon.
(3) Mass would be less on the Moon and its weight would be the same.
(4) Mass and weight would both be the same on the Moon.

5 ____

6 An object is moving with constant speed in a circular path. The object's centripetal acceleration remains constant in

(1) magnitude, only
(2) direction, only
(3) both magnitude and direction
(4) neither magnitude nor direction

6 ____

7 As shown in the diagram below, a rope attached to a 500.-kilogram crate is used to exert a force of 45 newtons at an angle of 65 degrees above the horizontal.

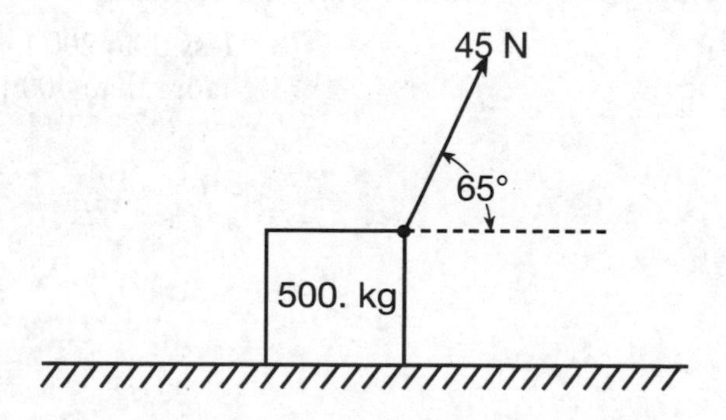

The horizontal component of the force acting on the crate is

(1) 19 N
(2) 41 N
(3) 210 N
(4) 450 N

7 ____

8 A spring with a spring constant of 68 newtons per meter hangs from a ceiling. When a 12-newton downward force is applied to the free end of the spring, the spring stretches a total distance of

(1) 0.18 m
(2) 0.59 m
(3) 5.7 m
(4) 820 m

8 ____

9 As a student walks downhill at constant speed, his gravitational potential energy

(1) increases and his kinetic energy increases
(2) increases and his kinetic energy remains the same
(3) decreases and his kinetic energy increases
(4) decreases and his kinetic energy remains the same

9 ____

10 When 150 joules of work is done on a system by an external force of 15 newtons in 20. seconds, the total energy of that system increases by

(1) 1.5×10^2 J
(2) 2.0×10^2 J
(3) 3.0×10^2 J
(4) 2.3×10^3 J

10 ____

11 A person on a ledge throws a ball vertically downward, striking the ground below the ledge with 200 joules of kinetic energy. The person then throws an identical ball vertically upward at the same initial speed from the same point. What is the kinetic energy of the second ball when it hits the ground? [Neglect friction.]

(1) 200 J
(2) 400 J
(3) less than 200 J
(4) more than 400 J

11 ____

12 Two construction cranes are used to lift identical 1200-kilogram loads of bricks the same vertical distance. The first crane lifts the bricks in 20. seconds and the second crane lifts the bricks in 40. seconds. Compared to the power developed by the first crane, the power developed by the second crane is

(1) the same
(2) twice as great
(3) half as great
(4) four times as great

12 ____

13 An ionized calcium atom has a charge of +2 elementary charges. If this ion is accelerated through a potential difference of 2.0×10^3 volts, the ion's change in kinetic energy will be

(1) 1.0×10^3 eV
(2) 2.0×10^3 eV
(3) 3.0×10^3 eV
(4) 4.0×10^3 eV

13 ____

14 A total charge of 100. coulombs flows past a fixed point in a circuit every 500. seconds. What is the current at this point in the circuit?

(1) 0.200 A
(2) 5.00 A
(3) 5.00×10^4 A
(4) 1.25×10^{18} A

14 ____

15 An aluminum wire of length 1.0 meter has a resistance of 9.0×10^{-3} ohm. If the wire were cut into two equal lengths, each length would have a resistance of

(1) 2.8×10^{-8} Ω
(2) 4.5×10^{-3} Ω
(3) 9.0×10^{-3} Ω
(4) 1.8×10^{-2} Ω

15 ____

16 In an operating electrical circuit, the source of potential difference could be

(1) a voltmeter
(2) a battery
(3) an ammeter
(4) a resistor

16 ____

17 A lightbulb with a resistance of 2.9 ohms is operated using a 1.5-volt battery. At what rate is electrical energy transformed in the lightbulb?

(1) 0.52 W
(2) 0.78 W
(3) 4.4 W
(4) 6.5 W

17 ____

18 A 40.0-kilogram child exerts a 100.-newton force on a 50.0-kilogram object. The magnitude of the force that the object exerts on the child is

(1) 0.0 N
(2) 80.0 N
(3) 100. N
(4) 125 N

18 _____

19 Two identical stationary bar magnets are arranged as shown in the diagram below.

N S

• P

S

N

What is the direction of the magnetic field at point P?

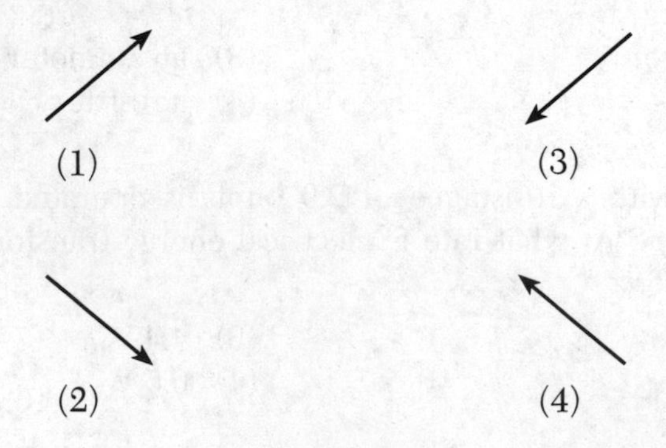

19 _____

20 A student claps his hands once to produce a sudden loud sound that travels through the air. This sound is classified as a

(1) longitudinal mechanical wave
(2) longitudinal electromagnetic wave
(3) transverse mechanical wave
(4) transverse electromagnetic wave

20 _____

21 A student generates water waves in a pool of water. In order to increase the energy carried by the waves, the student should generate waves with a

(1) greater amplitude
(2) higher frequency
(3) greater wavelength
(4) longer period

21 _____

22 A wave generator produces straight, parallel wave fronts in a shallow tank of uniform-depth water. As the frequency of vibration of the generator increases, which characteristic of the wave will always decrease?

(1) amplitude
(2) phase
(3) wavelength
(4) speed

22 _____

23 A space probe produces a radio signal pulse. If the pulse reaches Earth 12.3 seconds after it is emitted by the probe, what is the distance from the probe to Earth?

(1) 3.71×10^2 m
(2) 4.07×10^3 m
(3) 4.10×10^8 m
(4) 3.69×10^9 m

23 _____

24 The diagram below represents a light ray reflecting from a plane mirror.

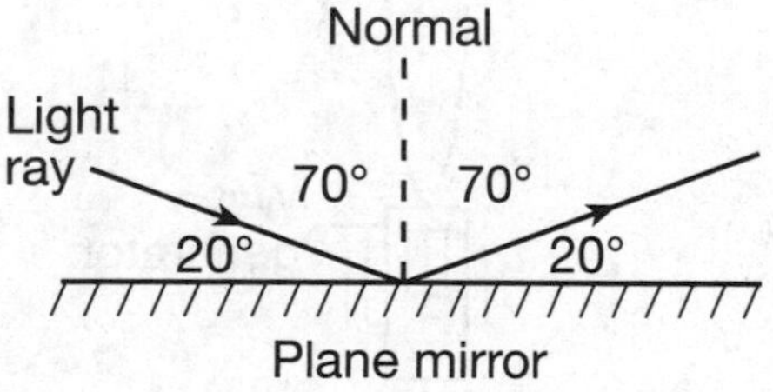

The angle of reflection for this light ray is

(1) 20°
(2) 70°
(3) 140°
(4) 160°

24 _____

25 A light wave travels from one medium into a second medium with a greater absolute index of refraction. Which characteristic of the wave can *not* change as the wave enters the second medium?

(1) frequency
(2) speed
(3) direction
(4) wavelength

25 ____

26 The speed of light ($f = 5.09 \times 10^{14}$ Hz) in glycerol is

(1) 1.70×10^6 m/s
(2) 2.04×10^8 m/s
(3) 3.00×10^8 m/s
(4) 4.41×10^8 m/s

26 ____

27 The diagram below represents a standing wave produced in a string by a vibrating wave generator.

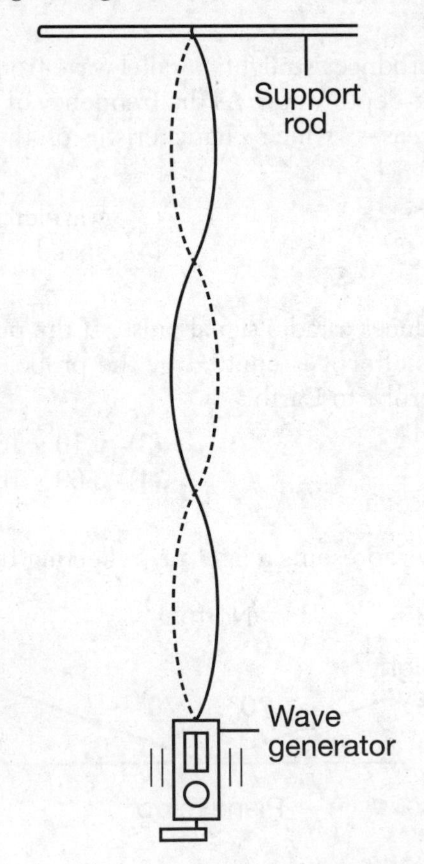

How many antinodes are shown in this standing wave?

(1) 6
(2) 2
(3) 3
(4) 4

27 ____

28 The Doppler effect is best described as the

(1) bending of waves as they pass by obstacles or through openings
(2) change in speed of a wave as the wave moves from one medium to another
(3) creation of a standing wave from two waves traveling in opposite directions in the same medium
(4) shift in the observed frequency and wavelength of a wave caused by the relative motion between the wave's source and an observer

28 ____

29 Which diagram represents diffraction of wave fronts as they encounter an obstacle?

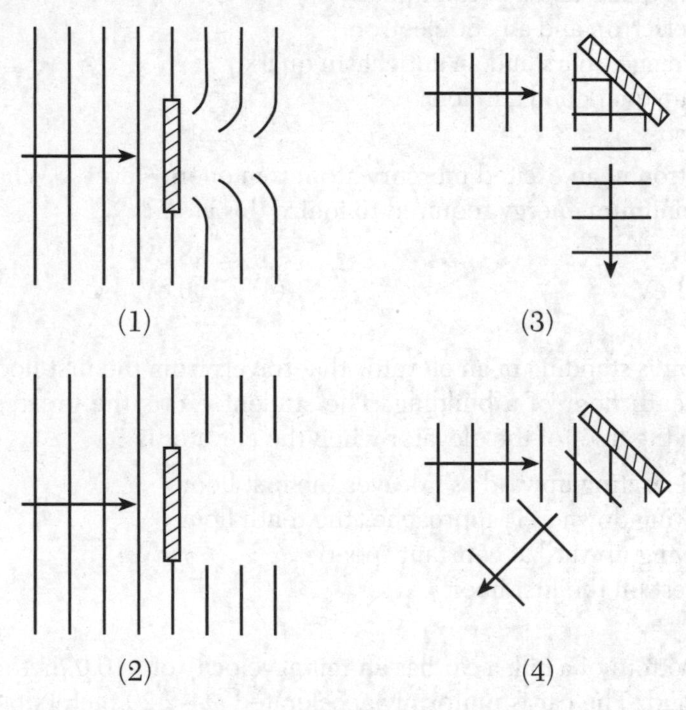

29 ____

30 Which types of forces exist between the two protons in a helium nucleus?

(1) a repulsive electrostatic force and a repulsive gravitational force
(2) a repulsive electrostatic force and an attractive strong nuclear force
(3) an attractive electrostatic force and an attractive gravitational force
(4) an attractive electrostatic force and an attractive strong nuclear force

30 ____

31 A meson could be composed of

(1) a top quark and a bottom quark
(2) an electron and an antielectron
(3) a strange quark and an anticharm quark
(4) an up quark and a muon

31 ____

32 An electron in an excited mercury atom is in energy level *g*. What is the minimum energy required to ionize this atom?

(1) 0.20 eV
(2) 0.91 eV
(3) 2.48 eV
(4) 7.90 eV

32 ____

33 A student is standing in an elevator that travels from the first floor to the tenth floor of a building. The student exerts the greatest force on the floor of the elevator when the elevator is

(1) accelerating upward as it leaves the first floor
(2) slowing down as it approaches the tenth floor
(3) moving upward at constant speed
(4) at rest on the first floor

33 ____

34 At the bottom of a hill, a car has an initial velocity of +16.0 meters per second. The car is uniformly accelerated at −2.20 meters per second squared for 5.00 seconds as it moves up the hill. How far does the car travel during this 5.00-second interval?

(1) 107 m
(2) 74.5 m
(3) 52.5 m
(4) 25.0 m

34 ____

35 A particle enters the electric field between two oppositely charged parallel plates, as represented in the diagram below.

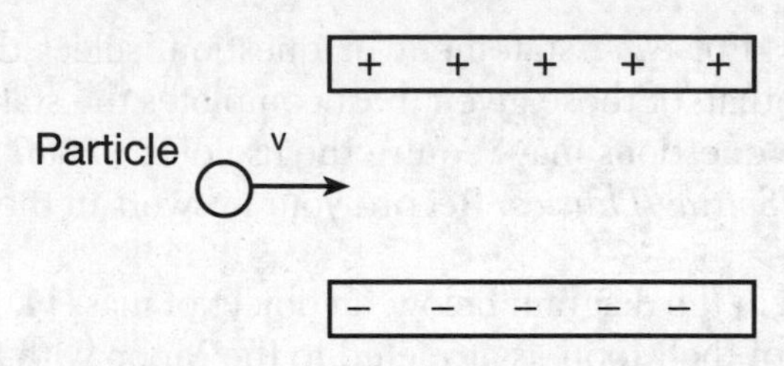

Which particle will be deflected toward the positive plate as it enters the electric field?

(1) photon
(2) proton
(3) electron
(4) neutrino

35 ____

PART B–1

Answer all questions in this part.

Directions (36–50): For *each* statement or question, select the *number* of the word or expression that, of those given, best completes the statement or answers the question. Some questions may require the use of the *2006 Edition Reference Tables for Physical Setting/Physics*. Record your answers in the spaces provided.

36 As represented in the diagram below, an object of mass m, located on the surface of the Moon, is attracted to the Moon with a gravitational force, F.

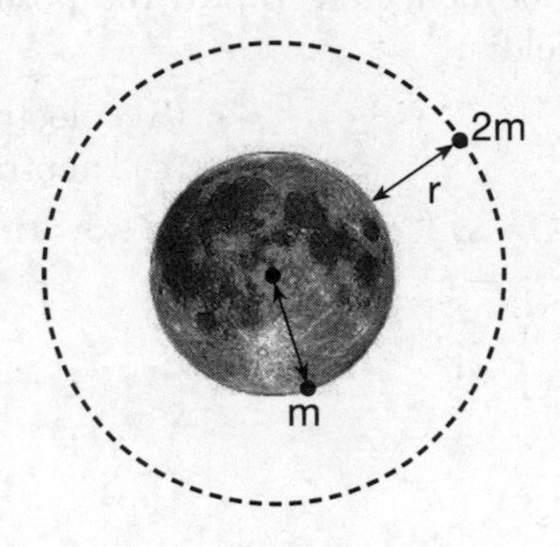

(Not drawn to scale)

An object of mass $2m$, at an altitude equal to the Moon's radius, r, above the surface of the Moon, is attracted to the Moon with a gravitational force of

(1) F
(2) $2F$
(3) $F/2$
(4) $F/4$

36 ____

37 The graph below represents the relationship between velocity and time for an object moving along a straight line.

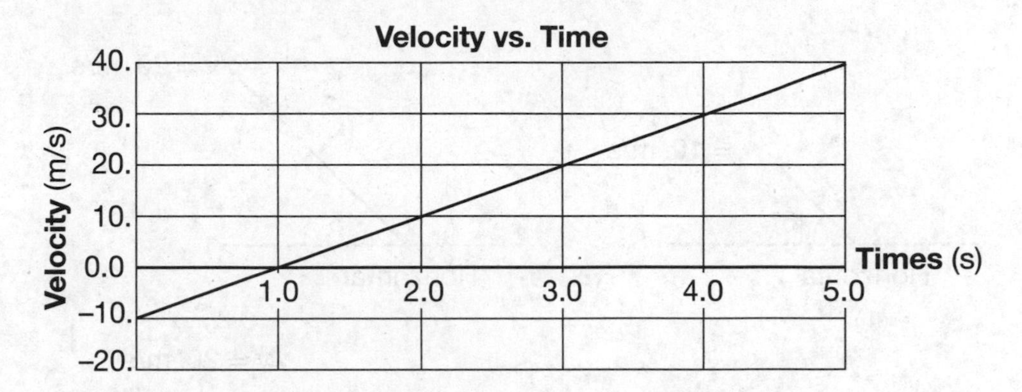

What is the magnitude of the object's acceleration?

(1) 5.0 m/s^2
(2) 8.0 m/s^2
(3) 10. m/s^2
(4) 20. m/s^2

37 _____

38 Two muons would have a combined charge of

(1) -3.2×10^{-19} C
(2) -1.6×10^{-19} C
(3) 0 C
(4) $+3.2 \times 10^{-19}$ C

38 _____

39 A 1.47-newton baseball is dropped from a height of 10.0 meters and falls through the air to the ground. The kinetic energy of the ball is 12.0 joules the instant before the ball strikes the ground. The maximum amount of mechanical energy converted to internal energy during the fall is

(1) 2.7 J
(2) 12.0 J
(3) 14.7 J
(4) 26.7 J

39 _____

40 A projectile lands at the same height from which it was launched. Which initial velocity will result in the greatest horizontal displacement of the projectile? [Neglect friction.]

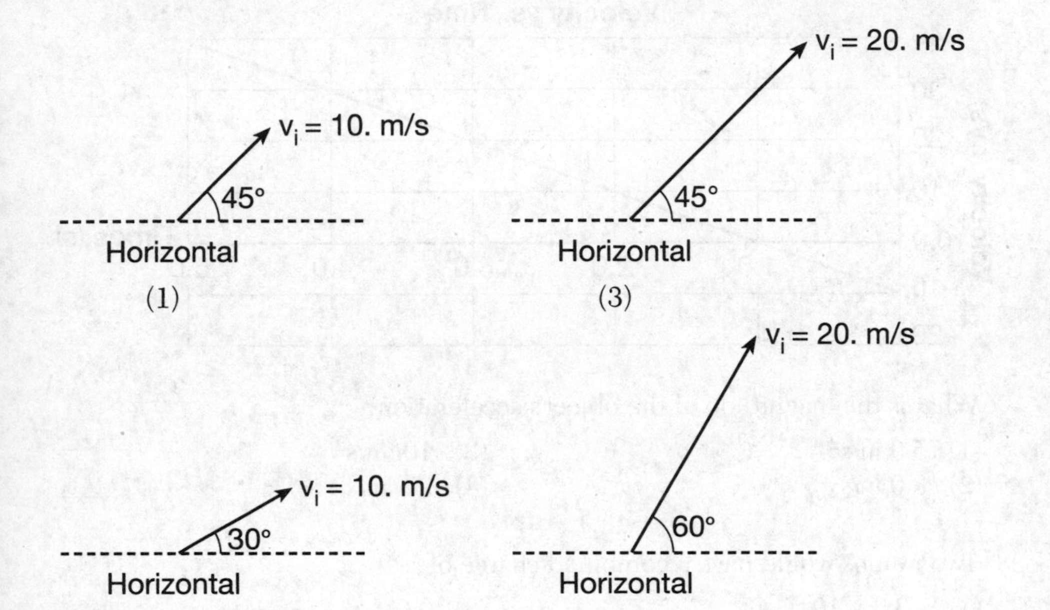

40 _____

41 A 5.0-kilogram box is sliding across a level floor. The box is acted upon by a force of 27 newtons east and a frictional force of 17 newtons west. What is the magnitude of the acceleration of the box?

(1) 0.50 m/s^2
(2) 2.0 m/s^2
(3) 8.8 m/s^2
(4) 10. m/s^2

41 _____

42 The diagram below represents a 2.0-kilogram toy car moving at a constant speed of 3.0 meters per second counterclockwise in a circular path with a radius of 2.0 meters.

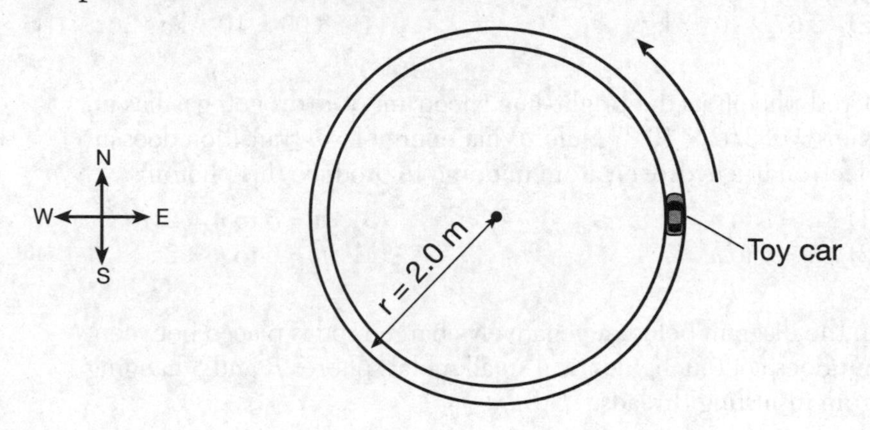

At the instant shown in the diagram, the centripetal force acting on the car is

(1) 4.5 N north
(2) 4.5 N west
(3) 9.0 N north
(4) 9.0 N west

42 ____

43 In which electric circuit would the voltmeter read 10 volts?

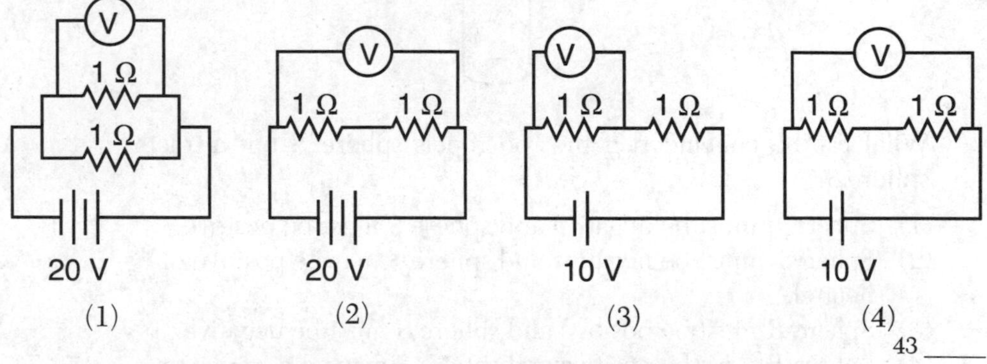

43 ____

44 The lambda baryon has the quark composition *uds*. Which particle has the same electric charge as the lambda baryon?

(1) neutron
(2) electron
(3) proton
(4) antimuon

44 ____

45 How many kilograms of matter would have to be converted into energy to produce 24.0 megajoules of energy?

(1) 2.67×10^{-16} kg
(2) 2.67×10^{-10} kg
(3) 8.00×10^{-8} kg
(4) 8.00×10^{-2} kg

45 ____

46 A red photon in the bright-line spectrum of hydrogen gas has an energy of 3.02×10^{-19} joule. What energy-level transition does an electron in a hydrogen atom undergo to produce this photon?

(1) $n = 3$ to $n = 2$
(2) $n = 4$ to $n = 2$
(3) $n = 5$ to $n = 2$
(4) $n = 6$ to $n = 2$

46 ____

47 In the diagram below, a negatively charged rod is placed between, but does not touch, identical small metal spheres *R* and *S* hanging from insulating threads.

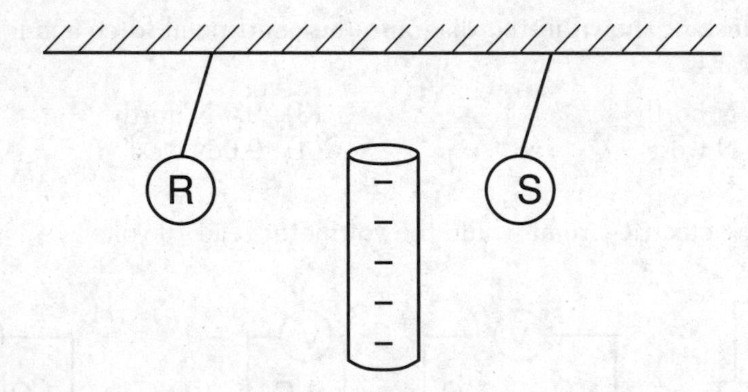

What can be concluded if the rod repels sphere *R* but attracts sphere *S*?

(1) Sphere *R* must be negative and sphere *S* must be positive.
(2) Sphere *R* must be negative and sphere *S* may be positive or neutral.
(3) Sphere *R* must be positive and sphere *S* must be negative.
(4) Sphere *R* must be positive and sphere *S* may be negative or neutral.

47 ____

48 The amount of electric energy consumed by a 60.0-watt lightbulb for 1.00 minute could lift a 10.0 newton object to a maximum vertical height of

(1) 6.00 m
(2) 36.7 m
(3) 360. m
(4) 600. m

48 ____

49 Microwaves can have a wavelength closest to the

(1) radius of Earth
(2) height of Mount Everest
(3) length of a football field
(4) length of a physics student's thumb

49 ____

50 Two pulses approach each other in a uniform medium, as represented in the diagram below.

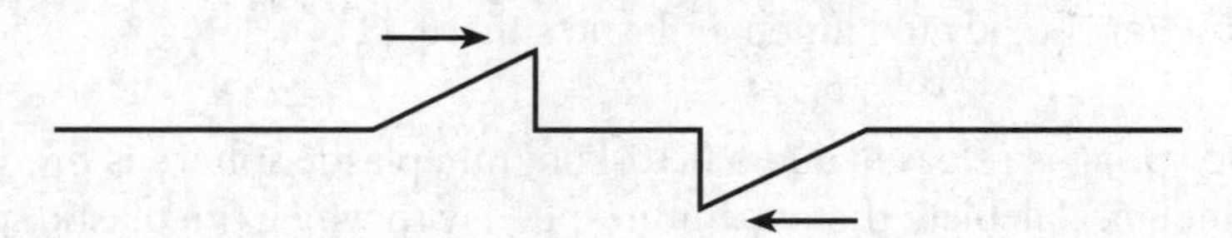

Which diagram best represents the superposition of the two pulses when the pulses overlap?

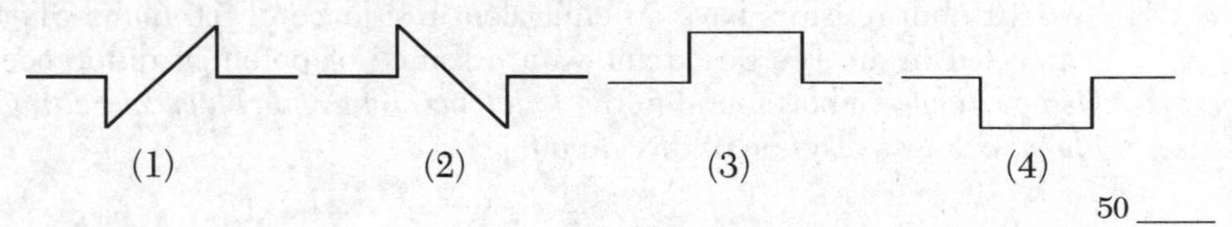

50 ____

PART B–2

Answer all questions in this part.

Directions (51–65): Record your answers on the answer sheet provided after the questions. Some questions may require the use of the *2006 Edition Reference Tables for Physical Setting/Physics*.

Base your answers to questions 51 through 53 on the information below and on your knowledge of physics.

A toy launcher that is used to launch small plastic spheres horizontally contains a spring with a spring constant of 50. newtons per meter. The spring is compressed a distance of 0.10 meter when the launcher is ready to launch a plastic sphere.

51 Determine the elastic potential energy stored in the spring when the launcher is ready to launch a plastic sphere. [1]

52–53 The spring is released and a 0.10-kilogram plastic sphere is fired from the launcher. Calculate the maximum speed with which the plastic sphere will be launched. [Neglect friction.] [Show all work, including the equation and substitution with units.] [2]

54 Two 10.-ohm resistors have an equivalent resistance of 5.0 ohms when connected in an electric circuit with a source of potential difference. Using circuit symbols found in the *Reference Tables for Physical Setting/Physics*, draw a diagram of this circuit. [1]

55 The graph below shows the relationship between distance, *d*, and time, *t*, for a moving object.

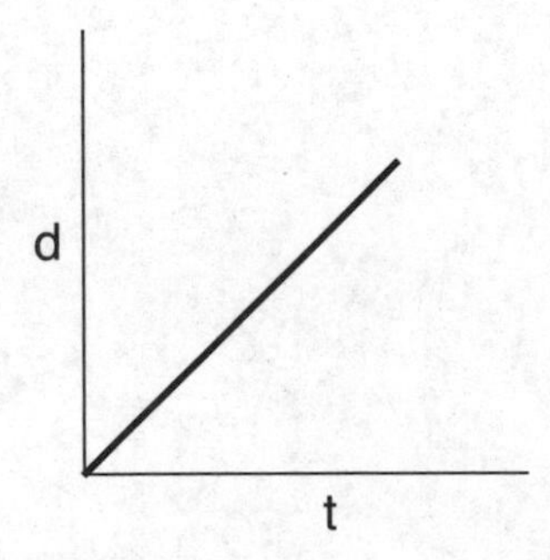

On the axes *on the answer sheet,* sketch the general shape of the graph that shows the relationship between the magnitude of the velocity, v, and time, t, for the moving object. [1]

Base your answers to questions 56 through 58 on the information and diagram below and on your knowledge of physics.

A ray of monochromatic light ($f = 5.09 \times 10^{14}$ Hz) passes from medium X into air. The angle of incidence of the ray in medium X is 25°, as shown.

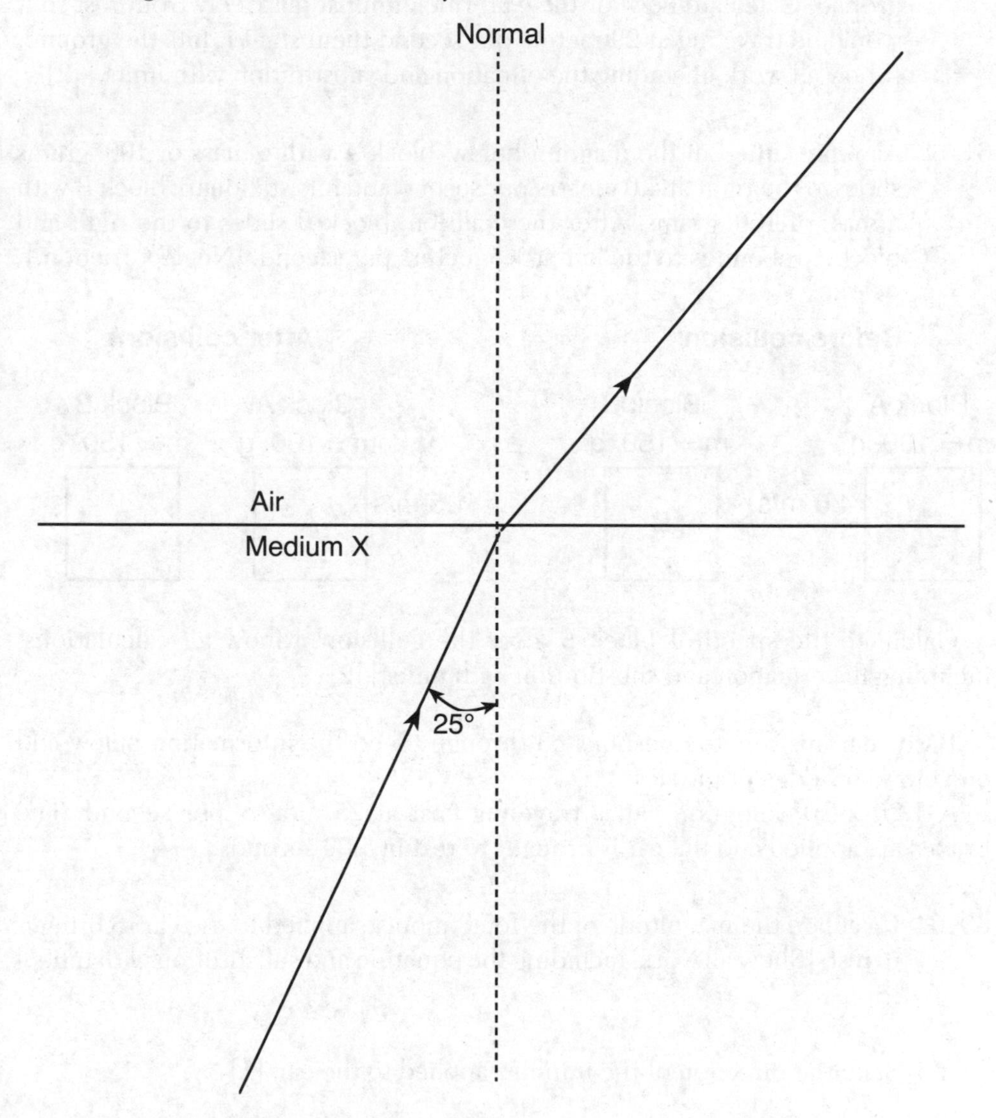

56 Using a protractor, measure and record the angle of refraction in the air, to the *nearest degree*. [1]

57–58 Calculate the absolute index of refraction of medium *X*. [Show all work, including the equation and substitution with units.] [2]

59–60 A student wishes to record a 7.5-kilogram watermelon colliding with the ground. Calculate how far the watermelon must fall freely from rest so it would be traveling at 29 meters per second the instant it hits the ground. [Show all work, including the equation and substitution with units.] [2]

61–62 As represented in the diagram below, block *A* with a mass of 100. grams slides to the right at 4.0 meters per second and hits stationary block *B* with a mass of 150. grams. After the collision, block *B* slides to the right and block *A* rebounds to the left at 1.5 meters per second. [Neglect friction.]

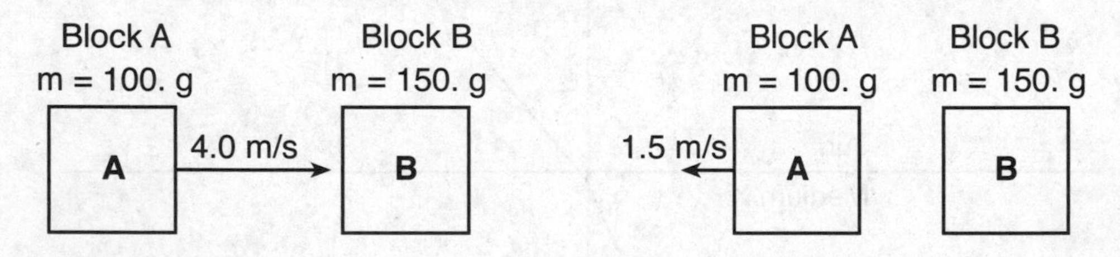

Calculate the speed of block *B* after the collision. [Show all calculations, including the equation and substitution with units.] [2]

Base your answers to questions 63 through 65 on the information below and on your knowledge of physics.

A 1.20×10^3-kilogram car is traveling east at 25 meters per second. The brakes are applied and the car is brought to rest in 5.00 seconds.

63–64 Calculate the magnitude of the total impulse applied to the car to bring it to rest. [Show all work, including the equation and substitution with units.] [2]

65 State the direction of the impulse applied to the car. [1]

PART C

Answer all questions in this part.

Directions (66–85): Record your answers on the answer sheet provided after the questions. Some questions may require the use of the 2006 *Edition Reference Tables for Physical Setting/Physics*.

Base your answers to questions 66 through 70 on the information and diagram below and on your knowledge of physics.

The diagram shows a negatively charged oil drop that is suspended motionless between two oppositely charged, parallel, horizontal metal plates. The electric field strength between the charged plates is 4.0×10^4 newtons per coulomb. The 1.96×10^{-15}-kilogram oil drop is being acted upon by a gravitational force, F_g, and an electrical force, F_e.

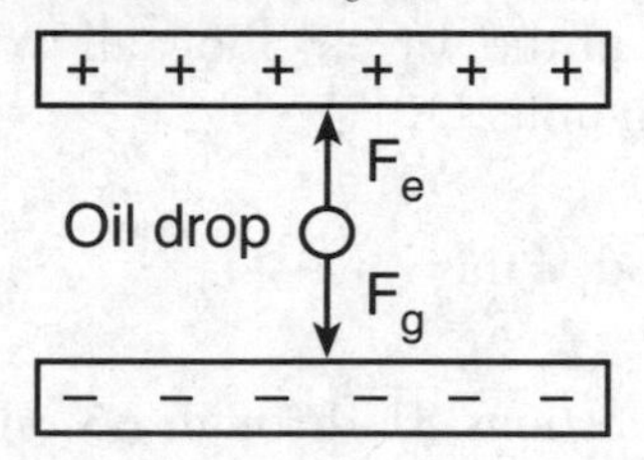

66–67 Calculate the magnitude of the gravitational force, F_g, acting on the oil drop. [Show all work, including the equation and substitution with units.] [2]

68 Determine the magnitude of the upward electrical force, F_e, acting on the oil drop suspended motionless between the charged metal plates. [1]

69–70 Calculate the net electric charge on the oil drop in coulombs. [Show all work, including the equation and substitution with units.] [2]

Base your answers to questions 71 through 75 on the information below and on your knowledge of physics.

In a circuit, a 100.-ohm resistor and a 200.-ohm resistor are connected in parallel to a 10.0-volt battery.

71–72 Calculate the equivalent resistance of the circuit. [Show all work, including the equation and substitution with units.] [2]

73–74 Calculate the current in the 200.-ohm resistor. [Show all work, including the equation and substitution with units.] [2]

75 Determine the power dissipated by the 100.-ohm resistor. [1]

Base your answers to questions 76 through 80 on the information below and on your knowledge of physics.

A wave traveling through a uniform medium has an amplitude of 0.20 meter, a wavelength of 0.40 meter, and a frequency of 10. hertz.

76–77 On the grid *on the answer sheet*, draw *one* complete cycle of the wave. [2]

78–79 Calculate the speed of the wave. [Show all work, including the equation and substitution with units.] [2]

80 Determine the period of this wave. [1]

Base your answers to questions 81 through 85 on the information and data table below and on your knowledge of physics.

In an experiment, the potential difference applied across an unmarked resistor was varied while the resistor was held at a constant temperature. The corresponding current through the resistor was measured. The data collected appear in the table below.

Potential Difference (volts)	**Current** (amperes)
1.5	0.0032
3.0	0.0059
6.0	0.0124
9.0	0.0177
12.0	0.0244

81 Mark an appropriate scale on the axis labeled “Current (A).” [1]

82 Plot the data points for current versus potential difference. [1]

83 Draw the line or curve of best fit. [1]

84–85 Using your graph, calculate the resistance of the resistor. [Show all work, including the equation and substitution with units.] [2]

Answer Sheet June 2019

Physics: The Physical Setting

PART B–2

51 ______________________ **J**

52–53

54

55

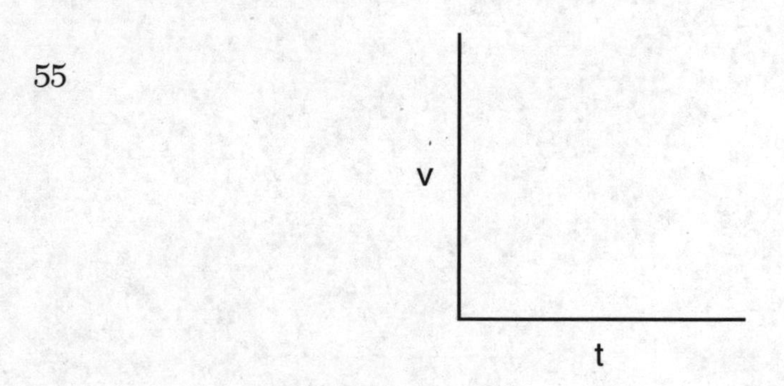

56 ______________________ °

57–58

59–60

61–62

63–64

65 ____________________

PART C

66–67

68 ______________________ **N**

69–70

71–72

73–74

75 ________________ **W**

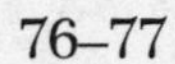
76–77

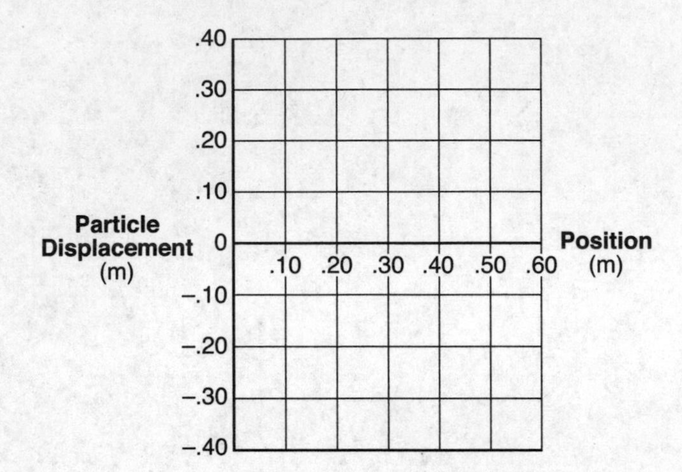
.40
.30
.20
.10
0
–.10
–.20
–.30
–.40
Particle
Displacement
(m)
.10
.20
.30
.40
.50
.60
Position
(m)

78–79

80 ________________ s

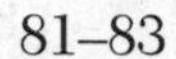
81–83

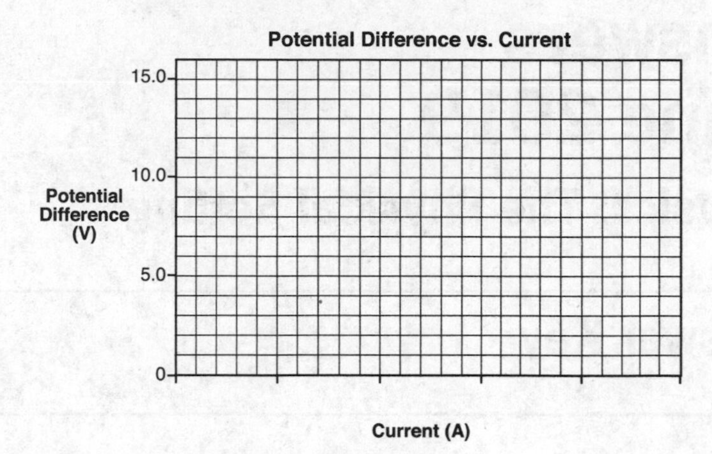
Potential Difference vs. Current
15.0
10.0
5.0
0
Potential Difference (V)
Current (A)

84–85

Answers June 2019

Physics: The Physical Setting

Answer Key

PART A

1. 4	**8.** 1	**15.** 2	**22.** 3	**29.** 1
2. 3	**9.** 4	**16.** 2	**23.** 4	**30.** 2
3. 2	**10.** 1	**17.** 2	**24.** 2	**31.** 3
4. 4	**11.** 1	**18.** 3	**25.** 1	**32.** 3
5. 2	**12.** 3	**19.** 3	**26.** 2	**33.** 1
6. 1	**13.** 4	**20.** 1	**27.** 3	**34.** 3
7. 1	**14.** 1	**21.** 1	**28.** 4	**35.** 3

PART B–1

36. 3	**39.** 1	**42.** 4	**45.** 2	**48.** 3
37. 3	**40.** 3	**43.** 4	**46.** 1	**49.** 4
38. 1	**41.** 2	**44.** 1	**47.** 2	**50.** 1

PART B–2 and **PART C**. *See* **Answers Explained**.

Answers Explained

PART A

1 **4** Scalar quantities have magnitude but no direction. Both energy and time are scalar quantities.

WRONG CHOICES EXPLAINED:

(1) Both displacement and velocity are vector quantities.
(2) Displacement is a vector quantity; time is a scalar quantity.
(3) Energy is a scalar quantity; velocity is a vector quantity.

2 **3** When dealing with a word problem requiring a mathematical solution, it is most helpful to translate the words into what quantities/variables are explicitly given and what constants or other variables can be deduced from the situation or words. We also write down the variable that needs to be solved for, or found. Then, it is much easier to determine which formula is required (refer to the appropriate section of the *Reference Tables for Physics*) and then use it to solve the problem. In this case we refer to the *Geometry and Trigonometry* section of *Reference Table K* for the correct equation.

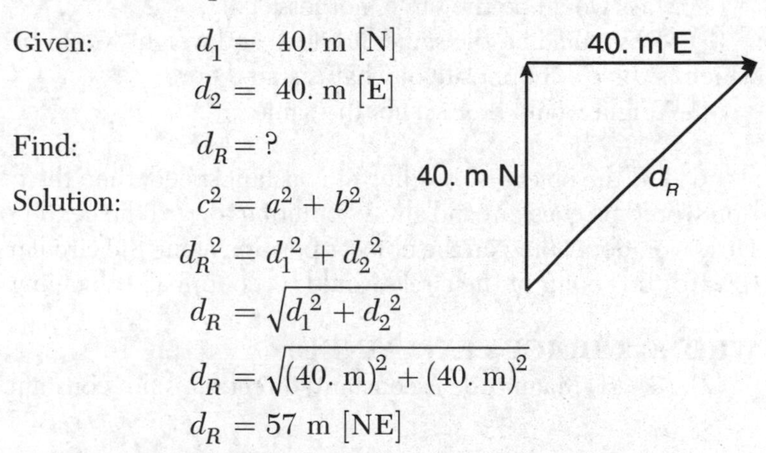

Given: $d_1 = 40.\text{ m [N]}$
$d_2 = 40.\text{ m [E]}$

Find: $d_R = ?$

Solution: $c^2 = a^2 + b^2$

$$d_R{}^2 = d_1{}^2 + d_2{}^2$$

$$d_R = \sqrt{d_1{}^2 + d_2{}^2}$$

$$d_R = \sqrt{(40.\text{ m})^2 + (40.\text{ m})^2}$$

$$d_R = 57\text{ m [NE]}$$

3 **2** As the ball travels upward, its velocity decreases; it decelerates at a rate of 9.81 m/s^2 (acceleration due to gravity) until its velocity is equal to zero at its highest point. Then its velocity increases; it accelerates downward at a rate of 9.81 m/s^2 (acceleration due to gravity). Acceleration due to gravity is constant throughout the entire path, including at the top or apex of the ball's flight.

WRONG CHOICES EXPLAINED:

(1) This choice is incorrect because acceleration is not zero.

(3) This choice is incorrect because velocity is zero and acceleration is nonzero, which is the exact opposite of what is stated here.

(4) This choice is incorrect because velocity is zero.

4 **4** Electrons carry a negative charge; therefore, electrons transfer from the hair to the comb in order for the comb to acquire a negative charge.

WRONG CHOICES EXPLAINED:

(1), (2) Protons do not transfer between objects.

(3) Electrons carry a negative charge; therefore, if they transferred from the comb to her hair, the hair would acquire a negative charge.

5 **2** The law of conservation of mass states that mass can neither be created nor destroyed, so the mass of the object would remain the same regardless of whether the object is on Earth or on the Moon. However, the acceleration due to gravity is less on the Moon than it is on Earth; so, the weight of the object, which varies depending on gravity ($F_g = mg$), would be less on the Moon than it is on Earth.

WRONG CHOICES EXPLAINED:

(1) Mass would be the same, not less.

(3) Mass would be the same, not less, and weight would be less, not the same, which is the exact opposite of what is stated here.

(4) Weight would be less, not the same.

6 **1** If the object is traveling at constant speed, then the centripetal acceleration would be constant and always pointing toward the center of the circular path. However, depending on the object's position along the circular path, the direction toward the center of the circle would be continually changing.

WRONG CHOICES EXPLAINED:

(2), (3), (4) Magnitude is constant; direction is not constant.

7 **1** Refer to the *Mechanics* or the *Geometry and Trigonometry* section of *Reference Table K.*

Given: $F = 45\text{ N}$

$\theta = 65°$

Find: $F_x = ?$

Solution:

$$A_x = A\cos\theta$$
$$F_x = F\cos\theta$$
$$F_x = (45\text{ N})\cos 65°$$
$$F_x = 19\text{ N}$$

or

$$\cos\theta = \frac{b}{c}$$
$$b = c\cos\theta$$
$$F_x = F\cos\theta$$
$$F_x = (45\text{ N})\cos 65°$$
$$F_x = 19\text{ N}$$

8 **1** Refer to the *Mechanics* section of *Reference Table K.*

Given: $k = 68\text{ N/m}$

$F_s = 12\text{ N}$

Find: $x = ?$

Solution:

$$F_s = kx$$
$$x = \frac{F_s}{k}$$
$$x = \frac{12\text{ N}}{68\text{ N/m}}$$
$$x = 0.18\text{ m}$$

9 **4** Refer to the *Mechanics* section of *Reference Table K.*

$$\Delta PE = mg\Delta h$$

Gravitational potential energy is proportional to height; therefore, as the student walks downhill, height decreases.

$$KE = \frac{1}{2}mv^2$$

Kinetic energy is proportional to speed squared. Since the student walks downhill at constant speed, there is no change in the student's kinetic energy. Choice (4) correctly states that gravitational potential energy decreases and kinetic energy remains the same.

10 **1** If 150 joules of work is done on a system, then the total energy of the system increases by 150 joules. Choice (1) correctly represents the amount in scientific notation of 150 joules.

11 **1** As the ball leaves the person's hand and travels upward, its speed decreases due to deceleration due to gravity until it reaches its highest point where its velocity will be equal to zero. The ball then begins to accelerate downward due to acceleration due to gravity and its speed increases until it reaches its starting point, where the magnitude of the velocity will be equal to the magnitude of the velocity at the start, only in the opposite direction (downward instead of upward). Given that the first ball was thrown downward with an equal velocity, and that resulted in the ball striking the ground below with 200 joules of kinetic energy, the second ball will also strike the ground with 200 joules of kinetic energy.

12 **3** Refer to the *Mechanics* section of *Reference Table K.*

Given:

$$m_1 = m_2 = 1200.\ \text{kg}$$
$$d_1 = d_2$$
$$t = 20.\ \text{s}$$
$$t = 40.\ \text{s}$$

Find:

$$P_2 : P_1?$$

Solution:

$$g = \frac{F_g}{m}$$

Since $m_1 = m_2$ then $F_{g1} = F_{g2}$

$$W = Fd$$

Since $d_1 = d_2$ and $F_{g1} = F_{g2}$ then $W_1 = W_2$

$$P = \frac{W}{t}$$
$$W = Pt$$
$$P_2 t_2 = P_1 t_1$$
$$P_2 = \frac{P_1 t_1}{t_2}$$
$$P_2 = \frac{P_1(20\ \text{s})}{(40\ \text{s})}$$
$$P_2 = \frac{1}{2} P_1$$

WRONG CHOICE EXPLAINED:

(2) You compared the power of the first crane to the second crane instead of the second crane to the first crane.

13 **4** Refer to *Reference Table A* and the *Electricity* section of *Reference Table K.*

Given:

$$q = +2e$$
$$e = 1.60 \times 10^{-19}\ \text{C}$$
$$q = 3.20 \times 10^{-19}\ \text{C}$$
$$V = 2.0 \times 10^{3}\ \text{V}$$
$$1\ \text{eV} = 1.60 \times 10^{-19}\ \text{J}$$

Find: $\Delta KE = ?$

Solution:

$$V = \frac{W}{q}$$
$$W = Vq$$
$$W = (2.0 \times 10^{3}\ \text{V})(3.20 \times 10^{-19}\ \text{C})$$
$$W = 6.4 \times 10^{-16}$$
$$W = \left(6.4 \times 10^{-16}\ \text{J}\right)\left(\frac{1\ \text{eV}}{1.6 \times 10^{-19}\ \text{J}}\right)$$
$$W = 4.0 \times 10^{3}\ \text{eV}$$

WRONG CHOICE EXPLAINED:

(1) This is equivalent to having completed the calculation using a +1 elementary charge instead of +2 elementary charges.

14 **1** Refer to the *Electricity* section of *Reference Table K*.

Given: $q = 100\ \text{C}$

$t = 500\ \text{s}$

Find: $I = ?$

Solution:

$$I = \frac{\Delta q}{t}$$

$$I = \frac{100\ \text{C}}{500\ \text{s}}$$

$$I = 0.200\ \text{A}$$

WRONG CHOICES EXPLAINED:

(2) You divided time by charge instead of charge by time.

(3) You multiplied the charge by time instead of dividing.

15 **2** Refer to the *Electricity* section of *Reference Table K*.

Given: $L_1 = 1.0\ \text{m}$

$R_1 = 9.0 \times 10^{-3}\ \Omega$

$L_2 = \frac{1}{2}L_1$

Find: $R_2 = ?$

Solution:

$$R = \frac{\rho L}{A}$$

Resistance is proportional to length. Halving the length would result in the resistance being halved as well, since both the resistivity and cross-sectional area have not changed.

WRONG CHOICE EXPLAINED:

(4) You doubled the resistance instead of halving it.

16 **2** A battery is a combination of two or more electric cells and is the source of potential difference in a circuit.

WRONG CHOICES EXPLAINED:

(1) A voltmeter is a device used to measure potential difference.
(3) An ammeter is a device used to measure electric current.
(4) A resistor is a device that supplies resistance to a circuit.

17 **2** Refer to the *Electricity* section of *Reference Table K.*

Given: $R = 2.9\ \Omega$

$V = 1.5\ \text{V}$

Find: $P = ?$

Solution:
$$P = \frac{V^2}{R}$$
$$P = \frac{(1.5\ \text{V})^2}{2.9\ \Omega}$$
$$P = 0.78\ \text{W}$$

WRONG CHOICE EXPLAINED:

(1) You forgot to square the voltage.

18 **3** Newton's Third Law states that if object A exerts a force on object B, then object B exerts an equal in magnitude but opposite in direction force on object A. Since the child exerts a 100. N force on the object, the object in return exerts a 100. N force on the child.

19 **3** Magnetic field lines can be used to visualize magnetic fields. By convention, field lines are drawn pointing away from north magnetic poles and toward south poles. At point *P* we could draw a field line horizontally from *P*, left toward the south pole of the magnet at the top of the diagram, and a vertical field line from *P* downward toward the south pole of the magnet below *P*. The net resultant magnetic force would be diagonally left and downward.

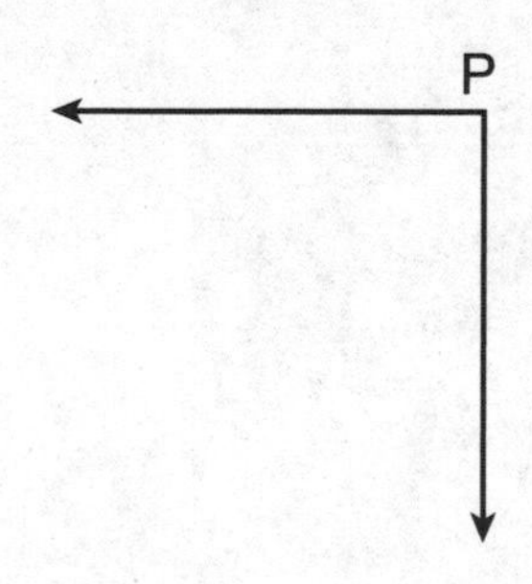

WRONG CHOICE EXPLAINED:

(1) You drew or assumed field lines away from south and toward north.

20 **1** Sound waves travel as longitudinal mechanical waves.

WRONG CHOICES EXPLAINED:

(3) Waves on a string is an example of transverse mechanical waves.
(4) Light travels as transverse electromagnetic waves.

21 **1** Amplitude is related to the energy carried by the wave. The greater the energy, the greater the amplitude of the wave.

WRONG CHOICES EXPLAINED:

(2) Frequency is the number of waves per unit time.
(3) Wavelength is the length of one complete wave cycle.
(4) Period is the time for one complete repetition of a cycle.

22 **3** Refer to the *Waves* section of *Reference Table K.*

$$v = f\lambda$$

The speed of a wave in a given medium is constant; therefore, frequency and wavelength are inversely proportional to each other. If frequency increases, wavelength will decrease and vice versa.

WRONG CHOICES EXPLAINED:

(1) Amplitude is related to the energy of the wave. Increasing the frequency that the wave is generated will not affect the amplitude.

(2) Points on a periodic wave that are at equal displacements from their rest positions and that are experiencing identical movements are said to be in phase.

(4) The speed of a mechanical wave is determined by the medium it is traveling in. The medium remains the same; therefore, the speed of the wave remains the same.

23 **4** Refer to the *Mechanics* section of *Reference Table K* and *Reference Table A*.

Given: $t = 12.3 \text{ s}$

$\overline{v} = c = 3.00 \times 10^8 \text{ m/s}$

Find: $d = ?$

Solution:

$$\overline{v} = \frac{d}{t}$$

$$d = \overline{v}t$$

$$d = \left(3.0 \times 10^8 \text{ m/s}\right)(12.3 \text{ s})$$

$$d = 3.69 \times 10^9 \text{ m}$$

WRONG CHOICE EXPLAINED:

(2) You calculated your answer using the speed of sound in air instead of the speed of light in a vacuum.

24 **2** Angles of incidence and reflection are calculated in reference to the normal. In this case, the angle that the reflected ray makes with the normal is 70°.

WRONG CHOICES EXPLAINED:

(1) This is the angle that the ray makes with the plane mirror.

(3) You added the angles of incidence and refraction to get the measure of the angle between the rays.

(4) You added all the angles except the angle that the reflected ray makes with the surface.

25 **1** The frequency of the light wave is determined by the source that produces it. As the light travels from one medium into another, the frequency does not change.

WRONG CHOICES EXPLAINED:

(2) As light travels from one medium to another with different absolute indices of refraction, its speed changes as it enters the new medium.

(3) Refraction, a change in direction of a wave when it passes obliquely from one medium into another, will occur due to the change in speed resulting from moving into a medium with a different absolute index of refraction.

(4) The speed of a wave is proportional to its frequency and wavelength. If frequency is constant, then, as the speed of a wave changes as it enters a new medium with a different absolute index of refraction, the wavelength will also proportionately change.

26 **2** Refer to the *Waves* section of *Reference Table K*, *Reference Table A*, and *Reference Table E*.

Given: $f = 5.09 \times 10^{10}$ Hz

$c = 3.00 \times 10^8$ m/s

$n = 1.47$

Find: $v = ?$

Solution:

$$n = \frac{c}{v}$$

$$v = \frac{c}{n}$$

$$v = \frac{3.0 \times 10^8 \text{ m/s}}{1.47}$$

$$v = 2.04 \times 10^8 \text{ m/s}$$

WRONG CHOICES EXPLAINED:

(3) You just used the speed of light in a vacuum without adjusting for change in speed in the new medium.

(4) You multiplied the speed of light by the absolute index of refraction instead of dividing.

27 **3** An antinode is the point or locus of points on an interference pattern that results in maximum constructive interference. In this diagram of a standing wave, there are 3 antinodes.

WRONG CHOICE EXPLAINED:

(4) These are the number of nodes, or points of maximum destructive interference.

28 **4** The Doppler effect is the apparent change in frequency that results when a wave source and an observer are in relative motion with respect to each other. As the distance between the source and the observer decreases, the frequency of the source, as perceived by the observer, is increased; as the distance increases, the apparent frequency is decreased.

WRONG CHOICE EXPLAINED:

(1) Diffraction is the bending of a wave around a barrier.

29 **1** Diffraction is the bending of a wave around a barrier.

WRONG CHOICES EXPLAINED:

(2) This diagram does not include the bending of the wave around the obstacle.
(3) This diagram depicts reflection of a wave on a surface.
(4) This diagram is similar to the reflection diagram; however, the reflected waves have changed wavelength. This would be what occurs if the wave were refracted as it traveled from one medium into another, but not from simple reflection.

30 **2** Protons carry a positive charge. Like charges repel; therefore, the protons experience a repulsive electrostatic force. The nuclear force is the attractive, short-range force responsible for binding protons and neutrons in the nucleus; therefore, the protons experience an attractive strong nuclear force. Choice (2) correctly indicates these two forces.

31 **3** Refer to *Reference Table G* and *Reference Table H*. A meson is composed of a quark and an antiquark. Choice (3) correctly lists a quark particle and an antiquark particle.

WRONG CHOICES EXPLAINED:

(1) These two particles are both quarks.
(2) These two particles are both leptons.
(4) One particle is a quark and the other is a lepton.

32 **3** Refer to *Reference Table F*. Energy level g is at −2.48 eV. In order to ionize the atom, 2.48 eV of energy needs to be provided.

WRONG CHOICES EXPLAINED:

(1) This is the difference between levels *f* and g.
(2) This is the difference between levels g and *h*.
(4) This is the difference between the ground state and g.

33 **1** Newton's Third Law states that if object A exerts a force on object B, then object B exerts an equal in magnitude but opposite in direction force on object A. Normally, the student's weight or force due to gravity is balanced by the upward normal force. In order for the elevator to accelerate upward, there has to be a

net upward force applied to the elevator and, thus, to the student inside the elevator. Therefore, at the time of acceleration, the student applies an equal downward force to the floor of the elevator.

WRONG CHOICES EXPLAINED:

(2) When the elevator decelerates, there would be a net downward force being applied to the elevator and, thus, to the student inside. Therefore, the net downward force that the student exerts on the elevator would be less than what it would be if the elevator were at rest.

(3), (4) Moving at constant speed would be the same as being at rest on the ground floor (no net force), so the student would only be exerting a downward force due to gravity or his or her weight.

34 **3** Refer to the *Mechanics* section of *Reference Table K.*

Given: $v_i = 16.0 \text{ m/s}$

$a = -2.20 \text{ m/s}^2$

$t = 5.00 \text{ s}$

Find: $d = ?$

Solution:

$$d = v_i t + \frac{1}{2}at^2$$

$$d = (16.0 \text{ m/s})(5.00 \text{ s}) + \frac{1}{2}(-2.20 \text{ m/s}^2)(5.00 \text{ s})^2$$

$$d = (80.0 \text{ m}) + (-27.5 \text{ m})$$

$$d = 52.5 \text{ m}$$

The negative acceleration or deceleration indicates that it is opposite to the direction of motion.

WRONG CHOICES EXPLAINED:

(1) You used positive 2.20 for the acceleration instead of −2.20, or you added the second half of the equation instead of subtracting it properly.

(2) You forgot to square the time in the second half of the equation.

(4) You forgot to multiply the second half of the equation by $\frac{1}{2}$.

35 **3** Electrons carry a negative charge. Like charges attract and opposites repel, so the electron would be deflected toward the positive plate and away from the negative one.

WRONG CHOICES EXPLAINED:

(1), (4) Photons and neutrinos do not carry charge and would not be deflected toward either plate.

(2) Protons carry a positive charge and a proton would be deflected toward the negative plate and away from the positive one.

PART B–1

36 **3** Refer to the *Mechanics* section of *Reference Table K.*

Given:

$$m_1 = m$$
$$r_1 = r$$
$$\text{mass of moon} = m_{moon}$$
$$F_{g1} = F$$
$$m_2 = 2m$$
$$r_2 = 2r$$

Find:

$$F_{g2} = ?$$

Solution:

$$F_{g1} = \frac{Gmm_{moon}}{r^2} = F$$

$$F_{g2} = \frac{G(2m)m_{moon}}{(2r)^2}$$

$$F_{g2} = \frac{2Gmm_{moon}}{4r^2}$$

$$F_{g2} = \frac{2}{4}\left(\frac{Gmm_{moon}}{r^2}\right)$$

$$F_{g2} = \frac{1}{2}F_{g1} \text{ or } \frac{1}{2}F$$

WRONG CHOICES EXPLAINED:

(1) You forgot to square the 2 in $2r$ in the denominator.

(2) You used r instead of $2r$ in the denominator.

(4) You forgot to use $2m$ instead of m in the numerator.

37 **3** The slope of the velocity versus time graph is given by the relationship $\Delta v/\Delta t$, the rate at which velocity has changed over the time interval—in other

words, the acceleration. We can pick any two points on the line in the graph and substitute them into the formula to calculate the acceleration.

Given: $v_1 = 0.0$ m/s

$v_2 = 10.$ m/s

$t_1 = 1.0$ s

$t_1 = 2.0$ s

Find: $a = ?$

Solution:
$$a = \frac{\Delta v}{\Delta t}$$
$$a = \frac{(v_2 - v_1)}{(t_2 - t_1)}$$
$$a = \frac{(10.0 \text{ m/s} - 0 \text{ m/s})}{(2.0 \text{ s} - 1.0 \text{ s})}$$
$$a = \frac{10.0 \text{ m/s}}{1.0 \text{ s}}$$
$$a = 10.0 \text{ m/s}^2$$

38 **1** Refer to *Reference Table H* and *Reference Table A*. A muon carries a −1 elementary charge. An elementary charge is equivalent to 1.6×10^{-19} C; therefore, 2 muons would carry a charge of $-2 \times 1.6 \times 10^{-19}$ C or -3.2×10^{-19} C.

WRONG CHOICES EXPLAINED:

(2) You forgot to multiply by 2.

(3) You most likely used the charge for a muon neutrino and not a muon.

(4) You multiplied by +2 instead of −2.

39 **1** Refer to the *Mechanics* section of *Reference Table K*.

Given:
$$F_g = 1.47\text{ N}$$
$$\Delta h = 10.0\text{ m}$$
$$KE_f = 12\text{ J}$$

Find:
$$\Delta Q = ?$$

Solution:
$$E_{T_f} = E_{T_i}$$
$$E_T = PE + KE + Q$$
$$PE_f + KE_f + Q_f = PE_i + KE_i + Q_i$$

The initial kinetic energy is zero since the ball is not moving at the start.

The final potential energy is zero since the ball is about to hit the ground (zero height).

$$PE_f + Q_f = KE_i + Q_i$$
$$Q_f - Q_i = PE_i - KE_f$$
$$\Delta Q = mg\Delta h - KE_f$$
$$g = \frac{F_g}{m}$$
$$F_g = mg$$
$$\Delta Q = F_g\Delta h - KE_f$$
$$\Delta Q = ((1.47\text{ N})(10.0\text{ m})) - 12\text{ J}$$
$$\Delta Q = 14.7\text{ J} - 12\text{ J}$$
$$\Delta Q = 2.7\text{ J}$$

40 **3** Horizontal displacement of the projectile depends on the horizontal component of the velocity and the amount of time the projectile travels.

Refer to the *Mechanics* section of *Reference Table K* for the equation to calculate the horizontal component of the velocity and the equation for calculating the horizontal distance traveled.

$$A_x = A\cos\theta$$
$$v_x = v_i \cos\theta$$
$$\overline{v} = \frac{d}{t}$$
$$d = \overline{v}t$$
$$d_x = \overline{v}_x t$$

The amount of time that the projectile travels in the horizontal direction is equal to the time it takes for the projectile to reach its highest point in the air, vertically, and come back down to the ground; in this case, it is the starting point, also known as the hang time. This time depends on the starting vertical component of the velocity and acceleration due to gravity.

Refer to the *Mechanics* section of *Reference Table K* for the equation to calculate the vertical component of the velocity and the hang time.

$$A_y = A\sin\theta$$
$$v_{iy} = v_i \sin\theta$$
$$v_f = v_i + at$$
$$v_{fy} = v_{iy} + a_y t$$

Once the total hang time is calculated, it can be substituted into the distance equation for the horizontal direction.

A smaller angle will result in a greater x component and smaller y component; i.e., it will be able to travel faster in the x direction but for less time. A larger angle will result in a smaller x component and a larger y component of velocity; i.e., it will be able to travel for longer in the x direction but at a slower speed. A 45 degree angle mathematically turns out to be the optimal angle for the greatest range—i.e., travel in the x direction for a given initial velocity.

Choice (1) and choice (2) both have a starting velocity of 10. m/s, but choice (1), which is launched at a 45 degree angle, will yield a longer range, so we can eliminate choice (2).

Choice (1) and choice (3) both have a launch angle of 45 degrees, but choice (3) will yield a longer range because it has a larger initial velocity, so we can eliminate choice (1).

Choice (3) and choice (4) both have the same initial velocity, but, since choice (3) is launched at a 45 degree angle and choice (4) is launched at a 60 degree angle, choice (3) will yield a larger range, so we can eliminate choice (4). Therefore, choice (3) will therefore result in the largest horizontal displacement.

41 **2** Refer to the *Mechanics* section of *Reference Table K.*

Given: $m = 5.0\ \text{kg}$

$F = 27\ \text{N east}$

$F_f = 17\ \text{N west}$

Find: $a = ?$

Solution:
$$F_{net} = F_{applied} - F_f$$
$$F_{net} = 27\ \text{N} - 17\ \text{N}$$
$$F_{net} = 10\ \text{N}$$
$$a = \frac{F_{net}}{m}$$
$$a = \frac{10\ \text{N}}{5.0\ \text{kg}}$$
$$a = 2.0\ \text{N/kg} = 2.0\ \text{m/s}^2$$

WRONG CHOICES EXPLAINED:

(1) You divided the mass by the net force instead of the other way around.

(3) You added the applied force and the frictional force instead of subtracting.

(4) You calculated the net force and didn't continue to solve for acceleration using mass.

42 **4** Refer to the *Mechanics* section of *Reference Table K.*

Given: $m = 2.0\ \text{kg}$

$v = 3.0\ \text{m/s}$

$r = 2.0\ \text{m}$

Find: $F_c = ?$

Solution:
$$F_c = ma_c$$
$$a_c = \frac{v^2}{r}$$
$$F_c = \frac{mv^2}{r}$$
$$F_c = \frac{(2.0\ \text{kg})(3.0\ \text{m/s})^2}{(2.0\ \text{m})}$$
$$F_c = 9.0\ \text{N}$$

Centripetal force and acceleration are always directed toward the center of the circular path, which in this diagram is west.

WRONG CHOICES EXPLAINED:

(1) You solved for centripetal acceleration instead of centripetal force and chose the direction of the instantaneous velocity instead of the centripetal force.

(2) You chose the correct direction for centripetal force but solved for centripetal acceleration instead of centripetal force.

(3) You calculated the correct magnitude but chose the direction of the instantaneous velocity instead of the centripetal force.

43 **4** Refer to the *Electricity* section of *Reference Table K.*

series circuit

$$V_T = V_1 + V_2$$

parallel circuit

$$V_T = V_1 = V_2$$

Choice (4) shows a series circuit with a 10 V battery and a voltmeter measuring the total potential difference across both resistors which, in a series circuit, is equal to the potential difference supplied by the power source, in this case 10 volts.

WRONG CHOICES EXPLAINED:

(1) This circuit has the voltmeter connected to one of the resistors, which is connected in parallel to the second resistor. In a parallel circuit, the potential difference in each resistor is equal to each other and to that of the power source, in this case 20 volts.

(2) This is a series circuit connected similarly to the correct choice (4). However, the power source is supplying 20 volts; therefore, the voltmeter connected across both resistors will also read 20 volts.

(3) This is a series circuit with the voltmeter connected to only one of the resistors. In a series circuit, the sum of the potential differences across each resistor will equal the total provided by the battery, which in this diagram is 10 V. Given that there are two equal resistors, each voltmeter will read 5.0 volts.

44 **1** Refer to *Reference Table H*.

Given: lambda baryon—made up of *uds* quarks

Find: particle with same electric charge as a lambda baryon

Solution:

$$\text{up quark} = +\frac{2}{3}e$$

$$\text{down quark} = -\frac{1}{3}e$$

$$\text{strange quark} = -\frac{1}{3}e$$

$$\frac{2}{3} + \left(-\frac{1}{3}\right) + \left(-\frac{1}{3}\right) = 0$$

The correct choice is (1); neutrons also carry zero electric charge.

WRONG CHOICES EXPLAINED:

(2) Electrons carry −1e charge.
(3) Protons carry +1e charge.
(4) Antimuons carry +1e charge.

45 **2** Refer to the *Modern Physics* section of *Reference Table K*, *Reference Table A*, and *Reference Table B*.

Given: $E = 24.0 \text{ MJ} = 24.0 \times 10^6 \text{ J} = 2.40 \times 10^7 \text{ J}$

Find: $m = ?$

Solution: $E = mc^2$

$$m = \frac{E}{c^2}$$

$$m = \frac{2.40 \times 10^7 \text{ J}}{\left(3.00 \times 10^8 \text{ m/s}\right)^2}$$

$$m = 2.67 \times 10^{-10} \text{ kg}$$

WRONG CHOICES EXPLAINED:

(1) You didn't convert the 24.0 megajoules into joules before substituting into the equation.

(3) You didn't convert the 24.0 megajoules into joules before substituting into the equation, and you forgot to square the speed of light in the denominator.

(4) You forgot to square the speed of light in the denominator.

46 **1** Refer to the *Modern Physics* section of *Reference Table K*, *Reference Table F*, and *Reference Table A*.

Given: $E_{photon} = 3.02 \times 10^{-19}$ J

Find: equivalent energy level transition—

$n = ?$ to $n = ?$

Solution:

$$E_{photon} = 3.02 \times 10^{-19}\text{J}\left(\frac{1\text{ eV}}{1.60 \times 10^{-19}\text{J}}\right)$$

$$E_{photon} = 1.89\text{ eV}$$

$$n_3 - n_2 = -1.51\text{ eV} - (-3.41\text{ eV}) = 1.89\text{ eV}$$

WRONG CHOICES EXPLAINED:

(2) This produces a photon with 2.55 eV of energy.

(3) This produces a photon with 2.86 eV of energy.

(4) This produces a photon with 3.02 eV of energy.

47 **2** A negatively charged rod brought near a neutral object causes a redistribution of charge within the object. It is still neutral but some of the charges have been separated; the negative electrons farther away from the rod on the far side of the object leave the side closest to the rod more positive. Given that like charges repel and opposite charges attract, if both objects were neutral, they would both be attracted to the rod. If an object were negative, it would be repelled by the rod. If an object were positive, it would be attracted to the rod. Since object R is repelled by the rod, we must conclude that it carries a negative charge. Since object S is attracted, it can either be positive or neutral.

48 **3** Refer to the *Electricity* and *Mechanics* sections of *Reference Table K*.

Given: $P = 60.0\text{ W}$

$t = 1.0\text{ min} = 60.0\text{ s}$

$F_g = 10.0\text{ N}$

Find: $d = ?$

Solution: $W = Pt$

$$W = (60.0\text{ W})(60.0\text{ s})$$

$$W = 3600.\text{ J}$$

$$W = Fd$$

$$d = \frac{W}{F}$$

$$d = \frac{3600.\text{ J}}{10.0\text{ N}}$$

$$d = 360.\text{ m}$$

WRONG CHOICES EXPLAINED:

(1) You forgot to convert 1 minute into 60 seconds before substituting into the equation.

(4) You just multiplied 60 watts by 10 newtons.

49 **4** Refer to the *Electromagnetic Spectrum* section of *Reference Table D*.

Microwaves have wavelengths between 10^{-4} and 10^{0} meters in length. This is an order of magnitude question. Ten to the −4 power is equivalent to 0.0001 meter or 0.004 inches. Ten to the zero power or one meter is approximately equal to 3.3 feet or 40 in. Answer choice (4), a student's thumb, falls within this range.

WRONG CHOICES EXPLAINED:

(1) The radius of Earth is on a 10^{6} order of magnitude.

(2) Mount Everest is on a 10^{4} order of magnitude at almost 10,000 meters high.

(3) A football field is on a 10^{2} order of magnitude at almost 100 meters wide.

50 **1** The effect produced when two or more wave pulses pass through the same point simultaneously is called interference. We add the amplitudes of each pulse together to get the combined amplitude pattern. Choice (1) correctly represents what the pulse would look like at the instant that they are superimposed at the same location in the medium.

WRONG CHOICES EXPLAINED:

(2) This is the mirror image of the correct answer choice.

(3) You flipped the bottom pulse to the top and combined the two into a rectangular shape.

(4) You flipped the top pulse to the bottom and combined the two into a rectangular shape.

PART B–2

51 Refer to the *Mechanics* section of *Reference Table K*.

Given: $k = 50.\ \text{N/m}$

$x = 0.10\ \text{m}$

Find: $PE_s = ?$

Solution:

$$PE_s = \frac{1}{2}kx^2$$

$$PE_s = \frac{1}{2}(50.\ \text{N/m})(0.10\ \text{m})^2$$

$$PE_s = 0.25\ \text{J}$$

Note: One credit is awarded for an answer of 0.25 J.

52–53 Refer to the *Mechanics* section of *Reference Table K*.

Given: $m = 0.10\text{ kg}$

Find: $v_{\text{max}} = ?$

Solution:

$$\Delta PE_s = \Delta KE$$

$$\Delta PE_s = \frac{1}{2}mv^2$$

$$v^2 = \frac{2(\Delta PE_s)}{m}$$

$$v = \sqrt{\frac{2(\Delta PE_s)}{m}}$$

$$v = \sqrt{\frac{2(0.25\text{ J})}{0.10\text{ kg}}}$$

$$v = 2.2\text{ m/s}$$

52 One credit is awarded for a correct equation and substitution of values with units *or* for an answer with units that is consistent with your answer to question 51.

53 One credit is awarded for the correct answer with units or for an answer, with units, that is consistent with your answer to question 52.

Note: Students will not be penalized more than 1 credit for errors in units in questions 52 and 53.

54

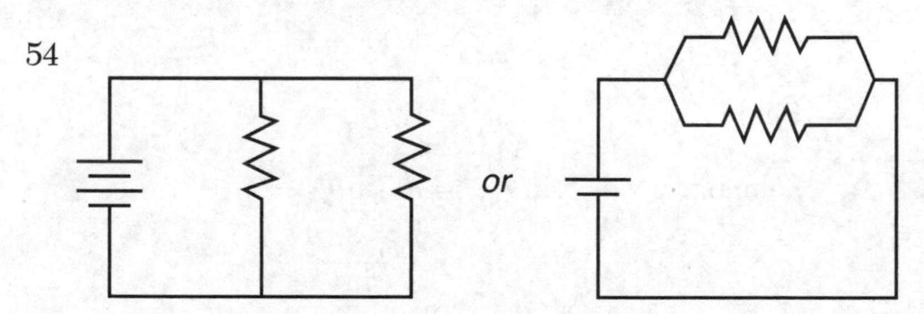

One credit is awarded for a circuit diagram showing two resistors connected in parallel with a cell or a battery.

Note: Credit is allowed for lines not touching the battery if the space between the lines and the battery is less than or equal to the distance between the battery symbol lines.

55 You are given a distance versus time graph for a moving object. The slope on a distance versus time graph is equal to the velocity of the object. This graph has a constant slope; hence, the velocity of this object is constant over time. You are asked to draw a graph for velocity versus time for this object. A horizontal line would represent a constant velocity over time.

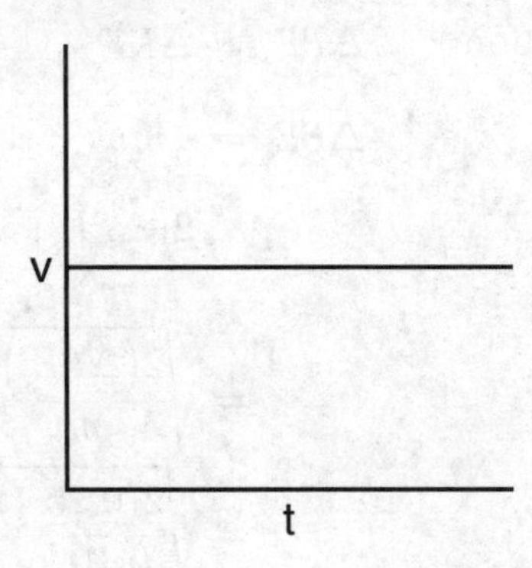

One credit is awarded for a line that approximates a horizontal line representing constant velocity.

56 You are asked to measure the angle of refraction with a protractor. Angles of refraction are measured with respect to the normal.
One credit is awarded for an answer of 40° ± 2°.

57–58 Refer to the *Waves* section of *Reference Table K.*

Given:

$$\theta_{air} = 40.0°$$
$$\theta_X = 25.0°$$
$$n_{air} = 1.00$$

Find:

$$n_x = ?$$

Solution:

$$n_{air} \sin\theta_{air} = n_x \sin\theta_x$$

$$n_x = \frac{n_{air} \sin\theta_{air}}{\sin\theta_x}$$

$$n_x = \frac{(1.00)\sin 40°}{\sin 25°}$$

$$n_x = 1.5$$

57 One credit is awarded for a correct equation and substitution of values with units *or* for an answer with units that is consistent with your answer to question 56.

58 One credit is awarded for the correct answer *or* for an answer, without units, that is consistent with your answer to question 57.

Note: Students will not be penalized more than 1 credit for errors in units in questions 57 and 58.

59–60 Refer to the *Mechanics* section of *Reference Table K* and *Reference Table A*.

Given:

$$m = 7.5\text{ kg}$$
$$v_i = 0.0\text{ m/s}$$
$$v_f = 29\text{ m/s}$$
$$a = 9.81\text{ m/s}^2$$

Find: $d = ?$ or $\Delta h = ?$

Solution:

$$v_f^2 = v_i^2 + 2ad$$
$$v_f^2 - v_i^2 = 2ad$$
$$d = \frac{v_f^2 - v_i^2}{2a}$$
$$d = \frac{(29\text{ m/s})^2 - (0\text{ m/s})^2}{2(9.81\text{ m/s}^2)}$$
$$d = 43\text{ m}$$

or

$$\Delta PE = \Delta KE$$
$$mg\Delta h = \frac{1}{2}mv^2$$
$$\Delta h = \frac{1}{2}\frac{mv^2}{mg}$$
$$\Delta h = \frac{1}{2}\frac{v^2}{g}$$
$$\Delta h = \frac{(29\text{ m/s})^2}{2(9.81\text{ m/s}^2)}$$
$$\Delta h = 43\text{ m}$$

59 One credit is awarded for a correct equation and substitution of values with units.

60 One credit is awarded for the correct answer with units *or* for an answer, with units, that is consistent with your answer to question 59.

Note: Students will not be penalized more than 1 credit for errors in units in questions 59 and 60.

61–62 Refer to the *Mechanics* section of *Reference Table K*.

Given:

$$m_A = 100.\text{ g}$$
$$v_{Ai} = 4.0\text{ m/s}$$
$$m_B = 150\text{ g}$$
$$v_{Bi} = 0\text{ m/s}$$
$$v_{Af} = -1.5\text{ m/s}$$

Find:

$$v_{Bf} = ?$$

Solution:

$$p_{before} = p_{after}$$
$$p = mv$$
$$m_1v_{1i} + m_2v_{2i} = m_1v_{1f} + m_2v_{2f}$$
$$m_1v_{1i} = m_1v_{1f} + m_2v_{2f}$$
$$v_{2f} = \frac{m_1v_{1i} - m_1v_{1f}}{m_2}$$
$$v_{2f} = \frac{(100.\text{ g})(4.0\text{ m/s}) - (100.\text{ g})(-1.5\text{ m/s})}{150\text{ g}}$$
$$v_{2f} = \frac{400\text{ gm/s} - (-150.\text{ gm/s})}{150\text{ g}}$$
$$v_{2f} = 3.7\text{ m/s}$$

61 One credit is awarded for a correct equation and substitution of values with units.

62 One credit is awarded for the correct answer with units or for an answer, with units, that is consistent with your answer to question 61.

Note: Students will not be penalized more than 1 credit for errors in units in questions 61 and 62.

63–65 Refer to the *Mechanics* section of *Reference Table K.*

Given: $m = 1.20 \times 10^3$ kg

$v_i = 25$ m/s

$t = 5.00$ s

$v_f = 0$ m/s

Find: $J = ?$

Solution: $p = mv$

$J = \Delta p$

$J = m\Delta v$

$J = m(v_f - v_i)$

$J = 1.20 \times 10^3 \text{ kg}(0 \text{ m/s} - 25 \text{ m/s})$

$J = -3.0 \times 10^4 \text{ kg} \bullet \text{m/s}$ or $3.0 \times 10^4 \text{ N} \bullet \text{s}$

63 One credit is awarded for a correct equation and substitution of values with units.

64 One credit is awarded for the correct answer or for an answer, with units, that is consistent with your answer to question 63.

Note: Students will not be penalized more than 1 credit for errors in units in questions 63 and 64. Credit is allowed for a correct answer with units that is positive or negative.

65 The car is traveling east. In order to stop the car's motion, the impulse, or force, over time would need to be applied opposite to the car's direction of motion, or in this case west.

Acceptable responses for credit include but are not limited to

- West
- Opposite to the direction of the car's motion
- Backward

PART C

66–67 Refer to the *Mechanics* section of *Reference Table K* and *Reference Table A*.

Given: $m = 1.56 \times 10^{-15}$ kg

$g = 9.81$ m/s^2

Find: $F_g = ?$

Solution:

$$g = \frac{F_g}{m}$$

$$F_g = mg$$

$$F_g = \left(1.96 \times 10^{-15}\ \text{kg}\right)\left(9.81\ \text{m/s}^2\right)$$

$$F_g = 1.92 \times 10^{-14}\ \text{N}$$

66 One credit is awarded for a correct equation and substitution of values with units.

67 One credit is awarded for the correct answer with units or for an answer, with units, that is consistent with your answer to question 66.

Note: Students will not be penalized more than 1 credit for errors in units in questions 66 and 67.

68 Given that the oil drop is motionless, the forces acting on the oil drop must be in equilibrium. We calculated the force due to gravity in question 67; therefore, the upward electrical force must be equal to 1.92×10^{-14} N.

One credit is awarded for an answer of 1.92×10^{-14} N *or* for an answer that is consistent with your answer to question 67.

69–70 Refer to the *Electricity* section of *Reference Table K*.

Given: $E = 4.0 \times 10^4$ N/C

$F_e = 1.92 \times 10^{-14}$ N

Find: $q = ?$

Solution:

$$E = \frac{F_e}{q}$$

$$q = \frac{F_e}{E}$$

$$q = \frac{1.92 \times 10^{-14}\ \text{N}}{4.0 \times 10^4\ \text{N/C}}$$

$$q = 4.8 \times 10^{-19}\ \text{C}$$

69 One credit is awarded for a correct equation and substitution of values with units or for an answer, with units, that is consistent with your responses to questions 67 and 68.

70 One credit is awarded for the correct answer with units *or* for an answer, with units, that is consistent with your answer to question 69.

Note: Students will not be penalized more than 1 credit for errors in units in questions 69 and 70.

71–72 Refer to the *Electricity* section of *Reference Table K.*

Given: $R_1 = 100.\ \Omega$

$R_2 = 200.\ \Omega$

$V_T = 10.0\ \text{V}$

Find: $R_{T_{parallel}} = ?$

Solution:

$$\frac{1}{R_{T_{parallel}}} = \frac{1}{R_1} + \frac{1}{R_2}$$

$$\frac{1}{R_{T_{parallel}}} = \frac{1}{100.\ \Omega} + \frac{1}{200.\ \Omega}$$

$$\frac{1}{R_{T_{parallel}}} = \frac{3}{200.\ \Omega}$$

$$R_{T_{parallel}} = \frac{200.\ \Omega}{3}$$

$$R_{T_{parallel}} = 66.7\ \Omega$$

71 One credit is awarded for a correct equation and substitution of values with units.

72 One credit is awarded for the correct answer with units or for an answer, with units, that is consistent with your answer to question 72.

Note: Students will not be penalized more than 1 credit for errors in units in questions 71 and 72.

73–74 Refer to the *Electricity* section of *Reference Table K.*

Given: $R_2 = 200.\ \Omega$

$V_T = 10.0\ \text{V} = V_2$

Find: $I = ?$

Solution:

$$R = \frac{V}{I}$$

$$I = \frac{V}{R}$$

$$I = \frac{10.0\ \text{V}}{200.\ \Omega}$$

$$I = 0.0500\ \text{A or } 5.00 \times 10^{-2}\ \text{A}$$

73 One credit is awarded for a correct equation and substitution of values with units.

74 One credit is awarded for the correct answer with units or for an answer, with units, that is consistent with your answer to question 73.

Note: Students will not be penalized more than 1 credit for errors in units in questions 73 and 74.

75 Refer to the *Electricity* section of *Reference Table K.*

Given: $R = 100.\ \Omega$

$V = 10.0\ \text{V}$

Find: $P = ?$

Solution: $P = \frac{V^2}{R}$

$$P = \frac{(10.0\ \text{V})^2}{100.\ \Omega}$$

$$P = 1.00\ \text{W}$$

One credit is awarded for an answer of 1.00 W.

76 One credit is awarded for at least one complete wave with a wavelength of 0.40 meter regardless of phase or shape.

77 One credit is awarded for at least one complete wave with an amplitude of 0.20 meter regardless of phase or shape.

Example of a 2-credit response for questions 76 and 77:

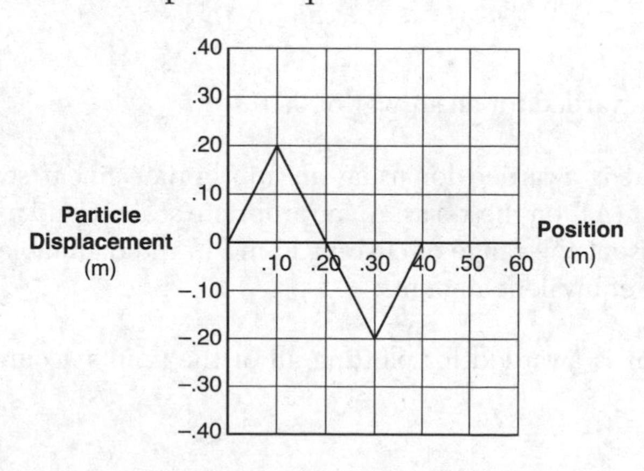

78–79 Refer to the *Waves* section of *Reference Table K.*

Given: $\lambda = 0.40\text{ m}$

$f = 10.\text{ Hz}$

Find: $v = ?$

Solution: $v = f\lambda$

$v = (10.\text{ Hz})(0.40\text{ m})$

$v = 4.0\text{ m/s}$

78 One credit is awarded for a correct equation and substitution of values with units.

79 One credit is awarded for the correct answer with units or for an answer, with units, that is consistent with your answer to question 78.

Note: Students will not be penalized more than 1 credit for errors in units in questions 78 and 79.

80 Refer to the *Waves* section of *Reference Table K.*

Given: $f = 10.\text{ Hz}$

Find: $T = ?$

Solution: $T = \frac{1}{f}$

$T = \frac{1}{10.\text{ Hz}}$

$T = 0.10\text{ s}$

One credit is awarded for an answer of 0.10 s.

81 One credit is awarded for using an appropriate linear scale on the axis marked "Current (A)" on the *x*-axis. An appropriate scale would use the length of the axis to represent the range of current found in the data table, and each box would represent equivalent amounts.

82 One credit is awarded for plotting all of the points accurately ±0.3 grid space.

83 One credit is awarded for drawing the best fit line or curve that is consistent with your responses to questions 81 and 82. A best fit line or curve is not a "connect the dots" but, rather, the **line** or curve that **best** represents the data on a scatter plot. This **line** may pass through some of the points, none of the points, or all of the points (ideally as many as possible that the line passes through), and there will be a similar number of points above or below at similar distances away from the line if the line is not passing through the points.

Example of a 3-credit response for questions 81 through 83:

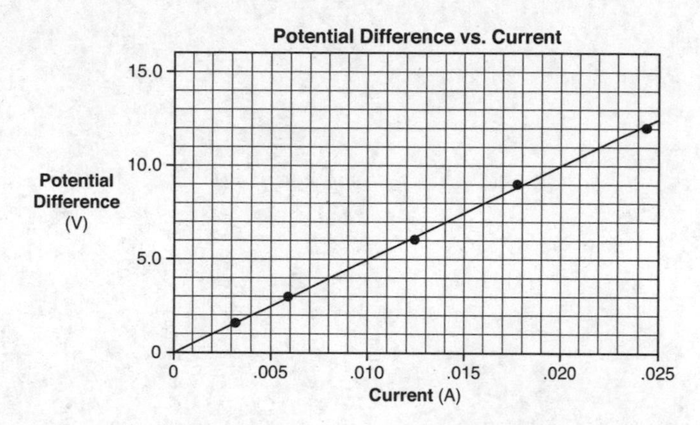

84–85 The slope of a potential difference versus current graph is given by the relationship $\Delta V/\Delta I$ or, in other words, the resistance. We can pick any two points on the line in the graph and substitute them into the formula to calculate the resistance.

Given: $V_1 = 0\text{ V}$

$V_2 = 3.0\text{ V}$

$I_1 = 0\text{ A}$

$I_2 = 0.006\text{ A}$

Find: $R = ?$

Solution:

$$R = \frac{\Delta V}{\Delta I}$$

$$R = \frac{(3.0\text{ V} - 0\text{ V})}{(0.006\text{ A} - 0\text{ A})}$$

$$R = 5.0 \times 10^2\ \Omega$$

84 One credit is awarded for the correct answer with units or for an answer, with units, that is consistent with your answers to questions 81–83.

85 One credit is awarded for the correct answer with units or for an answer, with units, that is consistent with your answer to question 84.

Note: Students will not be penalized more than 1 credit for errors in units in questions 84 and 85.

Topic	Question Numbers (total)	Wrong Answers (x)	Grade
Math Skills	2, 7, 8, 10, 12–18, 23, 26, 32, 34, 36–42, 45, 46, 48, 51–53, 56–64, 66–75, 78–85: (55)		$\frac{100(55-x)}{55} = \%$
Mechanics	1–3, 5–7, 11, 12, 18, 33, 34, 36, 37, 40–42, 55, 59–67: (26)		$\frac{100(26-x)}{26} = \%$
Energy	1, 8–11, 39, 48, 51–53: (10)		$\frac{100(10-x)}{10} = \%$
Electricity/ Magnetism	4, 13–17, 19, 35, 43, 47, 48, 54, 68–75, 84, 85: (22)		$\frac{100(22-x)}{22} = \%$
Waves	20–29, 50, 56–58, 76–80: (19)		$\frac{100(19-x)}{19} = \%$
Modern Physics	30–32, 38, 44–46, 49: (8)		$\frac{100(8-x)}{8} = \%$